Springer Climate

Series Editor
John Dodson, Institute of Earth Environment, Chinese Academy of Sciences, Xian, Shaanxi, China

Springer Climate is an interdisciplinary book series dedicated to climate research. This includes climatology, climate change impacts, climate change management, climate change policy, regional climate studies, climate monitoring and modeling, palaeoclimatology etc. The series publishes high quality research for scientists, researchers, students and policy makers. An author/editor questionnaire, instructions for authors and a book proposal form can be obtained from the Publishing Editor. **Now indexed in Scopus® !**

Neloy Khare
Editor

Science, Policies and Conflicts of Climate Change

An Indian Perspective

Springer

Editor
Neloy Khare
Ministry of Earth Sciences
Delhi, India

ISSN 2352-0698 ISSN 2352-0701 (electronic)
Springer Climate
ISBN 978-3-031-16253-4 ISBN 978-3-031-16254-1 (eBook)
https://doi.org/10.1007/978-3-031-16254-1

This Springer imprint is published by the registered company Springer Nature Switzerland AG
The registered company address is: Gewerbestrasse 11, 6330 Cham, Switzerland

Foreword

The subject of climate change has virtually engulfed the planet and is now seen as the dominant agenda in both national and international discourse. It transgresses disciplinary boundaries and constitutes a key element in informing the water–food–energy policy frameworks that underpin national and international governance structures required for the implementation of these policy frameworks. It is in this context that Dr. Neloy Khare has edited the present volume titled 'Science, Policies and Conflicts of Climate Change: An Indian Perspective'. In any global and planetary scale consideration, the Indian perspective matters simply because the subcontinent is home to about fifteen per cent of the global population.

This edited volume comprises thirteen thematic Chapters that address the issues of climate change related science, policies and the geopolitical issues that have arisen around the world and are the subject of planetary scale interrogation and conversation.

The enormous consequences of inevitable climate change could be minimized, to some extent, if the causes of such climate change are properly understood and attempts made to seed appropriate region-specific mitigation strategies. Awareness and climate literacy among the masses can encourage people to change their attitudes and behaviour; it also helps them to make informed decisions. Young budding minds can be ignited to tackle ongoing and emerging problems of global climate changes, including the impact of global warming and learn how to adapt to climate change. Undoubtedly, education empowers all people but especially motivates the young to act. It is, therefore, imperative that climate change awareness is spread to achieve sustainability in developing countries through the dissemination of data and the latest insights on the subject among the masses.

This book is a way forward in this direction to help understand the science behind climate change in time and space, the impacts of climate change on coupled environmental–social systems, and the required policy initiatives both in national and international arena, besides taking into account the associated geopolitical dimension of such global climate change.

The book begins with a chapter on climate change's science and policies focussing on the National Perspectives in Global context by *Ramesh* followed by a chapter that is dedicated to climate changes over the Indian subcontinent: Scenarios and impacts by *Mall et al.* Also, food security is at stake due to the ongoing rise in temperature and associated climate changes. *Praveen et al.* made a significant contribution to assess the historical and future relationship between climatic factors and the production of different crops in India.

To evaluate the geopolitical aspects of the ongoing climate change, *Jayaram and Sethi* detailed the Geopolitics of Climate Change and Water Security in the larger perspectives of South Asia, including India, with special reference to both conflict and cooperation.

One among many significant effects of global warming and climate change are perceived changes in the sea level, and therefore a study by *Kankara et al.* on the Vulnerability Analysis of Sea Level Rise (SLR) on a micro-tidal coast from Nagapattinam Coast has been included. To help understand climate changes in the geological past, *Humane et al.* studied the lacustrine records of anthropogenic change and precipitation in the monsoonal core zone of Central India; also, *Akhouri et al.*'s investigations to understand the anthropogenic interference and climatic changes influencing precipitation, sediment load and sand deposition by Phalgu River in and around Jahanabad district of Bihar provided additional insights from a fluvial landscape.

Mir assessed the impact of contrasting climates on the intensity of chemical weathering and maturity of sediments in different latitudes utilizing geochemical parameters, whereas *Yadav et al.* attempted to study the manganese mineralization in the manganiferous quartzite in the Boringpadar-Amath area, Odisha, India for their climate change implications. Interestingly, *Sabale* has provided archaeological evidence of climate change trends in the past. Besides, *Singh et al.* overviewed the scientific investigations conducted on Youngest Toba Tuff ash, artefacts and vertebrate remains that are preserved in the Quaternary alluvial sediments of India for their paleoclimatic implications. *Ghatak and Ghatak* studied active tectonics – climate interaction in the structural depo-centres of the Central Indian Tectonic Zone (CITZ) and *Singh and Kumar* put forth the significance and utility of cosmogenic radionuclides in the Quaternary Sciences using Accelerator Mass Spectrometry.

I am sure this book will receive the attention it deserves and emerge as a ready reference for obtaining recent insights and knowledge on the subject of climate change from the Indian perspective.

Indian Institute of Science Education and Research
Bhopal, India
July 2022

Prof. Sampat Kumar Tandon

Preface

It is often believed that climate change or global climate change is a more scientifically sound term than global warming. Such advocacy for the use of climate change instead of global warming is largely since changes to precipitation patterns and sea level are expected to have a more profound impact than the higher temperatures alone.

While global warming refers to the rise in global temperatures due mainly to the increasing concentrations of greenhouse gases in the atmosphere, climate change refers to the increasing changes in the measures of climate over a long period, including precipitation, temperature and wind patterns. Climate change can also impact human health by worsening air and water quality, increasing the spread of certain diseases, and altering the frequency or intensity of extreme weather events. The rising sea level threatens coastal communities besides the many adverse impacts mankind will face. Climate change may aggravate erosion, the decline in organic matter, salinization, soil biodiversity loss, landslides, desertification and flooding. The effect of climate change on soil carbon storage can be related to changing atmospheric CO_2 concentrations, increased temperatures and changing precipitation patterns. Temperatures are rising worldwide due to greenhouse gases trapping more heat in the atmosphere. Droughts are becoming longer and more extreme around the world. Tropical storms are becoming more severe due to warmer ocean water temperatures. The Earth's climate is influenced and changed through natural causes like volcanic eruptions, ocean currents, the Earth's orbital changes, solar variations and internal variability. The debate over climate change has become so contentious that many experts often opt to be limited in their claims as far as climate projections are concerned. Nevertheless, such predictions require a sound understanding of climate change science which helps evolve climate policies and its geopolitics globally. The larger the observational data set, climate data both in space and in time including geological past, the better the reliability of any projected climate model.

Nevertheless, the IPCC assesses the credibility of scientific research on global warming and uses that information to advise world leaders and policymakers on options for adapting to climate change, such as preventing the degradation of forests or reducing the consumption of gasoline and other fossil fuels. Climate change science for policy involves being responsive to policymakers. Such evidence-based scientific inputs on climate change will also help for expert judgment. Undoubtedly, everyone wants such state-of-the-art climate change assessments to help mitigate the devastating damages expected of changing climates and their extremes. Climate change refers to long-term shifts in temperatures and weather patterns. These shifts may be natural, but since the 1800s, human activities have been the main driver of climate change, primarily due to the burning of fossil fuels (e.g. coal, oil and gas), which produces heat-trapping gases. The sun heats the Earth, which then sends part of that heat back into space as infrared radiation. Certain gases—'greenhouse gases'—interact with this outgoing heat radiation, keeping the surface temperature higher than it would be without them. It remains an enigma whether the current climate change is a natural trend or an outcome of anthropogenic influence. It can only be understood when the climate database is enhanced covering larger geographical localities globally.

In view of the above, it is significant to bring out a detailed compilation of data on climate change, understand the science behind such changes in the climate and what could be expected in the geopolitical dimension of such global and regional climate changes envisaging. Therefore, the present book titled **'Science, Policies and Conflicts of Climate Change: An Indian Perspective'** aptly covers all these aspects from the perspectives of India.

The book begins with the Chapter by *Ramesh* who provides an exhaustive account on climate change science related policies with the national perspectives in the global context. While elaborating on this important topic he mentioned that India signed this United Nations Framework Convention on Climate Change (UNFCCC) in 1992 and ratified it in 1993. In general, UNFCCC efforts supported the developing countries in combating climate change by providing a platform for finance, technology transfers, discussions, global partnerships, etc., not seeking emission reduction commitments in recognition of their small contribution to global warming as well as low financial and technical capacities. UNFCCC initiatives helped create public awareness regarding climate change, which is much higher today than in the late 1990s. Although climate science in the late 1990s was certainly strong enough to negotiate an international treaty, it is hard to deny that the scientific understanding of the climate crisis has improved considerably over the past two decades through IPCC in which UNFCCC played a significant role. The Ministry of Environment and Forest and Climate Change (MOEF&CC) is the main organization responsible for climate change policy in India. Kyoto Protocol under the heads of Common but Differentiated Responsibilities (CBDR) says that the developed countries should take more responsibility and liability to take care of our degrading environment and make appropriate rules to save it for future generations by taking legally binding commitments. A major shortcoming of the Kyoto Protocol was the non-inclusion of dangerous microscopic environmental air pollutants that come from car engines and combustion-based power plants. Russia joined the Kyoto Protocol in 2005, but the USA did not join (and Canada dropped out of the Protocol subsequently). Now the world has reached almost 1.2 °C warming already, UNFCCC needs to deal with unsustainable commitments as the Paris 2015 NDC contributions are found to be inadequate to maintain 2 °C levels. Although low-lying countries that are facing unsustainable sea levels rises in a warming world argued for a tougher target of 1.5 °C, a largely unsatisfactory response was seen till the proceedings of the COP-26 held at Glasgow, UK, in 2021. The main challenge faced by UNFCCC is to ensure all rich nations pledged to maintain a $100bn a year funding beyond 2020 to contribute fully from now on at least, although several Pacific island countries advocated its inadequacy to fully meet the adverse impacts of climate change.

Mall et al. on the other hand focussed on the climate change exclusively in the Indian context, and they present shreds of evidence of the impact of climate change in past, present and future such as extreme events like heat waves, diurnal temperature range, shrinking of Himalayan glaciers, shifting of rainfall pattern, increased susceptibility to floods and droughts and its impact on some of the important sectors. Their findings show clear evidence of a decline in crop production and productivity of some of the important crops such as wheat, rice, sugarcane, maize, potato, tomato, etc. The recent studies established an increase in morbidity and mortality associated with extreme temperature and poor air quality associated with increased particulate matter (PM), NOx, SOx, O_3, black carbon and other ambient pollutants. They further pointed out that the important river basis of India like Gomti, Gandak, Vaigai, Mahi, Varuna and Ghaghra have shown increased susceptibility to flooding and drought events that are more likely to be frequent and severe in the future under different climate change scenarios owing to changes in erratic rainfall patterns and increasing temperature. This chapter also discusses the potential adaptation and mitigation strategies that would help policymakers to combat climate change amid the rising susceptible population.

While assessing the historical and future relationship between climatic factors and the production of different crops in India, *Praveen* explores the effects of climate variations over 15 different food and industrial crop productions in India using the econometric model and forecasts these crop outputs using the Artificial Neural Network (ANN) model up to 2030. The unit root test was used to inspect the stationarity in time series datasets in agricultural crop productions and climatic variables. The Dynamic Ordinary Least Square (DOLS) and Fully Modified Dynamic Ordinary Least Square (FMOLS) models were used for testing the accuracy of the outcomes. The results showed that land productivity declines if the annual mean temperature will increase, demonstrating the sensitivity of Indian agricultural crop production to the temperature variance. Meanwhile, results also unfold that if the temperature would increase, rainfall would become more unstable triggering the reduction of rice, wheat, tuar, cotton, and tea crop productions. major crop productions namely sesamum (til), gram, sugarcane, groundnut, jowar and bajra would experience a positive impact and subsequently increase slightly up to 2030. The outcomes of the study can aid in insights into sustainable agricultural crop productions in India, and livelihood patterns, particularly among minor and marginal agricultural families.

The book concludes with a devoted chapter by *Jayaram and Sethi* on the yet another important aspect of associated geopolitics with ongoing global climate changeswith special reference to South Asian region where climate vulnerabilities have direct and indirect implications for water security. While the effects of climate change on water access, safety and affordability in the region are becoming increasingly adverse, the transboundary nature of both climate change and water security, especially on account of shared water resources, complicates responses. The shared river basins such as the Indus, Ganges, Meghna and Brahmaputra, have been marred by geopolitical tensions, which renders transboundary climate cooperation to deal with water-related problems difficult. Although the region has seen different forms of cooperation in the form of river water sharing treaties/arrangements as well as regional organisations' role in advancing cooperation, these efforts rarely take cognizance of the effects of climate change on water security. In their chapter, they provide an analysis of the challenges posed by climate change to water security and geopolitics of transboundary conflicts and cooperation in South Asia, as well as explores opportunities for cooperation among the countries of the region.

The IPCC sixth assessment report shows that the world will probably reach or exceed 1.5 °C (2.7 °F) of warming within just the next two decades and the oceans will rise between 10 and 30 inches (26–77 centimetres) by 2100 (globally). The report highlights that the sea levels in South Asia could rise by 0.4 m in the low emissions scenario, about 0.5 m in the current pledge's scenario and 0.7 m in the high emissions scenario. Sea levels rise (SLR) has been rapidly increased which will lead to devastating effects on coastal habitats farther inland because it can cause destructive erosion, wetland flooding, aquifer and agricultural soil contamination with salt, and lost habitat for fish, birds and plants. Studies by Indian scientists reveal that the trends of sea level rise are estimated to be 1.3 mm/year along the Indian coasts during the past 40–50 years. Also, impacts of storm surges, tsunamis, coastal floods and coastal erosion in the low-lying coastal area cause gradual loss along with coastal land. *Kankara et al.* highlighted the impact of sea level rise on the sandy coast of the Nagapattinam sector and assessed the devastation incurred in both physical and socio-economic aspects. They further noted that the exposed vulnerable sectors to sea level rise are increasing with changing climate. The quantitative SLR impact assessment method suggested in the study forms the basic framework for coastal planners in disaster mitigation and building a climate-resilient community.

On the contrary, *Humane et al.* studied a sediment core from the Navegaon Bandh Lake (NBL) for diatom fossils and sediment geochemistry to investigate the anthropogenic impact and climatic variability over the past century. The declined geochemical concentrations show three major impacts on the NBL coinciding with the drought events in the foremost parts of India around ~1918, ~1975 and ~1965 and ~1991. The rise in the geochemical content of the NBL core was evident in the post dam construction period (i.e ~1917–1919) with the two major peaks around ~1975 and ~1998. These major peaks are indicators of the higher rate of soil erosion and the improved precipitation in the catchment of the NBL. Anthropogenic activities have brought five major changes in diatom assemblages and ecology in the NBL around ~1946, ~1956, ~1972, ~1978 and ~2006. The increased agricultural activities in the catchment of the NBL are evident since ~1906–1966 as the fertilizers (K, P) are mainly derived other than soil particles of the catchment. Similarly, deforestation in the region was evident during ~1967–1975 coinciding with the period of the highest soil erosion. Another major phase of soil erosion caused by anthropogenic activities was noticed during ~1992–1998. The hypolimnetic oxygen of the NBL has declined over the past ~80 years indicating increased eutrophication in the lake till the present. The nutrient input in response to climatic conditions and anthropogenic activities has played a vital role in the diatom shift of the NBL. *A. granulata* was predominant during ~1906–1914, ~1920–1922, ~1941–1943, ~1948–1956, ~1972–1976, and ~1973–1982 in the NBL along with increased soil erosion and flux, persistent wet period and increased nutrient levels. While the major period of the profusion of *R. musculus* in the NBL core was during ~1919–1921, ~1933–1940, ~1956–1972 and ~1978–2004, respectively, indicating mesotrophic, Meso euhyaline and alkalibiontic condition of the lake linked with the long dry period/less rainfall prevailed in the NBL watershed. The planktonic form *D. stelligera* was common during ~2007–2012 and indicates the existence of alkaline lake waters with moderate nutrients coinciding with the comparatively dry period/less rainfall and changing water level and deforestation in the catchment of the NBL. Overall, the investigation of the past evidence shows the importance of the study of fossil diatoms and geochemistry to understand the anthropogenic and precipitation changes along with the trophic status of the lake, which may be mainly used for the management of the tropical wetlands in India.

Similarly, *Akhouri et al.* dealt with arriving at the best possible balance resulting from environmental safety guideline procedures involved in mining, safeguarding the ecology of the river amidst challenges and ways to minimize the unprecedented effects of short-term climatic change. Their chapter also highlights aspects like drivers responsible for the reduced flow, erratic modes of sand deposition vis-a-vis undisciplined extraction and recommendations towards sustainable exploitation in tune with the expectation and guidelines served by the Hon'ble National Green Tribunal in this case along the Phalgu River, in Jehanabad district, Bihar.

Mir et al., on the other hand, attempted to study the impact of contrasting climates on the intensity of chemical weathering and maturity of sediments in different latitudes. In their chapter, chemical weathering and sediment maturity under contrasting climatic conditions from tropical peninsular and temperate extra peninsular parts of India are investigated. Rocks of peninsular India are moderate to intensively weathered while rocks of extra peninsular India are moderately weathered. High chemical weathering conditions in south India appear to result from climate forcing related to high rainfall and temperature. The distribution of C-values and Sr/Cu ratios collectively indicate a generally humid climate in peninsular India and a semi-humid to dry climate in extra peninsular India. There is an increase in humidity from higher to lower latitudes. The compositional maturity of the sediments is evaluated by using the ICV and $SiO2/Al2O3$ ratio. The results collectively indicate that the sediments in the peninsular area are mineralogically more mature than the sediments of the extra peninsular area. There is a decrease in sediment maturity from higher to lower latitudes meaning a high degree of clays and mineralogical maturity.

Further, *Yadav et al.* have brought to light an array of manganese mineralization with promising values, hosted in manganese ore and manganiferous quartzite and predict the responsible climatic environment for the formation of manganese based on field observations, ore petrography studies and geochemical inputs. Two bands of manganese ore and four bands of manganiferous quartzite are mapped to the southeast of Boringpadar and northeast of Amath area belonging to the Eastern Ghats Mobile Belt, India. The manganese ores are represented by bluish-black to brownish-black, massive to foliated, metallic luster, nodular and botryoidal forms which are mainly bedded and fragmental types. The ore bodies are strata bound occurring in between calc-silicate granulite and garnetiferous quartzite which defines a characteristic stratigraphic horizon. Manganiferous quartzite is hard, white to black, highly jointed, fractured and brecciated which occurs at the contact of khondalite. Braunite, bixbyite, jacobsite, manganite, pyrolusite and todorokite are primary minerals that are associated with host rock like quartzite and these minerals occur as inclusions and anhedral in shape. Secondary minerals are formed by the process of colloidal influx or metasomatism psilomelane-cryptomelane, pyrolusite and goethite are secondary minerals which occur in more quantities. A relatively high temperature metamorphic jacobsite-bixbyite-braunite assemblage, low temperature hydrothermal pyrolusite-psilomelane-cryptomelane assemblage and supergene pyrolusite-manganite assemblage are recorded which were formed by recycling of manganese in different stages of mineralization. Replacement and relict textures are common between braunite-manganite-pyrolusite, jacobsite-braunite-pyrolusite, pyrolusite-psilomelane-cryptomelane-goethite. A geophysical investigation has also been carried out in the area to know the subsurface continuity of manganese mineralization. In Boringpadar, the apparent resistivity values vary from 10 to 240 ohm-m which displays low resistivity and the apparent chargeability contour map shows high anomaly, indicative of a mineralized zone. In Amath, the apparent resistivity value varies from 20 to 140 ohm-m and the apparent chargeability contour map shows a high anomaly. The pseudo-depth sections were prepared over geophysical traverses to examine the subsurface distribution of apparent chargeability and apparent resistivity indicating mineralization prospects. This mineralized body starts almost from the surface and the contours of this anomaly zone are opening downward which is suggestive of the depth persistence of the causative source below 35 m. The analytical results of bedrock samples show high values of Mn, Fe, SiO_2, Al_2O_3, P, S and CaO. The values of Mn and Fe(T) ranged from 0.05 to 28.18 wt. % and 1.87 to 56.31 wt. %, respectively. In pitting/trenching samples, the Mn value ranges from 0.05 to 27.77 wt. % and the Fe(T) value ranges from 2.11 to 46.85 wt. %. The highest values of Mn come from manganese ore of the Amath area and maximum values of Fe(T) were recorded from manganese ore of the Boringpadar area. All the above-mentioned data indicate that the manganese ores and manganiferous quartzite might be formed in the freshwater environment.

It is a known fact that each river develops a typical type of topography according to her reaches because each flowing channel shapes the Earth's surface through the process of erosion, transportation and deposition of sediments. This is achieved by weathering and erosion of sediment from source areas and deposition of sediment in low-lying areas and streams modifying the Earth's surface. The old deposited sediments during the Quaternary era are immensely important from an archaeological point of view because human development has taken place contemporary to this period. Therefore, most of the contemporary cultural deposit is deposited in these sediments. In addition, such sediment is rich in flora and fauna which help to reconstruct the palaeoenvironment of those particular areas with reference to human development. In the present context, *Sabale* studied the details of the palaeoenvironment along with man–land observed in the Upper Reaches of the Sina by taking the quaternary trenches or excavations at suitable cultural settlement localities, in parts of south Ahmednagar district of Maharashtra.

A review of Youngest Toba Tuff (YTT) ash, artifacts and vertebrate remains uncovered in the alluvial deposits of various river basins of India has been attempted by *Singh et al.* in order to comprehend paleoclimate and paleoenvironmental conditions throughout the Quaternary period. The YTT eruption is regarded as the biggest eruption on the Earth's surface during the Pleistocene period, and hence is considered a catastrophe for ecosystem and mass populations. The ash of the same eruption has been found in laminated, sandy to clay alluvial columns in the form of light gray to yellowish brown, massive to structured, pockets, lenticles, and continuous to discontinues beds ranging in thickness from 0.2 to >3 m thick. Based on bedding geometries, preservational characteristics, and physical appearances, the ash is classified as primary and secondary. Geochemically and geochronologically, it overlaps elemental composition and isotopic age of the Oldest Toba Tuff (OTT) ash. Artifacts belonging to the Middle Paleolithic to Microlithic period are retained in the fluvial profiles that form pre- and post-YTT columns. Based on the typological and carving similarities, it has been linked to the Middle Paleolithic artifact industry of Africa, indicating migration of hominids in India. In addition, the majority of vertebrate remains reported mainly from both pre- and post-tephra successions of various alluviums reveal existence of herbaceous animal communities, implying the presence of small to tall grass vegetation and water pools to swamps environments under humid to temperate climates. Based on preservations of artifacts and vertebrate remains in both pre- and post-YTT ash strata, the study suggests existence of hominids and animals in the subcontinent prior to and after the eruption, indicating negligible effects of the eruption on the living media.

The Late Cenozoic period in the Central Indian Tectonic Zone (CITZ), marked by several episodes of crustal adjustments, resulted in the development of various landforms, drainage anomalies and pedosedimentary successions. The fault bound tectonic basins viz. Tapi, Purna and Narmada flanking the Satpura ranges hold telltale evidence of varying climatic and tectonic regimes. *Ghatak and Ghatak* in their chapter summarize data from various literature and the Active Fault Mapping programme in CITZ and study the fluvial archives to unravel the role of the tectonics-climate feedback system in shaping landscape architecture in and around the structurally controlled Quaternary River basins of Purna, Tapi and Narmada. The cumulative interactions between the deformational events and ongoing surface processes are best witnessed by the marker horizons, anomalous river courses, basin geometry, hill slopes and elevation characteristics and the way these have influenced its depositional environment. The study suggests that repeated reactivation of older crustal-scale faults under different stress regimes have been a guiding factor in landform evolution in CITZ. While tectonics holds the primary key to the overall landscape development and sedimentation, peninsular climatic or microclimatic variations have equally been significant as palaeo-climatic proxies mimic all India monsoon scenarios.

It is well known that the Quaternary Period begins at 2.58 Ma and was defined by the Global boundary Stratotype Section and Point (GSSP) in 2009. The Quaternary comprises only 0.4% of Earth's history but is crucial for humans because of glacial and interglacial events. The oxygen isotope studies in the microfossils of the deep marine sediments have revealed the glacial and interglacial cycles at 100 ka, 43 ka, 24 ka and 19 ka for the past 700 ka due to Milankovitch cycles. Therefore, it is important to reconstruct the past climate and environmental changes and model them to predict such events which may happen in future. Accelerator Mass Spectrometry (AMS) is an ultrasensitive technique to handle a very small sample size and provides isotopic ratios for many elements which help understand the evolution of Quaternary landforms and climate. *Singh and Kumar* in their chapter have discussed the AMS and different cosmogenic radionuclides (CRN) which help reconstruct past climate and tectonic events. Further, different geomorphological features and landforms such as loess, lakes, river terraces, ice cores, etc., are also discussed, where cosmogenic radionuclide proxy can be applied to understand the past changes in the climate on a global or regional scale. Denudation rates are an important proxy to understand the palaeo-humidity and temperature conditions and the role of CRNs to estimate accurate palaeo-denudation rates has also been discussed in their chapter.

Thus, altogether the present book emphasizes scientifically significant aspects of ongoing climate change and addresses specific issues of climate change, its science, policies and associated geopolitics largely from India's point of view. It is hoped that this book will serve its purpose and act as a ready reckoner to all policymakers, researchers and the scientific community.

New Delhi, India
July 2022

Neloy Khare

Acknowledgements

It is my great pleasure to express my gratitude and deep appreciation to all contributing authors. Without their valuable inputs on various facets of climate variability over the Antarctic and surrounding Southern Ocean region, the book would not have been possible. Various learned experts who have reviewed different Chapters are graciously acknowledged for their timely, constructive and critical reviews.

I sincerely thank Dr M. Ravichandran Secretary Ministry of Earth Sciences, Government of India, New Delhi (India), for their support, and encouragement. Prof. Govardhan Mehta, FRS has always been a source of inspiration and is acknowledged for his kind support.

This book has received significant support from Akshat Khare and Ashmit Khare, who have helped me during the book preparation. Dr Rajni Khare has unconditionally supported enormously during various stages of this book. Shri Hari Dass Sharma from the Ministry of Earth Sciences, New Delhi (India) has helped immensely in formatting the text and figures of this book and bringing it to its present form. Springer Nature has done a commendable job and is sincerely acknowledged.

New Delhi, India | Neloy Khare
July 2022

Contents

Part I Policy and Scenarios

1 **Climate Change Science, Policies, National Perspectives in the Global Context** 3
K. J. Ramesh

2 **Climate Changes over the Indian Subcontinent: Scenarios and Impacts** 27
R. K. Mall, Nidhi Singh, Subhi Patel, Saumya Singh, Aman Arora, R. Bhatla, R. S. Singh, and P. K. Srivastava

3 **Assessing the Historical and Future Relationship Between Climatic Factors and the Production of Different Crops over India** 53
Bushra Praveen, Kanak Singh, Pritee Sharma, Shaghla Parveen, and Swapan Talukdar

Part II Geopolitics and Conflicts

4 **Geopolitics of Climate Change and Water Security in South Asia: Conflict and Cooperation** 77
Dhanasree Jayaram and Gaurica Sethi

Part III Water (Water/Sea-Level Rise/Precipitation)

5 **Vulnerability Analysis of Sea Level Rise (SLR) on a Micro-tidal Coast: A Case Study of Nagapattinam Coast** 91
R S Kankara, Padmini Gunasekaran, and Dhanalakshmi Silamban

6 **Lacustrine Records of Anthropogenic Change and Precipitation in the Monsoonal Core Zone of Central India since the Last Century** 109
Samaya S. Humane, Sumedh K. Humane, and Snehal Juare

7 Anthropogenic Interference and Climatic Changes Influencing Precipitation, Sediment Load and Sand Deposition by the Phalgu River in and Around Jahanabad District, Bihar: A Case Study . . . 131
Akhouri Bishwapriya, Abhishek Kumar Chaurasia, and V. Aneesh Kumar

Part IV Land (Geology, Archeology/Paleo/Tectonics)

8 Impact of Contrasting Climate on the Intensity of Chemical Weathering and Maturity of Sediments in Different Latitudes: A Geochemical Study from Peninsular and Extra Peninsular India . . . 175
Ishfaq Ahmad Mir

9 Manganese Mineralization in Manganiferous Quartzite in the Boringpadar-Amath Area, Eastern Ghats Mobile Belt, Odisha, India: Implication for Climatic Changes . . . 195
Pawan Kumar Yadav, Manorama Das, and Sradhanjali Subhadarshini

10 Archaeological Evidence of Climate Change: Potential Source to Understand the Past Climate Trend . . . 221
P. D. Sabale

11 A Review Study on Youngest Toba Tuff Ash, Artifacts and Vertebrate Remains Preserved in the Quaternary Alluvial Sediments of India and Their Significances in Climate and Paleoenvironment Interpretations . . . 237
Ajab Singh, Ashok K. Srivastava, Nura Abdulmumini Yelwa, and Neloy Khare

12 Active Tectonics–Climate Interaction in the Structural Depo-Centres of Central Indian Tectonic Zone (CITZ) . . . 265
Snigdha Ghatak and Mriganka Ghatak

13 Application of Cosmogenic Radionuclides in the Quaternary Sciences Using Accelerator Mass Spectrometry . . . 297
Atul Kumar Singh and Pankaj Kumar

Glossary . . . 313

Index . . . 321

Contributors

V. Aneesh Kumar Central Ground Water Board, Patna, India

Aman Arora DST-Mahamana Centre of Excellence in Climate Change Research, Institute of Environment and Sustainable Development, Banaras Hindu University, Varanasi, India

R. Bhatla Department of Geophysics, Banaras Hindu University, Varanasi, India

Akhouri Bishwapriya Geological Survey of India, Patna, India

Abhishek Kumar Chaurasia Geological Survey of India, Patna, India

Manorama Das Geological Survey of India, SU: Bihar, Patna, India

Mriganka Ghatak Geological Survey of India, Kolkata, West Bengal, India

Snigdha Ghatak Geological Survey of India, Kolkata, West Bengal, India

Padmini Gunasekaran National Centre for Coastal Research (NCCR), Ministry of Earth Sciences, Government of India, NIOT-Campus, Chennai, India

Samaya S. Humane Department of Geology, Rashtrasant Tukadoji Maharaj Nagpur University, Law College Square, Nagpur, Maharashtra, India

Sumedh K. Humane Department of Geology, Rashtrasant Tukadoji Maharaj Nagpur University, Law College Square, Nagpur, Maharashtra, India

Dhanasree Jayaram Manipal Academy of Higher Education, Manipal, India
Alexander von Humboldt Foundation, International Climate Protection Fellow, Berlin, Germany

Snehal Juare Department of Geology, Yashwantrao Chawhan Art, Commerce and Science College, Lakhandur, Maharashtra, India

R S Kankara National Centre for Coastal Research (NCCR), Ministry of Earth Sciences, Government of India, NIOT-Campus, Chennai, India

Neloy Khare Ministry of Earth Science, New Delhi, India

Pankaj Kumar Inter-University Accelerator Centre, New Delhi, India

R. K. Mall DST-Mahamana Centre of Excellence in Climate Change Research, Institute of Environment and Sustainable Development, Banaras Hindu University, Varanasi, India

Ishfaq Ahmad Mir Geological Survey of India, State Unit: Karnataka and Goa, Bengaluru, Karnataka, India

Shaghla Parveen Department of Geography, Aligarh Muslim University, Aligarh, India

Subhi Patel DST-Mahamana Centre of Excellence in Climate Change Research, Institute of Environment and Sustainable Development, Banaras Hindu University, Varanasi, India

Bushra Praveen School of Humanities and Social Sciences, Indian Institute of Technology Indore, Indore, India

K. J. Ramesh Former Director General, India Meteorological Department, Bengaluru, India

P. D. Sabale Department of Archaeology, Deccan College Deemed University, Pune, India

Gaurica Sethi Department of Geopolitics and International Relations, Manipal Academy of Higher Education, Manipal, India

Pritee Sharma School of Humanities and Social Sciences, Indian Institute of Technology Indore, Indore, India

Dhanalakshmi Silamban National Centre for Coastal Research (NCCR), Ministry of Earth Sciences, Government of India, NIOT-Campus, Chennai, India

Ajab Singh Department of Geology, Faculty of Science, University of Malaya, Kuala Lumpur, Malaysia

Atul Kumar Singh Inter-University Accelerator Centre, New Delhi, India

Kanak Singh School of Humanities and Social Sciences, Indian Institute of Technology Indore, Indore, India

Nidhi Singh DST-Mahamana Centre of Excellence in Climate Change Research, Institute of Environment and Sustainable Development, Banaras Hindu University, Varanasi, India

R. S. Singh Department of Geophysics, Banaras Hindu University, Varanasi, India

Saumya Singh DST-Mahamana Centre of Excellence in Climate Change Research, Institute of Environment and Sustainable Development, Banaras Hindu University, Varanasi, India

P. K. Srivastava DST-Mahamana Centre of Excellence in Climate Change Research, Institute of Environment and Sustainable Development, Banaras Hindu University, Varanasi, India

Sradhanjali Subhadarshini Geological Survey of India, SU: Odisha, Bhubaneswar, India

Swapan Talukdar Department of Geography, University of Gour Banga, Malda, India

Pawan Kumar Yadav Geological Survey of India, SU: Bihar, Patna, India

Nura Abdulmumini Yelwa Department of Geology, Faculty of Science, University of Malaya, Kuala Lumpur, Malaysia
Department of Geology, Usmanu Danfodiyo University, Sokoto, Nigeria

About the Editor

Neloy Khare presently Adviser/Scientist 'G' to the Government of India at MoES has a very distinctive acumen not only in administration but also in quality science and research in his areas of expertise covering a large spectrum of geographically distinct locations like the Antarctic, Arctic, Southern Ocean, Bay of Bengal, Arabian Sea, Indian Ocean, etc. Dr. Khare has almost 30 years of experience in the field of paleoclimate research using paleobiology/palaeontology/teaching/science management/administration/coordination for scientific programmes (including Indian Polar Programme), etc. Having completed his doctorate (PhD) in tropical marine regions and Doctor of Science (D.Sc) in Southern High latitude marine regions towards environmental/climatic implications using various proxies including foraminifera (micro-fossil), has made significant contributions in the field of Paleoclimatology of Southern high latitude regions (the Antarctic and the Southern Ocean) using Micropaleontology as a tool. These studies coupled with his paleoclimatic reconstructions from tropical regions helped understand causal linkages and teleconnections between the processes taking place in Southern high latitudes with that of climate variability occurring in tropical regions. Dr. Khare has been conferred Honorary Professor and Adjunct Professor by many Indian Universities.

He has a very impressive list of publications to his credit (125 research articles in National and International Scientific journals; 3 Special Issues of National Scientific Journals as Guest Editor; Edited Special Issue of Polar Science (Elsevier), Journal of Asian Earth Science (Elsevier), Quaternary International (Elsevier) and Frontiers in Marine Science as its Managing/Guest Editor. Authored/edited many books, 130 Abstracts have been contributed to various seminars; 23 Popular Science Articles; 5 Technical reports). The government of India and many professional bodies have bestowed upon him many prestigious Awards for his humble scientific contributions to Past climate changes/Oceanography/Polar Science and Southern Oceanography. The most coveted award is the Rajiv Gandhi National Award in 2013 conferred by the Honourable President of India Others include the ISCA Young Scientist Award, BOYSCAST Fellowship, CIES French Fellowship, Krishnan Gold Medal, Best Scientist Award, Eminent Scientist Award, ISCA Platinum Jubilee

Lecture, IGU Fellowship, besides many more. Dr. Khare has made tremendous efforts to popularize ocean science and polar science across the country by way of delivering many invited lectures, radio talks and publishing popular science articles.

Dr. Khare has also sailed in the Arctic Ocean as a part of 'Science PUB' in 2008 during the International Polar Year campaign for scientific exploration and became the first Indian to sail in the Arctic Ocean.

Part I
Policy and Scenarios

Chapter 1
Climate Change Science, Policies, National Perspectives in the Global Context

K. J. Ramesh

Abstract India signed the United Nations Framework Convention on Climate Change (UNFCCC) in 1992 and ratified it in 1993. In general, UNFCCC efforts supported the developing countries in combating climate change by providing a platform for finance, technology transfers, discussions, global partnerships, etc., not seeking emission reduction commitments in recognition of their small contribution to global warming as well as low financial and technical capacities. UNFCCC initiatives helped creating public awareness regarding climate change, which is much higher today than in the late 1990s. Although climate science in the late 1990s was certainly strong enough to negotiate an international treaty, it is hard to deny that the scientific understanding of the climate crisis has improved considerably over the past two decades through IPCC in which UNFCCC played a significant role. The Ministry of Environment and Forest and Climate Change (MOEF & CC) is the main organization responsible for climate change policy in India.

The Kyoto Protocol under the heads of Common but Differentiated Responsibilities (CBDR) says that the developed countries should take more responsibility and liability to take care of our degrading environment and make appropriate rules to save it for future generations by taking legally binding commitments. A major shortcoming of the Kyoto Protocol was the non-inclusion of dangerous microscopic environmental air pollutants that come from car engines and combustion-based power plants. Russia joined the Kyoto Protocol in 2005 but the USA was never a part of it (and Canada subsequently dropped out of the Protocol).

Now the world has reached almost 1.2 °C warming already, and UNFCCC needs to deal with unsustainable commitments as the Paris 2015 NDC contributions are found to be inadequate to maintain 2 °C levels. Although low-lying countries that are facing unsustainable sea level rises in a warming world argued for a tougher target of 1.5 °C, a largely unsatisfactory response was seen till the proceedings of the COP-26 held at Glasgow, UK, in 2021. The main challenge faced by UNFCCC is to ensure all rich nations that pledged to maintain a $100bn a year funding beyond 2020

K. J. Ramesh (✉)
Former Director General, India Meteorological Department, Bengaluru, India

N. Khare (ed.), *Science, Policies and Conflicts of Climate Change*, Springer Climate,
https://doi.org/10.1007/978-3-031-16254-1_1

contribute fully from now on at least, although several Pacific island countries advocated its inadequacy to fully meet the adverse impacts of climate change.

Keywords Climate change policy shifts · Kyoto Protocol · Paris Agreement · Environmental laws · Montreal Protocol · Nationally determined contributions · Climate finance

Sustainable management and preservation of the natural environment are now becoming highly critical for human existence on the planet Earth. The United Nations recognized this as a paramount issue of global concern and convened the Stockholm Conference way back in 1972 to inspire and guide global actions holistically. Since then, several global conferences and summits have been held and many multilateral agreements/treaties/declarations adopted to promote sustainable use and management of natural resources and environmental protection globally by making the environment a major issue on the international development agenda and involving developed and developing countries alike as far as the global collective action is concerned. Consequently, countries across the globe have set up ministries or departments of the environment to build their respective national development agenda emphasizing environmental protection. Gradually, countless national level environmental laws and regulations have been enacted by member states of the United Nations for shaping the international environmental legal regimes to reverse environmental degradation on our planet.

The developing countries have always asserted that major environmental issues of serious concern identified by the Stockholm Conference such as industrial pollution, uncontrolled urban development, and wilderness degradation, were attributable to unrestrained industrialization in the United States, Europe and the Soviet Union. The developing countries have always been concerned about the possible impacts of industrial pollution on the developed world upon over and above the legacy issues such as access to clean water, malnutrition, proper sanitation and economic development. The Stockholm Conference gave birth to the concept of sustainable development, commonly known as 'eco-development' in those days, as a way of reconciling the competing claims of economic development and environmental protection. The hallmark outcome of the Stockholm Conference is the establishment of the United Nations Environmental Program (UNEP) within the United Nations system. UNEP has since served as an institutional catalytic agency for the promotion and coordination of global environmental activities. The world owes much to the Stockholm Conference because it was the catalysis for the new multilateral environmental cooperation and treaty-making era that followed. The Stockholm Declaration established international political goals and legal principles that have underpinned environmental discourse and lawmaking. By stressing that environmental issues are inherently political—not just scientific and technical, as many policymakers previously thought—and therefore emphasized the need for political negotiations and decision-making, the Stockholm Conference demonstrated how global cooperation could take place, even amid Cold War tensions.

Development of Environmental Law in India Post-Stockholm Declaration

This Stockholm Declaration is considered to be the Magna Carta of environmental law and is a significant parallel to that of the Universal Declaration on Human Rights of 1948. The Stockholm conference was attended by the then Prime Minister of India, Mrs. Indira Gandhi and she was greatly inspired by the discussions held, issues raised and developments that took place during that Conference. The results have reflected through quick developments that occurred immediately after the Stockholm Conference. The 42nd Amendment to the Indian Constitution, which was introduced after the Stockholm Conference of 1972, brought major changes to the Constitution of India and is considered to be the *mini constitution.* This Amendment brought changes to various subjects, but when we talk about the changes to the Environmental law, it introduced responsibility on part of both states and citizens to protect and improve the environment. Recognition of the right to a pollution free environment as part of a fundamental right under Article 21 of the Indian constitution, various rulings and observations of the Supreme Court of India on various matters on environmental pollution and degradation have also led to the introduction of various enactments for the protection of various subjects regarding the environment and air quality; all of which brought needful changes in the governance to protect the rights of the citizens at large.

To a great extent, in a broader perspective in India, the environmental legislation enacted after the Stockholm Conference was found to be mere rhetoric and ineffective in reality due to weak enforcement of laws perceived to inhibit industrial growth (Chandra 2015). Indeed, in India, even basic laws concerning urban planning and fundamental hygiene and public health were not given due priority and focus in the overall governance.

Significant United Nations engagements with environmental sustainability in the post-Stockholm era until the establishment of an exclusive global negotiating UN platform for planetary climate include

- The first World Climate Conference (WCC) held in Geneva in 1979 by World Meteorological Organization (WMO) led to the establishment of the World Climate Research Program (WCRP) in 1980 to promote global research on important climate issues, including ozone depletion and global warming.
- The Vienna Convention for the Protection of the Ozone Layer (1985) provided a global framework to protect the ozone layer from the illeffects of human activities that subsequently led to the adoption of the Montreal Protocol for global action in 1987 to reduce the production and consumption of ozone depleting substances.
- The Inter-Governmental Panel on Climate Change (IPCC) was established in 1989 with a functional Secretariat at WMO to build focused activities on scientific assessment of all aspects of climate change and their adverse environmental and socio-economic impacts to provide robust guidance for the formulation of realistic response strategies.

- The Second WCC held in 1990 was an important step towards a global climate treaty (Zillman 2009). The Conference was held at a crucial time in the climate treaty negotiation process. IPCC Published the First Assessment Report (FAR) in 1990 that warned of the impending threat of global warming and its adverse impacts on natural and human systems. The report consisted of three volumes and an overview, namely (1) Climate Change: The IPCC Scientific Assessment, (2) Climate Change: The IPCC Impacts Assessment, (3) Climate Change: The IPCC Response Strategies, and (4) FAR Overview. The FAR of IPCC had provided for the first time all necessary critical inputs for the first session of the International Negotiating Committee for a Framework Convention on Climate Change (INC)
- UN Conference on Environment and Development (UNCED) (known popularly as the Earth Summit/Rio Summit) was held in 1992 in Rio de Janeiro, Brazil. The United Nations Framework Convention on Climate Change (UNFCCC) established an international environmental treaty to combat 'dangerous human interference with the climate system', in part by stabilizing greenhouse gas concentrations in the atmosphere along with many other agreements such as a legally binding multilateral agreement on Convention on Bio-Diversity (CBD). UNFCCC established a Secretariat headquartered in Bonn and came into existence on 21 March 1994.
- The United Nations Convention on the Law of the Sea (UNCLOS) came into force in 1994 and aimed to prevent, reduce and control pollution of the marine environment, and protect fish stocks from depletion.
- IPCC released its Second Assessment Report (SAR) in 1995 comprising (1) Working Group I: The Science of Climate Change, (2) Working Group II: Impacts, Adaptations and Mitigation of Climate Change: Scientific-Technical Analyses, (3) Working Group III: Economic and Social Dimensions of Climate Change, and (4) IPCC Second Assessment. The essential message of the SAR was that carbon dioxide remains the most important contributor to global climate change with a warning that uncontrolled human activities led to greenhouse gas (GHGs) emissions which threatened to alter the Earth's climate to an unprecedented extent in human history with destructive impacts such as sea level rise, extreme events, etc.

Perspectives of the Global Climate Change Negotiation Agenda Discourse

While climate change was recognized by the UN General Assembly as the 'common concern of mankind' in 1988, the UNFCCC specifically acknowledges this 'common concern' and placed it at the centre of the international climate change negotiation regime. It promotes a precautionary approach to addressing climate change and articulating the principle of inter-generational equity by paying special attention to protecting the global climate system in the best interests of future generations. UNFCCC also incorporates Principle 7 of the Rio Declaration on the Environment and Development on the principle of 'common but differentiated responsibilities' in addressing climate change.

The basis of the UNFCCC negotiated Convention of 1992 was precisely to note that the developed countries have historically been responsible for the harmful greenhouse gas (GHG) emissions and remain their main emitters. Always on a per capita basis, the primary responsibility devolves on them to take the tough measures needed to address the climate change issue using their strong financial resources and high technological capacities.

The negotiated Convention envisaged that all of the GHG stabilization in the environment be achieved within a specific time frame so that the new adapted environment would not disturb the production of food grains and thereby ensure the growth of humans and our environment together in a sustainable manner. While negotiating the convention, although it was very difficult to prove the existence of ongoing environmental degrading activities all over the world due to lack of scientific evidence, UNFCCC took inspiration from the Montreal Protocol of 1987 and convinced the member countries to work towards the human safety and protection that was only possible when our environment is safe from all the polluting activities. Thus, the Convention maintained a differentiation between the countries according to their development and their contribution to the GHG emissions by categorizing them as developed, developing and underdeveloped.

- The Annex 1 parties include the industrialized countries that were members of the Organisation for Economic Co-operation and Development (OECD) in 1992, plus countries with economies in transition (the EIT Parties), including the Russian Federation, the Baltic States, and several Central and Eastern European States. All of these countries were included in Annex 1 because they are developed and contributed maximum towards the GHG emissions. Annex 1 countries must make National Policies for their own countries so that they can mitigate climate change and then reduce the amount of greenhouse gases in the atmosphere.
- Annex 2 countries include part of the OECD but not part of EIT. These countries are supposed to provide financial resources and support to the other developing countries so that they can make their own GHG emission reduction plans and policies and then help them to adapt to the effects of climate change.
- Then there are the Non-Annex countries that are financially or technically poor and they cannot handle the effects of climate change; thus, they have to be given financial support from other countries to bring in new policies for themselves.

India was placed under the category Non-Annex country according to the UNFCCC. All 197 countries that ratified the 1992 Convention were called Parties to the Convention. The members of this Convention formed themselves into a Conference of Parties (herein referred to as COP) so that they can meet every year and discuss adopting new practices and bringing up new rules so that they can achieve the goals and objectives of UNFCCC. The COP, the supreme decision-making body of the Convention, had built in a mechanism to assess the progress of their actions towards climate change. Annually, all State Parties to the Convention that were represented at the COP meet and review the implementation of the Convention and any other legal instruments that the COP adopts and take decisions necessary to promote the effective implementation of the Convention, including institutional and administrative arrangements.

The UNFCCC framework continues to undertake assumed responsibility to reduce GHG emissions and to build consensus to work collectively to avoid dangerous anthropogenic interference with the climate system. So far, the UNFCCC has been able to provide significantly acceptable guiding principles for world leaders to make progress through effective solutions for addressing the adverse impact of climate change.

The expectation at the UNFCCC platform has always been that the developed world provide new and additional financial resources and technological cooperation on preferential terms to the developing world; those that are seen to face the adverse impacts of a global warning generated by historical GHG emissions, to enable many of them to more effectively respond to climate change. Although the science of climate change was evolving and expected to improve the assessment of impacts at global, regional and national scales by successive IPCC Assessment Reports, the developing/underdeveloped world was seeking due guidance from the subsidiary bodies of UNFCCC viz. Subsidiary Body for Science & Technology Advice (SBSTA) and Subsidiary Body for Implementation (SBI), for building appropriate adaptation measures to minimize the adverse impacts of climate change. At the same time, it was emphasized that the developing world cannot accept binding mitigation commitments, although they could consider entering into contractual agreements to implement mitigation actions provided incremental costs are met by the developed world so that the requirement to divert scarce resources from their overriding priorities of poverty eradication and economic development gets avoided.

International climate change negotiations broadly relied on how the world addresses environmental concerns and shares the costs and benefits of enhanced environmental protection. Such negotiations have always sought to build an international consensus on

(a) identification of GHG emission sources responsible for climate change, both historically and currently;
(b) steps need to be taken under a negotiated umbrella to tackle the problem; and
(c) workout ways and means of bearing the burden of corrective actions between financially sound, technologically advanced, the industrialized world and not so well off developing/underdeveloped world.

COP-2 held at Geneva in 1996 saw an important shift with the USA supporting a 'realistic, verifiable and binding medium term GHG emission target' as long as market-based flexible mechanisms would be included in the future climate treaty. Geneva Ministerial Declaration of COP-2 endorsed the IPCC's SAR and called upon Parties to accelerate negotiations for a legally binding treaty that would set GHG reduction targets and timetable for industrialized (Annex I) countries.

However, momentum towards an agreement that set binding emissions limits for individual countries soon became a hurdle over the argument put forward by most developing countries that rich countries need to shoulder the entire burden of reducing emissions. In the late 1990s, such a continued divide prevented the United States from ratifying the Kyoto Protocol, the main mechanism for reducing global emissions. The U.S. Senate reasoned that unless China and other large developing countries agreed to limit their emissions, an international agreement would be meaningless.

Despite the above, to boost the effectiveness of the UNFCCC, the Kyoto Protocol was adopted in December 1997 at COP-3 in Kyoto, Japan. It commits industrialized countries and countries in transition (Central and Eastern Europe and the former Soviet Union) to a market economy to achieve quantified emissions reduction targets for a basket of six GHGs by an average of 5% below 1990 levels in 2008–2012. The Kyoto Protocol entered into force on 16 February 2005 and has 192 parties. Its first commitment period took place from 2008 to 2012.

Kyoto Protocol a Climate Change Success or a Global Warming Failure?

The Kyoto Protocol was opened for signature in 1998, but it wouldn't come into force until at least 55 parties ratified the treaty and also needed to account for at least 55% of the total CO_2 emissions of the Annex I countries. This political process took 7 years, and it was the ratification of Russia that finally brought the treaty into force in 2005. However, the factual analysis makes everyone believe that the 2005 ratification of the Kyoto Protocol was a calculated political strategy to ensure the success of the Kyoto Protocol before it had even begun.

It is clear from the below chart that emissions came down rapidly after 1990; in fact, by 1994 (4 years before the Kyoto Protocol was even open for signatures), members of the designated group had already reduced emissions by 11.2%, which was largely attributed to the rapid collapse of energy consumption after 1990. This was already well ahead of the Kyoto Protocol target of 4.7% CO_2 reduction by 2012 (Fig. 1.1). The rapid decline of CO_2 emissions during the early 1990s was due to the collapse of the Soviet Union and associated rapid decline in heavy manufacturing industries across Russia and the newly formed independent states. The two largest energy consumers in this group, Russia and Ukraine, had a particularly significant decline in energy consumption. Further, analysis suggests that the collapse of the Soviet Union appears to be the key reason for the decline in energy consumption of the Kyoto Protocol parties as evident from the chart below:

Strikingly, it becomes evident that without Russia and Ukraine (including Canada which pulled out in 2011) the Kyoto Protocol Parties would only have reduced their emissions by a mere 2.7%. This is well below their originally envisaged target. It is only the fact that carbon emissions reduced by Russia + Ukraine together amount to 32.4% that made the headline Kyoto Protocol figures down to a 12.5% reduction (Fig. 1.2).

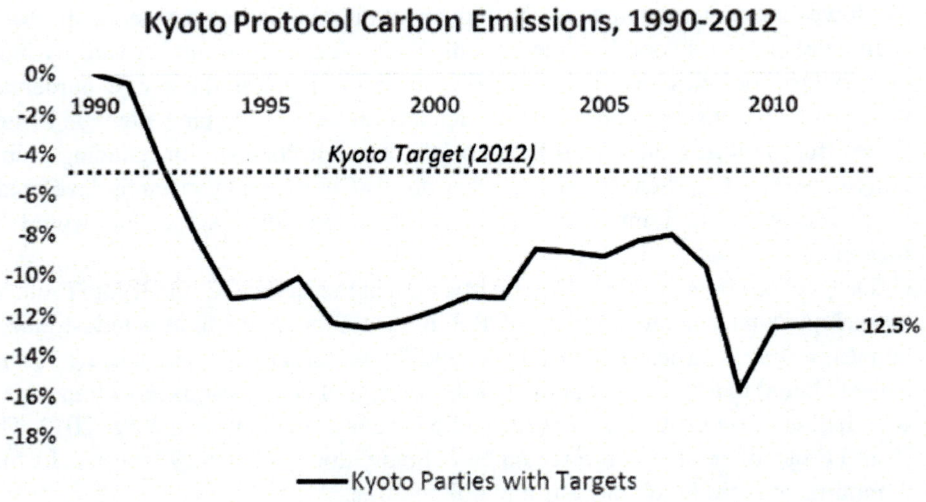

Fig. 1.1 Carbon Emissions during 1990–2012 as per Kyoto Protocol Target (Circular Ecology 2015)

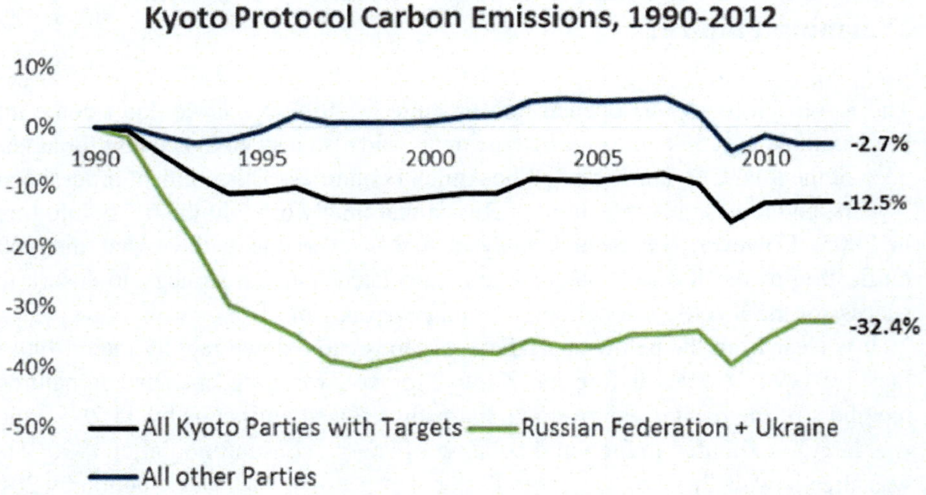

Fig. 1.2 Carbon Emissions Targets during 1990–2012 by All Kyoto Protocol Parties, Russia & Ukraine and All Other Parties

Hence, general headline figures that make the Kyoto Protocol seem like a huge climate change success and a valuable contribution to environmental sustainability ultimately appear to be a global warming failure.

The 2012 Doha Amendment established the second commitment period from 2013 to 2020. The Amendment strengthened quantified emission reduction commitments for developed countries and set a goal of reducing GHG emissions by 18% compared to 1990 levels. The assessment (Kyoto 2020) of the latest information received from Parties with commitments under the Doha Amendment (Annex B

Parties), based on data for the period 1990–2018, shows that total aggregate GHG emissions in 2018 were 25.3% lower than in 1990. It is to be noted that the above assessment only applies to a group of some 37 countries that agreed to emission reduction targets under the Doha Amendment of the Kyoto Protocol. The above outcome demonstrates the potential of consistently implementing climate change policies and actions at the national level.

The continued dissociation of the US towards binding emission reduction commitments was finally bridged in a big way during President Obama's visit to China in early November, when the world's two largest emitters, the United States and China, announced a bilateral agreement in which Washington pledged to reduce emissions by one quarter below 2005 levels by 2025, and Beijing committed to preventing its emissions from growing after 2030. The Sino-American agreement came at a critical time for the global climate negotiations: the Kyoto Protocol, despite America's refusal to ratify it, remained the world's only binding international agreement to reduce emissions. And it was scheduled to expire in 2020. Most climate experts believed that a replacement agreement needed to be formulated by the end of 2015 to allow enough time for it to come into force before 2020, giving the world just 1 year to overcome decades of stalemate in global climate negotiations. The European Union has launched its own attempt to jump-start climate negotiations, announcing it will pursue emissions reductions of 40% below 1990 levels by 2040. Among the world's large economies—and big emitters—that left India as the biggest wild card going into the Lima negotiations of COP-20 in 2014 and the 2015 climate conference COP-21, scheduled to be held in Paris.

In December 2015, parties adopted the Paris Agreement. Under the terms of the Agreement, all countries submit Nationally Determined Contributions (NDCs) and will review the aggregate progress on mitigation, adaptation, and means of implementation every 5 years through a Global Stocktake. The Paris Agreement entered into force on 4 November 2016, and, to date, 193 parties have ratified the Agreement. Through the NDC process, countries have the opportunity to further advance climate policies and actions and ratchet them up over time.

The Gist of Key Turning Points under UNFCCC Negotiations and Associated Climate Policy Shifts

(i) Long-term negotiations during 2005–2009: Convening in Montreal, Canada, in 2005, CMP 1 established the Ad Hoc Working Group on Annex I Parties'. Further Commitments under the Kyoto Protocol (AWG-KP) which mandated consideration of Annex I parties' further commitments at least 7 years before the end of the first commitment period. In December 2007, COP 13 and CMP 3 in Bali, Indonesia, resulted in an agreement on the Bali Roadmap on long-term issues. COP 13 adopted the Bali Action Plan (BAP) and established the Ad Hoc Working Group on Long-term Cooperative Action under the

Convention (AWG-LCA), with a mandate to focus on mitigation, adaptation, finance, technology, capacity building and a shared vision for long-term cooperative action. Negotiations on Annex I parties' further commitments continued under the AWGKP. The deadline for concluding the two-track negotiations was 2009 in Copenhagen, Denmark.

(ii) The UN Climate Change Conference in December 2009 was held in Copenhagen. The event was marked by disputes over transparency and process. After a lengthy debate, delegates ultimately agreed to 'take note' of the Copenhagen Accord and to extend the mandates of the negotiating groups until COP 16 and CMP 6 to be held in 2010. In 2010, over 140 countries indicated support for the Accord and over 80 countries provided information on their national mitigation targets or actions.

(iii) The UN Climate Change Conference convened in December 2010 at Cancun, Mexico, adopted the Cancun Agreements and agreed to consider the adequacy of the long-term global goal during the 2013–2015 review. The Cancun Agreements established several new institutions and processes, including the Green Climate Fund (GCF), the Cancun Adaptation Framework, the Adaptation Committee and the Technology Mechanism, which includes the Technology Executive Committee (TEC) and the Climate Technology Centre and Network (CTCN).

(iv) The negotiating mandate for the Paris Agreement was adopted at the UN Climate Change Conference in 2011 in Durban, South Africa. Parties agreed to launch the Ad Hoc Working Group on the Durban Platform for Enhanced Action (ADP) with a mandate 'to develop a protocol, another legal instrument or an agreed outcome with legal force under the Convention applicable to all Parties' no later than 2015, to enter into force in 2020. In addition, the ADP was mandated to explore actions to close the pre-2020 ambition gap concerning the 2 °C targets.

(v) The UN Climate Change Conference held in 2014 in Lima, Peru, adopted the 'Lima Call for Climate Action', which furthered progress on the negotiations towards the Paris Agreement. It elaborated the elements of a draft negotiating text and the process for submitting and synthesizing intended nationally determined contributions (INDCs), while also addressing pre-2020 ambition.

(vi) The UN Climate Change Conference convened in 2015 in Paris, France, adopted the Paris Agreement on 12 December 2015. The Agreement includes the goal of limiting the global average temperature increase to well below 2 °C above pre-industrial levels and pursuing efforts to limit it to 1.5 °C. It also aims to increase parties' ability to adapt to the adverse impacts of climate change and make financial flows consistent with a pathway towards low GHG emissions and climate-resilient development. The Agreement will be implemented to reflect equity and the principle of common but differentiated responsibilities and respective capabilities, in light of different national circumstances. Under the Paris Agreement, each party shall communicate, at five-year intervals, successively more ambitious NDCs. The Paris Agreement also includes a transparency framework and a process known as the Global

Stocktake. Beginning in 2023, parties will convene this process at five-year intervals to review collective progress on mitigation, adaptation and means of implementation. The Agreement also includes provisions on adaptation, finance, technology, loss and damage, and compliance. When adopting the Paris Agreement, parties launched the Paris Agreement Work Programme (PAWP) to develop the Agreement's operational details. Parties also agreed on the need to mobilize stronger and more ambitious climate action by all parties and non-party stakeholders to achieve the Paris Agreement's goals. Several non-party stakeholders made unilateral mitigation pledges in Paris, with more than 10,000 registered actions.

The Paris Agreement entered into force on 4 November 2016, which was 30 days after the dual entry into force requirement of ratification by at least 55 countries representing at least 55% of global GHG emissions was met. As of 19 November 2016, 111 countries have ratified the agreement.

(vii) The UN Climate Change Conference took place in 2016 at Marrakech and adopted several decisions related to the PAWP, the terms of reference for the Paris Committee on the Capacity building; and initiated a process to identify the information to be provided following Paris Agreement related to ex-ante biennial finance communications by developed countries. Other decisions adopted included approving the five-year work plan of the Warsaw International Mechanism on Loss & Damage (WIM) associated with Climate Change Impacts, enhancing the Technology Mechanism, and continuing and enhancing the Lima work programme on gender.

(viii) The Fiji/Bonn Climate Change Conference convened in 2017 at Bonn, Germany, under the COP Presidency of Fiji. The COP launched the Talanoa Dialogue, a facilitative dialogue to take stock of collective progress towards the Paris Agreement's long-term goals. The COP also established the 'Fiji Momentum for Implementation', a decision giving prominence to pre-2020 implementation and ambition. Parties also guided the completion of the PAWP and decided that the Adaptation Fund shall serve the Paris Agreement.

(ix) The Katowice Climate Change Conference convened in 2018 at Katowice, Poland, concluded a busy and hectic year of negotiations that featured an additional negotiation session to advance work on the Paris Agreement Work Programme (PAWP). Parties adopted the Katowice Climate Package, which finalized nearly all of the PAWP, including decisions to facilitate common interpretation and implementation of the Paris Agreement on the mitigation section of NDCs, adaptation communications, transparency framework, Global Stocktake, and financial transparency, among others. The COP was unable to agree on whether to 'welcome' or 'note' the Intergovernmental Panel on Climate Change's (IPCC) Special Report on 1.5 °C of Global Warming.

(x) The Chile/Madrid Climate Change Conference convened in 2019 in Madrid, under the COP Presidency of Chile. Decisions were adopted on the review of the WIM and some finance-related issues, such as guidance to the Global Environment Facility (GEF) and Green Climate Fund (GCF). Parties also adopted three decisions each named the Chile/Madrid Time for Action.

(xi) The global COVID-19 pandemic disrupted the normal meeting cycle. Online sessions were held in June and November 2020 to hear updates from the constituted bodies and hold mandated events. The Climate Ambition Summit in December 2020 served as a platform for countries to put forward new NDCs and net zero pledges. A subsequent online summit was convened in April 2021 by US President Biden.

India's Stand at UNFCCC

India consistently highlighted issues relating to equity, historical responsibilities and per capita emissions as the basis for a differentiated approach to the collective arrangements being considered. The notions of fairness, justice and equity underlying its differentiation between developed and developing countries in terms of responsibilities and capabilities remain as relevant today as when first agreed to at the time of negotiation of the Convention as summarized below (Table 1.1) following Mohammad Mohnish (2016).

Climate Change Centric Environment Policy of India: Post COP-26 Focus

India continued to emphasize that developed nations have not only failed to meet the US$ 100 billion goals annually of support to developing nations since 2009 but also continue to present it as the ceiling of their ambition to 2025. India is pursuing the principle of 'One Sun, One World, One Grid' (OSOWOG) intending to harness solar energy wherever the sun is shining to ensure that generated electricity flows to areas that need it most. India also pressed the issue of emission reduction to combat climate change at the 16th G20 Summit held in Naples, where it urged the Group to bring down per capita emissions to the global average by 2030 given the fast-depleting available carbon space. Several initiatives are taken to tackle the menace of toxic air through the National Clean Air Programme (NCAP).

India entered into Climate Action and Finance Mobilization Dialogue (CAFMD) with the US under the India–US Climate Clean Energy Agenda 2030 that aimed to provide both countries an opportunity to renew collaborations on climate change while addressing the financing aspects. Various climate change actions should address all anticipated adverse impacts on India due to threats emanating from the higher rate of Indian Ocean warming than other oceans, increased heat waves, heavy rainfall and flooding. India ratified the Kigali Amendment for phasing out hydrofluorocarbons (HFCs) under the Montreal Protocol on Substances that Deplete the Ozone Layer.

Table 1.1 Major outcome and India's during different meetings of Conference of Parties (CoPs)

Conference of Parties (COP)	Key outcomes	India's stance
COP 1 (Berlin, 1995)	COP-1 initiated a two-year process to negotiate another legal instrument to galvanize 'appropriate action for the period beyond 2000 by establishing a special negotiating body, the Ad Hoc Group on the Berlin Mandate (AGBM)	India's stand was based on the pragmatic view that the emission reduction must be initiated by the industrialized countries to provide necessary space for the developing countries to increase their emission India pressed for a 20% cut in GHG emissions of the industrialized countries by the year 2005
COP 2 (Geneva, 1996)	The Geneva Ministerial Declaration was noted, but not adopted A decision on guidelines for the national communications to be prepared by developing countries was adopted Also discussed, Are quantified Emissions Limitation and Reduction Objectives (QELROs) (Zillman 2009) for different Parties and an acceleration of the Berlin Mandate talks so that commitments could be adopted at COP-3 To evaluate the progress made since COP-1 Oil producing countries & industrialized groups were opposed throughout the meeting	India along with other developing countries has pursued industrialized countries' emission cuts and associated targets
COP 3 (Kyoto, 1997)	The Kyoto Protocol (Chandra 2015) was adopted by consensus The Kyoto Protocol includes legally binding emission targets for a developed country, (Annex I) Parties for the six major greenhouse gases, are to be reached by the period 2008–2012 Most industrialized countries and some central European economies in transition (all defined as Annex II countries) agreed to legally binding reductions in greenhouse gas emissions of an average of 6–8% below 1990 levels between the years 2008–2012, defined as the first emissions budget period	India endorsed the Kyoto Protocol as an Annex II member
COP 4 (Buenos Aires, 1998)	Parties adopted a 2-year 'Plan of Action' (IPCC 2021) to advance efforts and to devise mechanisms for implementing the Kyoto Protocol, to be completed by 2000	India echoed to support global efforts to protect the environment without hindering India's development pathways

(continued)

Table 1.1 (continued)

Conference of Parties (COP)	Key outcomes	India's stance
COP 5 (Bonn, 1999)	Primarily it was a technical meet A focus on the adoption of the guidelines for the preparation of national communications by Annex I countries, capacity-building (IPCC 2018), transfer of technology and flexible mechanisms	India pointed out that the United Nations Framework Convention on Climate Change (UNFCCC) (Circular Ecology 2015) too acknowledges that development and poverty eradication are the first and the overriding priorities of the developing countries, consequent to which their greenhouse gas (GHG) emissions would grow Further, it was pointed out that the UNFCCC explicitly underlined the conspicuous North-South disparities by recognizing the 'common but differentiated responsibilities' (CBDR) and respective capabilities. Thus, while underpinning the principle of Equity, the Convention differentiated the levels of commitment for the developed and the developing countries
COP 6 (The Hague, 2000) Part II of the sixth COP (Bonn, 2000)	The consensus was finally reached on the so-called Bonn Agreement Work was also completed on a number of detailed decisions of the Bonn Agreements, including capacity-building for developing countries and countries with economies in transition	India supported the resurrection of the Kyoto Protocol in part II of the COP
COP 7 (Marrakech, 2001)	Parties agreed on a package deal, with key features including rules for ensuring compliance with commitments, consideration of removal of emissions on account of land use, land-use change and forestry (LULUCF) Principles in reporting of such data and limited banking of units generated by sinks under the Clean Development Mechanism (CDM) The meeting also adopted the Marrakech Ministerial Declaration as an input into the World Summit on Sustainable Development in Johannesburg	India played a major role in the codification of the Kyoto Protocol at this meeting for the implementation of concrete measures to safeguard the climate. Further highlighted the fact that the impacts of climate change will affect the developing countries more adversely than the developed countries and called for greater attention to be paid globally to targeting the adaptation needs of the developing countries
COP 8 (New Delhi, 2002)	The Delhi Ministerial Declaration on Climate Change and Sustainable Development reiterated the need to build on the outcomes of the World	India emphasized the need for financial resources to help developing countries to adopt to the adverse impact of climate change

(continued)

Table 1.1 (continued)

Conference of Parties (COP)	Key outcomes	India's stance
	Summit Delegates agreed on principles for the financing of a fund to help the poorest nations cope with the adverse effects of climate change The parties approved a mechanism to incorporate forest protection into the effects of the international community to combat climate change	India Pointed out that climate change and sustainable development are interlinked and it was necessary to focus on poverty, land degradation, access to water and food and human health to effectively address climate change concerns As always, India remained firm in its rejection of emission commitment for developing countries India fully embraced CDM
COP 9 (Milan, 2003)	Adopted decisions focused on the institutions and procedures of the Kyoto Protocol The formal decisions adopted by the Conference intend to strengthen the institutional framework of both the Convention and the Kyoto Protocol New emission reporting guidelines based on the good-practice guidance by the Intergovernmental Panel on Climate Change (IPCC) were adopted to provide a sound and reliable foundation for reporting on changes in carbon concentrations resulting from land-use changes and forestry Two funds, the Special Climate Change Fund and the Least Developed Countries Fund were developed to support technology transfer, adaptation projects and other activities	India called for the removal of obstacles – financial & Intellectual Property Rights (IPRs), for the successful adaptation of emerging technology options in energy efficiency, fuel switching & renewable energy India strongly advocated that all facets of sustainable development must contribute positively to the dominant challenge of our times of poverty alleviation
COP 10 (Buenos Aires, 2004)	Discussed the progress made since the first Conference of the Parties, till then over a decade, and its future challenges with special emphasis on climate change mitigation and adaptation The Buenos Aires Plan of Action was adopted to promote developing countries' better adapt to climate change Discussions were made on the Post-Kyoto mechanisms on how to allocate emission reduction obligations following 2012 when the first commitment period ends	India stated that the immediate priority must be to put the Kyoto mechanism fully into operation at the soonest possible Observed that transfer of environment-friendly technologies to developing countries remained a major concern Further suggested that technologies for mitigation of GHGs & adaption to the impact of climate change be made freely accessible for use by developing countries

(continued)

Table 1.1 (continued)

Conference of Parties (COP)	Key outcomes	India's stance
COP 11 (Montreal, 2005)	The COP agreed on a process for considering future action beyond 2012 under the UNFCCC The Montreal Protocol was established to extend the life of the Kyoto Protocol beyond its 2012 expiration date and negotiate deeper cuts in greenhouse-gas emissions	India played an important role in drafting the decision providing further guidance relating to the CDM
COP 12 (Nairobi, 2006)	A wide range of decisions was adopted at COP-12 designed to mitigate climate change and help countries adapt to the impacted adverse effects There was agreement on the activities for the next few years under the 'NairobiworkprogramonImpacts, VulnerabilityandAdaptation', as well as on the management of the Adaptation Fund under the Kyoto Protocol Parties welcomed the 'Nairobi Frame work' which will provide additional support to developing countries in successfully developing projects for the CDM Parties in Nairobi also adopted rules of procedure for the Kyoto Protocol's Compliance Committee, making it fully operational	India and other major developing countries joined in vigorously challenging any dialogue on taking on binding commitments and therefore were blamed for halting reckonable advancement on the road to new agreements on international action beyond 2012 when the Kyoto commitments were planned to expire
COP 13 (Bali, 2007)	This COP resulted in the Adaptation of the Bali Action Plan It calls for a shared vision for long-term cooperative action on four key elements: Mitigation, Adaptation, Finance, and Technology The Ad Hoc Working Group on Long-term Cooperative Action under the Convention (AWG-LCA) was established as a new subsidiary body to conduct the negotiations aimed at urgently enhancing the implementation of the Convention up to and beyond 2012	India demanded that the rich countries compensate developing countries for the afforestation drive & for avoiding deforestation, this demand was accepted by the key decision making Contact Group of the conference Further, India highlighted the poor record of rich countries in emission cuts of GHGs, which was backed by the UN data
COP 14 (Poznan, 2008)	Adaptation Fund was launched under the Kyoto Protocol, to be filled by a 2% levy on projects under the CDM Parties agreed that the Adaptation Fund Board should have the legal capacity to grant direct access to developing countries	India's stand was that developed countries must sharply reduce their emission to release atmospheric space for the development of poorer countries in a manner that is consistent with the achievements of stabilization of GHGs concentration in the atmosphere

(continued)

Table 1.1 (continued)

Conference of Parties (COP)	Key outcomes	India's stance
	Also saw Parties endorse an intensified negotiating schedule for 2009	On technology issue, India had called for strategic support for the transfer of technology that can accelerate carbon mitigation in developing countries and compensation for all additional costs related to the accelerated transfer of technology to promote mitigation & adaptation
COP 15 (Copenhagen, 2009)	It produced the Copenhagen Accord which was supported by a majority of countries A number of developing countries agreed to communicate their efforts to limit GHG emissions every 2 years On long-term finance, developed countries agreed to support a goal of mobilizing US$100 billion a year by 2020 to address the needs of developing countries	The Environment Minister of India stated at the Conference that India's entire approach is anchored in the sanctity of the troika; the UNFCCC, the Kyoto Protocol and the Bali Action Plan. India believes that the well-known and widely accepted principles of common but differentiated responsibilities (CBDR) and historical responsibilities are sacrosanct
COP 16 (Cancun, 2010)	Produced the Cancun Agreements Parties agreed to: commit to a maximum temperature rise of 2 °C above pre-industrial levels; make a fully operational technology mechanism by 2012 to boost the development and spread of new climate-friendly technologies; establish a Green Climate Fund Also agreed on a new Cancun Adaptation Framework, which included setting up an Adaptation Committee to promote strong, cohesive action on adaptation	India along with South Africa and other Like-minded countries submitted a proposal calling for Annex I parties to agree to at least a 40% emission reduction commitment by 2020 as compared to 1990 levels Further advocated for the strengthening of clean development mechanism (CDM)
COP 17 (Durban, 2011)	At COP 17, Parties decided to adopt a universal climate agreement by 2015, with work beginning under a new group called the Ad Hoc Working Group on the Durban Platform for Enhanced Action (ADP) Parties also agreed to a second commitment period of the Kyoto Protocol from 1 January 2013. A significantly advanced framework for the reporting of emission reductions for both developed and developing countries was also agreed upon, taking into consideration the principle of common but differentiated responsibilities	India proposed three agenda items for consideration by COP. These were issues of equity, unilateral actions & technology related IPRs Played a key role in establishing the second commitment period under the Kyoto Protocol As it was also decided to begin a process for developing legal arrangements for enhancing the actions of all parties under the Convention. India ensured that the new arrangements, which have to be decided by 2015 and implemented from 2020 are established under the

(continued)

Table 1.1 (continued)

Conference of Parties (COP)	Key outcomes	India's stance
		Convention India highlighted the issues of equity and CBDR in the climate change negotiations With the support of India, the Green Climate Fund was also established
COP 18 (Doha, 2012)	Parties set out a timetable to adopt a universal climate agreement by 2015, to come into effect in 2020 Parties also completed the work under the Bali Action Plan to concentrate on new work towards a 2015 agreement under a single negotiating stream, the ADP Emphasized the need to increase their ambition to cut greenhouse gases and to help vulnerable countries to adapt Also saw the launch of a second commitment period under the Kyoto Protocol, from 1 January 2013 to 31 December 2020, with the adoption of the Doha Amendment to the Kyoto Protocol	India pursued the strategy of working together with the Group of 77 & China to protect the overall interest of developing countries Also raised the issue of equity in climate change related action & commitments, technology-related, IPRs & unilateral measures taken by some countries in the name of climate change Further succeeded in having these issues included in the ongoing work of various bodies of the Convention
COP 19 (Warsaw, 2013)	Governments advanced the timeline for the development of the 2015 agreement It was decided to either begin or intensify domestic preparations for their Intended Nationally Determined Contributions (INDC) towards the agreement so that they are ready well before December 2015 and ideally by the first quarter of 2015. This is an important part of the timeline of the negotiations It was also decided that nationally determined contributions would be put forward clearly and transparently. Developed country governments were urged to provide support to developing countries for this important domestic process Developed countries were keen on ensuring that climate negotiations focus on reducing emissions in the agricultural sector	India called for CBDR. India, countered by pointing out that emission reduction efforts in the agricultural sector would affect farmers who constitute a large percentage of the population, and are often the poorest, in the developing world Also argued that the effort to reduce emissions should be the focus on fossil-fuel-based activities that spew out carbon dioxide, the greatest contributor to global warming by far Indian intervention supported by many other countries ensured that the talks remained focused on adaptation and only a report on this specific matter is produced for the countries to discuss in future

(continued)

Table 1.1 (continued)

Conference of Parties (COP)	Key outcomes	India's stance
COP 20 (Lima 2014)	For the first time, an agreement was reached in which all countries will specify their objectives, and they will submit their CO^2 emissions information by March 2015 (INDC) The first Multilateral Assessment was held in Lima, providing greater transparency for actions by developed countries, as they can compare their degree of compliance with the emission reduction goals The Lima Conference agreed that the contribution of countries has to be more than their current commitments	India's stand was guided by the principle of Equity and CBDR, which is the bedrock principle of the UNFCCC Regarding submission of INDCs, India was of the view that the INDCs would be mitigation centric and that after countries submit their INDCs, these would be aggregated to ascertain whether the total contributions are adequate to achieve the global goal of containing temperature rise to below 2 °C by the end of the century from pre- industrial levels. Any gap between the two could mean pressure on countries to re-submit their INDCs or enhance their contributions However, India and many other countries of the developing world were not in favour of such an externally imposed review as it would compromise the sovereignty of Parties in determining their targets as per their national circumstances.
COP 21 (Paris, 2015)	The COP agreed to a set of decisions with immediate effect to accelerate climate action and to prepare for the implementation of the Paris Agreement once it enters into force Reaffirmed the goal of limiting global temperature increase well below 2 °C, while urging efforts to limit the increase to 1.5 °C Established binding commitments by all parties to make 'nationally determined contributions' (NDCs), and to pursue domestic measures aimed at achieving them Reaffirmed the binding obligation of developed countries under the UNFCCC to support the efforts of developing countries, while for the first time encouraging voluntary contribution by developing countries too The Agreement also explicitly recognizes that the principles of equity and CBDR and respective capabilities, in the light of different national circumstances will be respected	India was able to secure its interest and that of developing countries in the Paris Agreement The Purpose of the Agreement notes that the Agreement is to enhance the implementation of the Convention. This was a key demand of India so that the Agreement remains under the Convention and does not create a completely new regime India called for climate justice (a fair share of the carbon budget); its attempt to accommodate 'differentiation' in the agreed outcome and its reluctance to give in on its coal production goals had already created an atmosphere of disbelief in terms of attaining a strong legally binding agreement

(continued)

Table 1.1 (continued)

Conference of Parties (COP)	Key outcomes	India's stance
COP-22 (Marrakech, 2016)	Parties adopted 35 decisions, several related to the Paris Agreement work programme, including that such work should conclude by 2018; that the Adaptation Fund should serve the Paris Agreement; the terms of reference for the Paris Committee on Capacity-building; and initiating a process to identify the information to be provided in accordance with Agreement Article 9.5 (biennial finance communications by developed countries). COP 22 also adopted decisions, including approving the five-year work plan of the WIM for Loss and Damage, enhancing the Technology Mechanism, and continuing and enhancing the Lima work programme on gender	India stressed that the Convention, with its principles of equity and common but differentiated responsibilities, continues to be the political and legal basis for parties to enhance climate action and international cooperation in the post-2020 period India underscored the need to identify sources within and outside the UNFCCC and called for reviewing the ToR of the 'financial bodies.' The allocation of financial resources should be based on criteria of justice and fairness India, on behalf of many developing countries and coalitions, stressed that mechanisms developed under International Civil Aviation Organization (ICAO) and International Maritime Organization (IMO) should align with the principles of the Convention and COP decisions
COP-23 (Fiji/Bonn)	COP-23 decided that the Adaptation Fund shall serve the Paris Agreement; assess the technical examination process on mitigation and adaptation; take work forward on long-term finance; and conclude reviews of the Standing Committee on Finance, the Adaptation Fund, capacity building in countries with economies in transition, and developing countries; and give guidance to the Executive Committee of the Warsaw International Mechanism for Loss and Damage associated with Climate Change Impacts	India along with Brazil, South Africa and China, expressed concern about developed countries unilaterally creating new criteria for funding under the GCF, stressing that this practice has no legal basis
COP-24 (Katowice, 2018)	This meeting focused on completing work on the Paris Agreement Work Programme (PAWP), a set of decisions meant to operationalize the Paris Agreement. To this end, parties adopted the Katowice Climate Package, which includes decisions on nearly all of the issues mandated as part of the PAWP, including:	India underscored that the PAWP should support enhanced ambition without 'back-sliding' on rules. Also, emphasized that public finance is 'at the heart' of climate action in developing countries

(continued)

Table 1.1 (continued)

Conference of Parties (COP)	Key outcomes	India's stance
	on mitigation: further guidance to nationally determined contributions (NDCs), common time frames, and modalities, work programmes, and functions under the Paris Agreement of the forum on the impact of the implementation of response measures; on adaptation: further guidance on adaptation communication; on finance: identification of the information to be provided by parties by Agreement Article 9.5 (*ex-ante* finance transparency), matters relating to the Adaptation Fund, and setting a new collective quantified goal on finance; on technology: scope of and modalities for the periodic assessment of the Technology Mechanism, and the technology framework the modalities, procedures, and guidelines for the transparency framework for action and support; the global stocktake; and modalities and procedures for the effective operation of the committee to facilitate implementation and promote compliance	
COP-25 (Madrid, 2019)	The COP had relatively modest ambitions in the continuum of the UNFCCC process: its main task was to finalize the Paris rulebook in anticipation of 2020 when the Paris Agreement comes into effect. At the same time, the conference was a time to reflect on progress made, and assess if there are the necessary mechanisms for all countries to enter into the post-2020 period. Despite some advances, its outcomes were met with largely negative reactions, with UN Secretary-General António Guterres himself declaring that 'the international community lost an important opportunity...to tackle the climate crisis.'	India called for developed countries to honour existing financial commitments, scale up financial support, and increase the predictability of support
COP-26 (Glasgow, 2021)	The Glasgow Climate Change Conference convened after a year-long postponement due to the global COVID-19 pandemic. Parties adopted the Glasgow Climate Pact: a series of three overarching cover decisions that provide an overall political narrative of the	India announced five new targets: (i) increasing non-fossil fuel energy capacity to 500 Giga Watts by 2030; (ii) fulfilling 50% of energy requirements from renewable sources by 2030;

(continued)

Table 1.1 (continued)

Conference of Parties (COP)	Key outcomes	India's stance
	Conference of the Parties (COP). For the first time in the UNFCCC process, there is a reference to phasing down unabated coal power and phasing out inefficient fossil fuel subsidies. Some of the substantive aspects of the decisions include calls for: developed countries to double their adaptation finance from 2019 levels, by 2025; and parties that have not yet communicated new or updated nationally determined contributions (NDCs) to do so before the next COP. The Pact also establishes: an annual high-level ministerial roundtable on pre-2030 ambition; the Glasgow Dialogue between parties on loss and damage, to convene from 2022 to 2024; and an annual dialogue to strengthen ocean-based action.	(iii) reducing one billion metric tons of carbon emissions between now and 2030; (iv) reducing carbon intensity by 45% by 2030; and (v) Achieving net zero emissions by 2070. Citing the need to match finance with increasing ambition, India urged developed countries to provide USD 1 trillion of climate finance as soon as possible and proposed to set up a tracking system of climate finance to put pressure on countries that have not lived up to their financial commitments

Analysing Post-COP 26 Possibilities of Limiting the Global Warming Below 2.0 °C by 2100

Prior to the COP-26, 154 Parties have formally placed 2030 mitigation goals in their nationally determined contributions and 76 of them have even placed longer-term pledges. Analysis of the pledges before COP26 suggests a less than 50% possibility of limiting global warming below 2 °C by 2100. A new perspective (Meinshausen et al. 2022) is presented on the possibilities of limiting global warming below 2 °C by 2100 provided all conditional and unconditional pledges are implemented fully.

Peak warming could be limited to 1.9–2.0 °C (5–95% range 1.4–2.8 °C) in the full implementation case – building on a probabilistic characterization of Earth system uncertainties in line with the Working Group I Contribution to the Sixth Assessment Report of the Intergovernmental Panel on Climate Change (IPCC) (IPCC 2021). By retrospective projection of twenty-first-century warming, it has become clear that the aggregate level of ambition changed from 2015 to 2021. Further, the perspective relied on the extrapolation of time-limited targets beyond 2030 or 2050, characteristics of the IPCC 1.5 °C Special Report (SR1.5) (IPCC 2018) scenario database and the full implementation of pledges.

Further independent emissions modelling framework suggests that peak warming of 1.8 °C, supporting the finding that realized pledges could limit warming to just below 2 °C. However, limiting warming not only to 'just below' but to 'well below' 2 °C or 1.5 °C calls for immediate climate actions/policies to bring about steep emission reductions during the current decade and fully getting aligned with global net-zero CO_2 emission targets.

References

Chandra M (2015) Environmental concerns in India: problems and solutions. J Int Bus Law 15(1): Article 1. Available at: http://scholarlycommons.law.hofstra.edu/jibl/vol15/iss1/1

Blog, Carbon Footprint, Energy, News, Sustainability-The Kyoto Protocol: Climate Change Success or Global Warming Failure? February 4, 2015 by Circular Ecology

IPCC-Special-Report-on-Global Warming of 1.5 °C (2018)

IPCC AR6 Working Group I Report (2021)

Kyoto's Second Phase Emission Reductions Achievable-But Greater Ambition Needed, UNFCCC (17 June, 2020)

Meinshausen M, Lewis J, McGlade C, Gütschow J, Nicholls Z, Burdon R, Cozzi L, Hackmann B (2022) Realization of Paris Agreement pledges may limit warming just below 2°C. Nature 604(7905):304–309. Available at: https://www.researchgate.net/publication/359936322_Realization_of_Paris_Agreement_pledges_may_limit_warming_just_below_2_C?msclkid=6b9d9568c2ba11ecafd695b675c4e4c9

Mohnish M (2016) Outcome and Indian stance in COPs, 1–21. Available at: https://www.icwa.in/view points

Zillman J (2009) A history of climate activities. World Meteorol Org Bull 58(3):141–150. Available at: http://public.wmo.int/en/bulletin/historyclimate-activities

Chapter 2
Climate Changes over the Indian Subcontinent: Scenarios and Impacts

R. K. Mall, Nidhi Singh, Subhi Patel, Saumya Singh, Aman Arora, R. Bhatla, R. S. Singh, and P. K. Srivastava

Abstract It has now been well established that the rise in global mercury has driven climate change phenomena that have led to extreme temperature events, sea level rise, change in the hydrological cycle, frequent droughts and floods, and cyclones and forest fires and caused a myriad of adverse impacts on vital worldwide sectors such as agriculture, water and health. The impact of climate change is anticipated to be more adverse for destitute and socioeconomically deprived populations from developing and underdeveloped nations owing to poor adaptive capacity and higher sensitivity. The present chapter focuses on the Indian context, where it presents shreds of evidence of the impact of climate change in the past, present and future such as extreme events like heat waves, diurnal temperature range, shrinking of Himalayan glaciers, shifting of rainfall patterns, increased susceptibility to floods and droughts, and its impact on some of the important sectors. The chapter shows clear evidence of a decline in crop production and productivity of some of the important crops such as wheat, rice, sugarcane, maize, potato, tomato, etc. The recent studies established an increase in morbidity and mortality associated with extreme temperature and poor air quality associated with increased particulate matter (PM), NOx, SOx, O_3, black carbon and other ambient pollutants. In addition, important river basins of India, such as Gomti, Gandak, Vaigai, Mahi, Varuna and Ghaghra, have shown increased susceptibility to flooding and drought events that are more likely to be frequent and severe in the future under different climate change scenarios owing to changes in erratic rainfall patterns and increasing temperature. The chapter also discusses the potential adaptation and mitigation strategies that would help policymakers to combat climate change amid the rising susceptible population.

R. K. Mall (✉) · N. Singh · S. Patel · S. Singh · A. Arora · P. K. Srivastava
DST-Mahamana Centre of Excellence in Climate Change Research, Institute of Environment and Sustainable Development, Banaras Hindu University, Varanasi, India
e-mail: rkmall@bhu.ac.in

R. Bhatla · R. S. Singh
Department of Geophysics, Banaras Hindu University, Varanasi, India

N. Khare (ed.), *Science, Policies and Conflicts of Climate Change*, Springer Climate,
https://doi.org/10.1007/978-3-031-16254-1_2

Keywords Climate change · Impact assessment · Adaptation · Mitigation · Climate impacts

General Introduction

The unequivocal change in climate has been observed throughout the world with an average increase in human-induced global surface temperature in 2010–2019 relative to 1850–1900 of 1.07 [0.8–1.3] °C (IPCC 2021). The increase in global mercury has contributed to sea level rise, extreme events, and caused adverse impacts on vital sectors such as agriculture, water and health (Mall et al. 2014; Ghosh et al. 2016; Mall et al. 2017; Singh et al. 2018). Extreme climate events, in particular heat waves, floods, droughts, cyclones and wildfires, have put both man-made and natural ecosystems at risk (Mall et al. 2011; Field et al. 2014; Hari et al. 2021). With the present trend in warming, many multimodel projections have reported that the global mean surface temperature is likely to rise by 1.5 °C by the end of the twenty-first century from 1850–1900 (IPCC 2014; Singh et al. 2018) which may lead to more severe and frequent extreme events (high confidence) and may thus cause a significant global impact (Vittal et al. 2013; Field et al. 2014; Singh et al. 2021a, b, c).

India too is expected to witness a rise in global surface temperature by 0.4–2.6 °C in the next 50 years with more intense and frequent heat waves under 4.5 and 8.5 scenarios (IPCC 2014; Sonkar et al. 2019). Researchers have already found a shift in the spatial-temporal occurrence of heat waves (HW) in India with three new hotspots (northwestern, central, and south-central India) and an overall increase in HW during the past three decades (Sharma and Mujumdar 2017; Singh et al. 2021a). Besides, a significant decrease in the diurnal temperature range (DTR) in India was observed in the past three decades owing to an increase in minimum temperature and a decline in solar radiation which is more prominent over north-west India, parts of Gangetic plain, north-east and central India. This might have serious repercussions on human health and agricultural production (Mall et al. 2021).

High-resolution climate models are often used to study climate change phenomena (Bhatla et al. 2019), the performance of which is further enhanced using different convective parameterization schemes (CPSs) and bias correction methods (Choudhary and Dimri 2019; Jaiswal et al. 2022). In one such study, the CPSs viz. Grell and Mix99 (Grell- > Land & Emanuel- > Ocean schemes were found to best simulate the Indian Summer Monsoon (ISM) rainfall over the Indian subcontinent for RegCM-4.3 (Bhatla et al. 2019). Also, Grell followed by Kuo CPS performed best to simulate the Indian summer monsoon (ISM) surface air temperature and EL_GO performed well in simulating extreme rainfall events. Further, Bhatla et al. 2020 using six Conformal-Cubic Atmospheric Models (CCAM) over India, found that distribution mapping performed best to correct surface temperature. The inter-comparison of various climate models to study climate change and associated events is being undertaken to find the best-performing models (ensemble) to efficiently cover the spatial variability, magnitude and heterogeneity in various climate variables (Bhatla et al. 2020a; Mall et al. 2021; Singh et al. 2021d). Climate change is anticipated to cause substantial alterations in morphometric parameters that may lead to hydro-meteorological hazards such as an increase in the intensity, duration, and severity of droughts and floods, along with a shift in the seasonal and inter-annual precipitation patterns that put water resources and economic and human welfare at risk (Bhatla et al. 2020b; Mishra et al. 2021; Pandey et al. 2021), particularly to Indian agriculture that is highly governed by the southwest monsoon (Parthasarathy et al. 1992).

Climate change, a result of global warming, has accelerated present rates of sea level rise, inundating many low-lying coastal and inter tidal locations (Uk et al. 2007). In response to the 0.8 °C global warming, the pace of sea level rise has almost tripled throughout the twentieth century (Galbraith et al. 2002). The North Indian Ocean Sea level has risen significantly in recent decades due to the weakening of the Indian summer monsoon. (Swapna et al. 2017). Another indicator of climate change is glaciers. Based on the upward migration of the glacier, due to temperature changes, it appears that the Himalayan glaciers in India have been steadily shrinking since 1980 (Kumar et al. 2021). Mall et al. (2021) has shown an increase in the diurnal temperature range (DTR) with 0.038 °C per decade over different agroclimatic regions of India. Also, a significant increase in minimum temperature over the Himalayan region and Gangetic plain was observed.

Climate change has severely challenged the agricultural system through climate variability, extreme weather events, increasing water scarcity, issues related to crop management practices and soil fertility and increased pollution levels (Aggarwal and Mall 2002; Mall et al. 2006, 2018; Gupta et al. 2017; Sonkar et al. 2019) that have ultimately raised uncertainties in crop yield and global food security (Burney and Ramanathan 2014). The important staple crops such as wheat and rice have shown a significant declining trend in India and other parts of the world which is mainly attributed to a rise in minimum temperature and declining solar radiation (Pathak et al. 2003; Challinor et al. 2014; Asseng et al. 2017; Sonkar et al. 2019). Other important crops such as sugarcane, pigeon pea, and maize, have also observed the impact of changing climate (Mishra et al. 2017; Tigchelaa et al. 2018; Sonkar et al. 2020). The impact on crop production may further be complicated due to high concentrations of atmospheric aerosols that may influence many crop physiological behaviours (Myhre et al. 2013; Gupta et al. 2017).

As per The Lancet Climate Commission, anthropogenic activities driven by climate change have challenged public health in the past 50 years (Watts et al. 2017) and children may bear 88% of the disease burden, disproportionately affecting the poorest (Zhang et al. 2007; Singh et al. 2021d) as typical childhood diseases such as diarrhoea, pneumonia, malaria, malnutrition, are sensitive to climate. The adverse impact of extreme temperatures like heatwaves and cold spells on human health in terms of mortality has been studied in different parts of India such as Varanasi, Ahmedabad and North-West India (Azhar et al. 2014; Nori-Sarma et al. 2019; Singh et al. 2019). Most of the studies have shown that excess mortality during extreme temperature events is mainly attributed to cardiovascular diseases (CVD), respiratory diseases (RD) and cerebrovascular diseases (Kilbourne 1999; Azhar et al. 2014), especially for people with pre-existing medical conditions. Driven by climate change, the rising temperature may further deteriorate the air quality and cause a plethora of environmental issues, including an increase in public health issues (Srivastava et al. 2019; Singh et al. 2020a, 2021c).

Other population characteristics such as demographics, socio-economic resilience, vulnerability and physiologic acclimatization may further aggravate the climate change impacts (Singh et al. 2021d). Therefore, it is imperative to strengthen the institutional capacity for better mitigation and response strategies, improving overall resilience (Mall et al. 2019).

An Outlook of Historical Climate Trends and Projected Climate Change

The enormous increase in greenhouse gas concentrations owing to human activity, particularly industrialization, fossil fuel burning, and land use/land cover changes is considered to alter the climate, cause global warming and global energy imbalance (Huang et al. 2011; Mahmood et al. 2019). Since the mid-twentieth century, human influence on climate has been the leading driver of observed warming, with the global average surface temperature rising by 0.85 °C between 1880 and 2012 (Stocker et al. 2018). On account of the changing climatic conditions, several studies have been completed for the projection of climate change and ongoing research are continuing using Representative Concentration Pathways (RCPs) where the concentration of greenhouse gases (GHGs) emissions is one of the prime parameters along with aerosols and other chemically active gases, as well as land use/land cover (Moss et al. 2008). Different RCPs scenarios, ranging from RCP 2.6 to RCP 8.5, suggest low (1.0 °C) to extreme temperature (3.7 °C) increase in future, which would lead to major imbalance in climatic conditions leading to extreme weather situations. In order to overcome the impact of such disasters, the projection of climatic conditions & their unpredictable nature in the form of climate extremes are being studied worldwide (Lewis and King 2017; Tien Bui et al. 2019). In recent years, the climate models, crop simulation models, hydrological models and weather forecast models have been developed and used extensively to study climate change and its impacts. For example, studies have been undertaken to analyse and predict the impact of heat waves on Europe using the EURO-Coordinated Regional Climate Downscaling Experiments (CORDEX) project (Vautard et al. 2013), estimate future heat wave mortality under global climate change (Peng et al. 2011), evaluate CORDEX products to project heat waves over India (Singh et al. 2021b), and so on. Similarly, the research focusing on the impacts such as rainfall extremes (flood/droughts), temperature extremes (heat/cold waves), diurnal temperature change, dust storms, etc., on a broader scale are also apprehended by utilizing climate big data, geo big data, robust climate models such as Global Climate Models (GCMs), Regional Climate Models (RegCM), and other physical as well as statistical models (Asadieh and Krakauer 2017; DeBeer et al. 2016; Kumar et al. 2017;

Lavers et al. 2013; Mall et al. 2021; Singh et al. 2019; Srivastava et al. 2021; Chaubey et al. 2022). With the advent of Artificial Intelligence (AI) and Machine Learning (ML) techniques in the modern technology era, mainly in solving the problems related to classification, regression, and clustering for complicated data structures henceforth providing the best solution, it has been studied in different fields and applications of Science and Technology (Das et al. 2015; Greene et al. 2019; Riedl 2019; Thrall et al. 2018; Ullah et al. 2020; Arora et al. 2020; Dey et al. 2021a, b). The use of AI/ML in climate change simulations and impact studies has only recently begun (Huntingford et al. 2019; Milojevic-Dupont and Creutzig 2020; Stein 2020; Sanchez-pi et al. 2021), but it is expected that with the increased use of high-performance computers, improved skill and good data sets, the use of simulation models by professionals, policymakers and decision-makers will increase in the coming decades for projection of climate change.

Climate Change and Extreme Events: Past Observations and Future Projections

Globally, a wide range of natural hazards are occurring as extreme events instigated by climatological, geophysical, and hydro-meteorological events, causing casualties, injuries, economic loss, property and infrastructure damages, etc. (Lindell 2013). In the Indian weather chronology, high-impact weather phenomena such as cyclonic storms (synoptic-scale weather disturbances that last a few days), thunderstorms (that last less than a day), and short-lived cloudbursts (that last a few hours) that can produce intense rainfall amounts are classified as severe or extreme weather events (Ghosh and Krishnan 2020). Natural climatic variability (including events like El Niño) is responsible for many kinds of weather and climate extremes, and multi-decadal climate oscillations provide the backdrop for anthropogenic climate change (Seneviratne et al. 2012). Extreme climatic occurrences, in particular, highlight the relevance of climate to humanity and highlight associated vulnerabilities. The number of climate extreme events has steadily increased from 1900 to 2015 (CRED 2016; Mall et al. 2019). According to IPCC (2014) reports, the evidence suggests that the frequency and severity of high-temperature extremes are increasing globally, along with more frequent and severe heavy precipitation events, especially in tropical regions. Multiple studies through future projections modelling confirm that the number of extreme events would be increased in the coming future (Alexander and Arblaster 2009; CCSP 2008; Orlowsky and Seneviratne 2012). Using multiple RCMs, a study was steered to determine how heat waves, heavy precipitation, drought, wind storms and storm surges change between the present (1961–1990) and future (2071–2100) climates; the results show that by the end of the twentieth century, Europe will experience more hot days, and heavy winter precipitation will increase over north-eastern Europe (Beniston et al. 2007). Fischer and Schär (2010) predict that the frequency of 'heatwave days' in the Iberian

Peninsula and the Mediterranean will increase from an average of around two days every summer from 1961 to 1990 to around 13 days for 2021 to 2050 (and 40 days for 2071–2100). Similarly, Bhatla et al. (2020a, b) conducted a study to monitor the temperature changes in India from 1901 to 2010, the temperature increased nearly double in the 50 years since 1950, and the mean temperature from the 2000s onward show a continuous warming pattern over the Himalayan region, northern, southwest and central parts of India. Using data from CMIP5 GCMs and observations, a study shows that if the global mean temperature rises above 1.5° (2.0°), 3-hourly precipitation maxima at 100-year return intervals are anticipated to increase by 20% (25%) and the short-term precipitation extremes are also expected to increase more in India (Ali and Mishra 2018). Likewise, several other studies, based on country-specific areas of interest, focusing on the changes in precipitation patterns and their intensity in prospects have also been quantified worldwide (Endo et al. 2017; Zarekarizi et al. 2018; Cooper 2019; Duan et al. 2019). A study looked at the temporal and geographical changes in precipitation extremes for the period 2020–2100 over 9 river basins in China, based on the results of 18 GCMs with 3 RCPs from the CMIP5 (Xu et al. 2019). The results indicate the amount and intensity of precipitation in high-latitude and high-elevation regions of China will increase, whereas precipitation frequency will increase in southeastern and southern China. A combination of 10 statistically GCMs and 03 dynamically downscaled GCMs have been utilized to project past and future climate projections in the Pacific Northwest USA; analysis showed high values of extreme precipitation clustered in western & northern parts of the Columbia River Basin for the historical period whereas the northern part is experiencing a higher degree of change in the indices for future scenario (Zarekarizi et al. 2018). Another study used five regional climate simulations under CORDEX-SA to look at changes in precipitation extremes over India in the near (2020–2049) and far (2070–2099) future under RCPs 4.5 and 8.5, and found a slight increase in the frequency of wet precipitation extremes over central India in the near (2020–2049) and far (2070–2099) future, as well as a consistent increase throughout the twenty-first century (Rai et al. 2019).

Climate Change and Air Pollution: A Reciprocal Relationship

The relationship between climate change and air quality is interchangeable, both can impact each other. The changing climate can lead to modifications in local air quality. For example, associated warming of the atmosphere due to climate change has the potential to escalate ground-level ozone and related impacts (EPA 2020). Climate change driven by global warming, on the other hand, is escalated by the emission of air pollutants mainly greenhouse gases such as carbon dioxide (CO_2), methane, nitrous oxide, water vapour and synthetic fluorinated gases (Chauhan et al. 2014). Thus, the present warming of the globe is directly attributable to anthropogenic activities, primarily the burning of fossil fuels such as coal, oil, gasoline and natural gas. Though various efforts have been made to curb emissions and use alternative sources of energy, a lot more efforts are needed to comply with the commitments made in the Paris Climate Agreement-2015. It is suggested that to avoid accelerating climate change, a reduction of about 40% in global carbon emissions is required by 2030 (NRDC 2021).

Global warming is already evident in the melting of glaciers and diminishing snow covers, warming of the oceans, and sea level rise (Singh and Singh 2012; Mishra et al. 2014; Jyoti et al. 2019). Global sea levels are projected to rise by at least 44–74 cm by the end of the century relative to the mid-1990s, excluding the risk of ice-sheet collapse. Sea levels along the Indian coast are not forecast to rise quite as much as the average, increasing by 20–30 cm compared with current levels (Swapna et al. 2020). Coastal communities also face more severe storms: cyclones in the Bay of Bengal are projected to nearly double by 2070–2100, compared to the baseline period of 1961–1990 (Sarthi et al. 2014). These storms will be characterized by not only faster wind speeds but also greater storm surges due to higher sea levels. Low-income urban households are particularly susceptible, as they often live in dense settlements that lack basic services and infrastructure that could reduce risk: piped water, storm water drainage, paved roads or decent housing. Many households also live on hazardous sites such as steep slopes, floodplains and low-lying coastal areas, where the cost of land is cheaper and/or formal development is prohibited (Satterthwaite et al. 2020).

Aerosols influence climate in several ways by both scattering and/or absorbing radiation through interaction with cloud microphysics and other cloud properties (Srivastava et al. 2019). Atmospheric aerosols are not only a local issue as they can travel across transboundary to other countries, oceans or even continents, for example, atmospheric brown clouds (ABCs), from East Asia, spread across the Pacific and likewise, pollution from North America spreads across the Atlantic (Ramanathan 2007). Importantly, air pollutants are responsible for global dimming, whereas greenhouse gases are responsible for global warming. Thus, utmost care is required in curbing the air pollution that may unmask the solar dimming effect and might enhance warming (Ramanathan 2007).

Atmospheric aerosols are also involved in producing brighter clouds that are responsible for less precipitation. Among other pollutants, black carbon aerosols play an important role in solar dimming. The ABCs might also be responsible for reduced precipitation during the Indian summer monsoon (Ramanathan et al. 2005). Overall, aerosol radiation and microphysical effects might be responsible for a weakened hydrological cycle and fresh water crisis, forming a major environmental concern of the twenty-first century (Ramanathan 2007).

The Entire Indo-Gangetic Plain (IGP) has been documented to have an excessive burden of air pollutants with many associated sources like biomass and waste burning, and automobile emissions that are primarily sensitive to changing climate (Mall et al. 2018; Sonkar et al. 2019; Banerjee et al. 2021) and pollution-related health impacts (Singh et al. 2021c). Several studies exist in this region about source apportionment, size segregation, chemical composition and distribution of particulate matter (Singh et al. 2017; Vinjamuri et al. 2020). In one such study, the vertical distribution of smoke aerosols across South Asia shows abundance over upper IGP and a strong diurnal variation with abundance at lower heights (<4 km) during the daytime, indicating primary contribution from the local emission sources may have greater implications on regional climate, air quality, smoke transport and AOD-particulate modelling (Vinjamuri et al. 2020). Murari et al. 2020 reported that fine particulates were observed to be dominant during the post-monsoon and winter. The molecular distributions of organics indicate both biogenic and anthropogenic emissions over IGP, whereas both crustal resuspensions and secondary aerosols were responsible emitters of PM $>$ 2.1 (Singh et al. 2021c).

Climate Change and its Impact on Various Sectors: In Indian Perspective

Climate Change-Extreme Events-Aerosols-Agriculture Nexus

Climate change poses serious concerns about the agricultural production and livelihood of stakeholders causing a change in quality production and, threatening food security (Fig. 2.2). The increasing population has urged the need to increase the availability and affordability of food using the limited land resource. Climate has a direct impact on agriculture as it affects crop development, quality of products and ultimately crop yields (Mall et al. 2006). Empirical evidence shows temperature has a negative association with crop yields while rainfall anomalies are also strongly linked with yields (Challinor et al. 2014). However, a higher level of CO_2 concentration offsets the negative effect of temperature marginalizing the yield losses (Mall et al. 2017). Weather extremes like drought and heat stress have caused global cereal production losses of 9–10% (Lesk et al. 2016). In India, the decline in summer monsoon rainfall in 2002–2003 led to a decline in food grain production that year (Mall et al. 2006). The influence of other atmospheric components like increased aerosols, cloud cover, ozone, greenhouse gas accumulation, etc., has also been accounted for. The effect of atmospheric brown clouds is less certain on agriculture. A study that couples a statistical rice model to an RCM has reported an increment in crop yield due to reductions in both carbon dioxide and atmospheric brown clouds (Auffhammer et al. 2006). Similarly, the vulnerability of wheat crops to rising temperature and aerosols over five wheat growing zones of India was negative and spatially inconsistent, particularly in IGP which accounts for India's major food production area and sustains around 60% of India's population ; thus, making them predominantly sensitive to climate change (Mall et al. 2018; Sonkar et al. 2019). In future, other factors such as increased surface ozone may also lead to losses in crops like wheat, maize and soybean (Avnery et al. 2011). Climate change's impact on the ground water table, quality of soil, nutrient content of the grain, soil evapotranspiration, pest and disease infestation, and weeds could also lead to a loss in crop yield.

It is projected that global warming will reduce crop production in India by 30–40% by 2080 (Cline 2007). Rice and wheat, the staple diet of the country will witness losses in yield due to heat stress at end of the twenty-first century (Soora et al. 2013; Ray et al. 2015; Yadav et al. 2015; Bhatt et al. 2019; Patel et al. 2022) and production of pulse, oilseed and vegetable crops will also be affected by climate change (Yadav et al. 2016, 2021). There will be a rise in the evapotranspiration of rice and wheat in 2040–2061 (Tyagi et al. 2019). Maize, the third most important cereal, is expecting losses of up to 17% in Mid Indo-Gangetic Plains while in the southern plateau the yield will reduce by up to 35% by 2080 (Byjesh and Kumar 2010). Cash crops like sugarcane are widely affected by temperature, precipitation, CO_2 and extreme weather events (Srivastava and Rai 2012; Mall et al. 2016; Verma et al. 2019; Pathak et al. 2019) and sugarcane production and sugar content is projected to decrease under future climate conditions (Ray et al. 2019; Sonkar et al. 2019).

Further, a change in the 2 °C rise in temperature and increase in precipitation can reduce 8.4% of the total revenue for India (Kumar and Parikh 2001). Kahn et al. (2019) predict that climate change could reduce India's GDP by around 2.6% by 2100 even if the global temperature increase is held below 2 °C; however, this increases up to 13.4% in a 4 °C scenario.

Studies have shown that such events can be minimized by the adoption of certain adaptation strategies at the genetic level as well as crop management levels, such as the development of heat-tolerant varieties, short growing season varieties, and improved protein assimilation at elevated temperatures. In addition, crop management practices like change in sowing dates to escape heat stress at critical crop growth stages, climate-smart agriculture, scheduling of irrigation, and fertilizer can help reduce the negative impacts.

Climate Change and Water Resources: The Costly Affair

With global warming, the hydrological cycle is predicted to become more intense, potentially increasing the severity of extreme precipitation events and the risk of flooding across all climate areas as water availability shifts from dry to wet places (Tabari 2020). The impact of climate change on water, changes in positive, flood, or, negative, drought, is being quantified by researchers and the assessment of the future climate change on it is also being practised worldwide (Dai et al. 2018; Mukherjee et al. 2018; Nagalapalli et al. 2019; Pandey and Srivastava 2019; Srivastava et al. 2020; Pandey et al. 2021; Wasko et al. 2021). Notably, the hydroclimatic predictions over the Indian Monsoon Region are also being affected by the land cover and land-use change and for which the study of coupled land-atmosphere interactions is required to understand the climate change impact on hydroclimatic predictions (Niyogi et al. 2018). Increasing variations in monsoon cycles and extreme rainfall have also been witnessed in several places nationwide. Extreme rainfall episodes across the whole central subcontinent increased threefold in the form of widespread extreme rain events over central India between 1950 and 2015 due to increasing variability of the low-level monsoon westerlies over the Arabian Sea (Roxy et al. 2017). A study systematically assessed the hydrological response to climate change and projected the future situation of global water resources using 03 global climate and 08 hydrological models (Hagemann et al. 2013). The findings suggest that accessible water resources are increasing in many locations, but in Central and Southern Europe, the Middle East, the Mississippi River basin, southern Africa, southern China and southeastern Australia, the available water resources are decreasing. On a local level, a study used CMIP5 GCMs under two RCPs 2.6 and 8.5 to analyse the influence of climate change on flood frequency and flood source area in the Talar River Basin, Iran (Maghsood et al. 2019). The study found that projected climate change will likely result in a decrease in average discharge in January, February and March for both RCPs and an increase in September and October for RCP 8.5, as well as an increase in maximum and minimum temperature and annual precipitation for all months in the near future. A study analysed the change of dry and wet circumstances between the present (1979–2003) and future climates (2075–2099) using MRI-AGCM3.2 S with a 20-km mesh global atmospheric model with four alternative sea surface temperature patterns and a 12-month comparative standardized precipitation index (SPI) (Hasegawa et al. 2016). The median of the 12-month cSPI moves to severely dry around the Mediterranean Sea to the Persian Gulf and to highly wet in the Tibetan Plateau, North and South India, and around the Yellow Sea, according to one of the study's findings. Jehanzaib et al. (2020) show the risk of meteorological drought becoming hydrological drought increased by 13% and 2%, respectively, under RCP 4.5 and 1.5% and 84%, respectively, under RCP 8.5 scenarios, demonstrating the impact of climate change from baseline period 1981–2010 for different future periods, 2011–2040, 2041–2070 and 2071–2099. Mishra et al. (2021) reported that flash droughts predominantly occur during the summer monsoon season (June–September) and are driven by the intra-seasonal variability of monsoon rainfall, and their occurrence is projected to rise by about fivefold in the twenty-first century in India. Recent studies using ML models for Karnataka state show that more than 50% of areas fall under very high vulnerability zones (Saha et al. 2021). Important river basins of India such as Gomti, Gandak, Vaigai, Mahi and Varuna have shown increased susceptibility to flooding and drought events that are more likely to be frequent and severe in the future under different climate change scenarios owing to changes in erratic rainfall patterns and increasing temperature (Bhatt et al. 2020; Chaubey et al. 2021; Maurya et al. 2021; Dey et al. 2021a, b; Bhatt et al. 2022). A recent study has illustrated a significant amount (12.5% and 27.35%) of flood hazard classes would be increased under RCP 4.5 & 8.5, respectively, in Mahanadi River Basin, India (Gusain et al. 2020).

Climate, Air Pollution, Extreme Events and Human Health

Climate change threatens to undermine the advances of improvement in global health and the attempts that have been made in reducing global mortality over the past two and a half decades (Philipsborn and Chan 2018). However, the impact will disproportionately affect the children and SES poor, thus widening the disparities in existing social determinates of health (Philipsborn and Chan 2018; Singh et al. 2019).

Heat and Cold Wave Related Health Effects

A drastic increase in heat waves over several parts of the world like Russia in 2003 and 2010 (Han et al. 2017; Luterbacher et al. 2004), Europe in 2003, 2007 and 2010 (Åström et al. 2013), in the USA in 2006 (Knowlton et al. 2009), and India in 2010, 2013 and 2015 (Mazdiyasni et al. 2017) had driven several epidemiological research studies to investigate the extent of association. The extreme events in terms of heat or cold spells were reported to cause cardiovascular (CVD), cerebrovascular and chronic respiratory diseases (RD, Hajat et al. 2014; Gosling et al. 2017). Thus, a better understanding of the effects of extremes temperature on morbidity and mortality is vital for India, because of its higher vulnerability and low population resilience to climate change.

Air Pollution and Related Health Effects

Climate change is also expected to influence the concentration and distribution of air pollutants (Akhtar and Palagiano 2018). Air pollution is already known to negatively influence lung development and acute or chronic respiratory diseases (Rajak and Chattopadhyay 2020). Chronic exposure to ambient ozone (O_3) may exacerbate asthma symptoms (Chakraborty et al. 2014). The burden of disease driven by climate change would likely be relatively high among developing countries that already suffer from a lack of adequate health infrastructure. Thus, a focused attempt on reducing these inequities and saving lives are key priorities. Each year the combined exposure to outdoor and indoor air pollutants kills about 3.7 million people worldwide and South-East Asia and the Western Pacific Regions witness the highest health burden owing to air pollution (Nyhan 2015). Particulate matter, mainly fine particulates (PM2.5), were more frequently used to estimate the association between health (mortality and morbidity); at global (Janssen et al. 2011; Cao et al. 2012) and local scale (Chowdhury et al. 2018; Saini and Sharma 2020). Apart from PM2.5, black carbon (BC) has also shown a high adverse impact that is often regarded as a more robust estimate of heath impact (Singh et al. 2021c). Other gaseous pollutants like NO_2, SO_2 and O_3 can also influence health impacts (Singh et al. 2021c). Thus, understanding the effects of air pollution-induced health risks is imperative owing to large SES vulnerability and low population resilience to changing climate (Fig. 2.1).

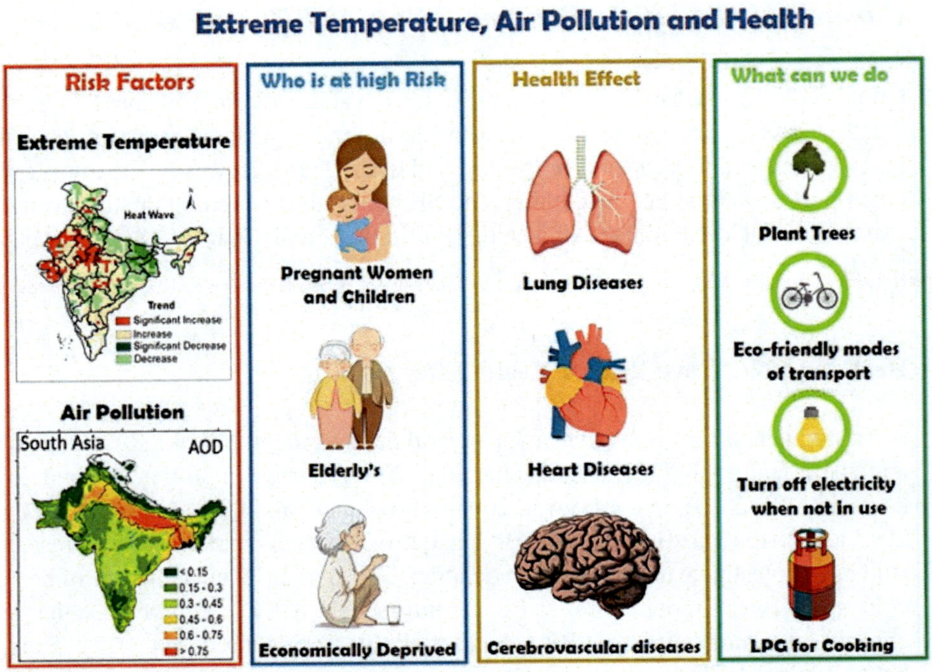

Fig. 2.1 Pictorial representation of the impact of extreme temperature and air pollution on human health and possible solutions to combat adverse effects

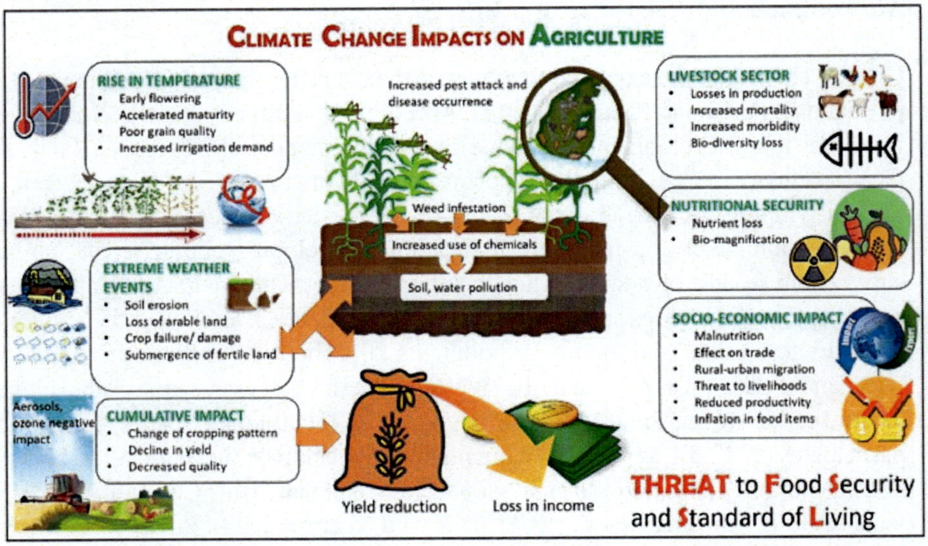

Fig. 2.2 Pictorial representation of the various impacts of climate change on agriculture

Gastrointestinal Diseases

Globally, GIT related illness is one of the leading causes of mortality and has a death toll of 450,000 annually (UNICEF 2020). Importantly, a high percentage of deaths occurred in children under 2 years of age. With the increase in occurrences of storms or floods, a drastic rise in water and food-borne illnesses such as gastroenteritis and infectious diarrhoea was observed immediately; children being the most susceptible due to their developing immune systems. Bacteria, viruses and parasites are considered the most common causal organism for GI infections (Mohapatra et al. 1989). Importantly, under the scenario of climate change, an additional 48,000 diarrheal deaths are expected by 2030, among children under 15 years of age, mostly in developing and underdeveloped nations (Philipsborn and Chan 2018).

Upper Respiratory Tract Infections

URTI is often linked with high morbidity and mortality cases. Influenza is attributed roughly to 250,000–500,000 global deaths annually (WHO 2007). Further, the warmer temperatures were found to have extended the length of the pollen season and thus increased asthma exacerbation (D'Amato et al. 2002). The warmer temperatures also enhance the ground level O_3 production which in turn damages the lungs and triggers asthma attacks in children and causes breathing problems in people with pre-existing chronic lung disease (Zhang et al. 2017, 2019). However, the current estimates of the association between URTI and climate are very limited for India (Awasthi and Pande 1997). Therefore, future research is needed in tropical and subtropical countries for a better understanding of the association with change in climate through a set of observational and experimental studies that shall be further useful to predict the future dynamics given global climate change.

Lower Respiratory Tract Infections

LRTI is considered the third foremost cause of death worldwide, including high child mortality in India (WHO 2017). The typical humid subtropical climate in the northern part of India presents unique pneumonia associated with mortality during summer between June and August (Farrar et al. 2019). Mycoplasma pneumonia and Respiratory syncytial virus are attributed as major causes of most pneumonia infections (Kumar et al. 2018). Due to the association of LRTI with large numbers of climate variables, the prediction of transmission and epidemics of associated viruses is a bit challenging. Climate change can modify both the inter-host and host characteristics. Thus, strong attempts must be made through widespread vaccination and efficient preventive mechanism to combat the increase in LRTI cases in future.

Vector-Borne Diseases

Of total disability and related illnesses occurred worldwide, about one-sixth are due to VBD, and currently, more than 50% of the world's population is at risk (Lozano et al. 2012). Much of the risk associated with VBD is typically high among the economically and socially deprived population with poor environmental and social conditions and improper medical interventions (Dhiman et al. 2010). VBD is the most researched disease due to its association with changing climate and owing to its high sensitivity to climatic factors (Campbell-Lendrum et al. 2015; Mall et al. 2017; Singh et al. 2020b). The climate change projections for VBDs show an overall favourable condition in future with large spatial heterogeneity (Astrom et al. 2012). Therefore, future trends in climate can be expected to amplify the VBD burden in association with non-climate factors.

Impacts on Other Sectors of Society – an Emerging Area of Interest

Ramifications of climate change impact are not limited to extreme events, health and agriculture but also include other sectors like tourism, migration, culture and gender. Climate change has been identified as one of the macro-economic drivers of tourism. Research done in the area of tourism climatology shows that tourism is a climate-dependent activity and hence affected by impacts of climate change, for example, changes in the aesthetic value of the destination, alterations in environmental conditions, effect sports tourism and thus influence the travel decisions and economy of the tourist spots (Scott et al. 2012; Gómez-Martín 2021). The Uttarakhand flood in 2013 and the Kerala flood in 2018 devastated tourist spots causing loss of lives and livelihoods. A substantial scope lies in comprehensive studies based on the impact of climate change on tourism in India specifically. Another dimension related to human movement is migration. As migration is an option to adapt to the negative impacts of climate change, a new term 'Climate migrant' has emerged which refers to people migrating from their homes due to climate change impacts. Labours from flood-affected, drought-affected regions migrate from their native region in search of livelihood opportunities and income generation. It has been observed that in India the weather extremes have increased rural–urban migration majorly among agricultural labourers, while a reduction is evident in rural–rural and international migration (Sedova and Kalkuhl 2020). The reason is that climate, agricultural yields and migration are interlinked and there is a negative relation between agricultural yields and out-migration of agricultural labours in India (Viswanathan 2015). This migration also leads to increased exposure to diseases and health issues like mental stress, heat stress of migrant workers, thus having serious implications on health (Semenza and Ebi 2019).

Focusing on the societal structural impacts of climate change, it has been observed that climate change is not gender-neutral and it has implications on the culture of society also. Climate change has a profound effect on culture or community with special emphasis on indigenous communities in which inter-dependence of humans and the environment is deeply intricate. Climate change impacts affect the cultural assets, the extent of adaptation, and the effectiveness of policy which needs to be taken care of while formulating policy and adaptation strategies like migration or shifting (Adger et al. 2012). Women in rural India are vulnerable to climate change and also play a pivotal role in the adaptation to climate change in climate-smart agriculture (Yadav and Lal 2018; Chanana-Nag and Aggarwal 2020). Both of the above-discussed dimensions have a significant impact on the building up of climate resilience. This emphasizes a comprehensive consideration of parameters influencing the adaptations for climate change while outlining and implementing suitable policies.

Mitigation and Adaptation Strategies

The two strategies for dealing with climate change are mitigation and adaptation. Mitigation is a strategy for reducing greenhouse gas emissions or increasing greenhouse gas sinks. Adaptation is defined as an 'adjustment in natural or human systems in response to present or anticipated climatic stimuli or their effects, which mitigates harm or maximizes benefits' (IPCC 2001). The difference between climate change mitigation and adaptation strategies is that mitigation focuses on addressing the causes of climate change and minimizing the potential impacts, whereas adaptation focuses on reducing the negative effects of climate change while also taking advantage of any opportunities that arise. As the issues related to climate change impact increase in multiple sectors, i.e. agriculture, infrastructure, health, water, etc. adaptation and mitigation strategies are essential to tackle unforeseen situations and maintain sustainability. Similarly, when these strategies are adopted in agriculture to mitigate the potential negative consequences of climate change, essential synergies must be discovered, as mitigation practices may compete with local agricultural practices targeted at sustaining production and income (Rosenzweig and Tubiello 2007). In comparison to the earlier studies which were focused on global climate change, the focus has now shifted to local climate change or in other words urban climate change. In a study, it is suggested that the use of roadside plantations can act as the first line of defence to reduce pollution impacts against vehicular emissions generated by the rapidly increasing number of vehicles, which harms natural ecosystems, including trees and physical properties, and human health (Singh et al. 2020a). Because cities are prime contributors with 70% of CO_2, the adaptation and mitigation plans for urban climate change are becoming more common as cities strive to improve climate resilience (Sharifi 2021). Owing to the increasing population and the changing urban climate scenarios, extreme events have also increased in the country. For example, flooding has severely affected metropolitan cities such as

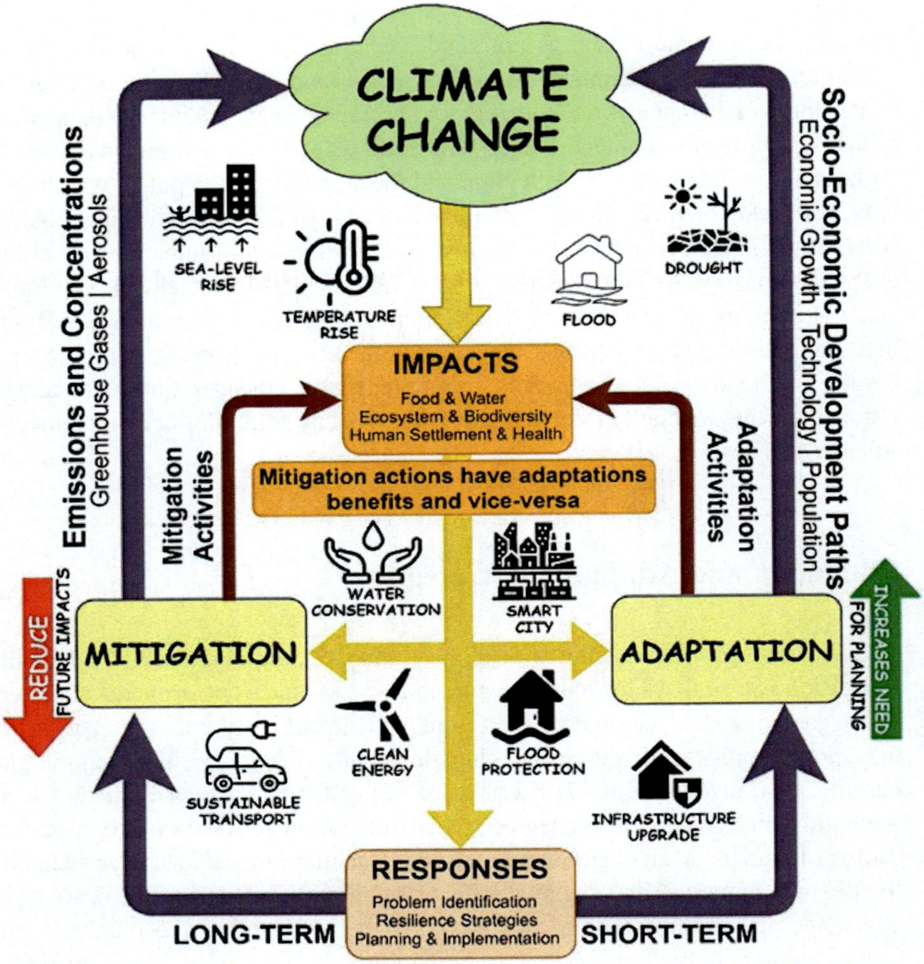

Fig. 2.3 Role of mitigation and adaptation strategies in climate change

Mumbai and Kolkata where the population has increased multifold in the past two to three decades. Furthermore, while it is technically difficult to re-plan the city's structure in the current situation to combat such hazards (tsunamis, storms and floods), it is possible to adapt to and mitigate the effects of natural hazards through appropriate planning and management, with the integrated cooperation and participation of citizens and government (Dhiman et al. 2019).

The flow diagram provided in Fig. 2.3 represents the best scenario for Mitigation and Adaptation strategies.

The Way Forward and Future Challenges

The climate change crisis in India is more than just an environmental concern and has emerged as the biggest developmental challenge over the years. In this regard, the government of India in 2008 launched the National Action Plan for Climate Change (NAPCC) programme outlining eight National Missions on climate change to mitigate and adapt to the adverse impact of climate change (CCP 2008). Under the initiative of the National Mission on Strategic Knowledge for Climate Change (NMSKCC) a series of initiatives have been taken, one of which was the establishment of centres for excellence in climate change research studies all over India. These initiatives will generate evidence on the past impact and future vulnerability of various sectors to combat some of the important social and global impacts of climate change and its devastating effects.

There are a range of options such as stabilizing and reducing GHG emissions by two thirds by 2050, investing in efficient alternate energy sources, strict environmental law and introducing new regulations for transportation and building codes, and realization of a carbon tax. Apart from such regulatory steps, the government institutions such as the India Meteorological Department, Ministry of Earth Science and State Agricultural Universities/Indian Council of Agricultural Research, etc., are actively engaged in projects such as the Gramin Krishi Mausam Sewa project (GKMS) and FASAL (Forecasting Agricultural output using Space, Agrometeorology and Land-based) observations for regularly generating weather and crop forecasts at the district/state/national level and educating and evaluating farmers perception about climate change issue to combat the losses due to future climate change along with developing an early warning system. Early warning systems for health hazards and meteorological hazards are being envisioned for direct societal development and to address the direct and indirect impacts and prevention and control of climate risks for an environmentally unsustainable economy in India by creating local, regional and national evidence of climate change and its impact and key suggestions to mitigate the future challenges.

Acknowledgements The authors wish to thank the Climate Change Programme, Department of Science and Technology, New Delhi for providing financial support (Award no.: DST/CCP/CoE/80/2017(G)).

References

Adger WN, Barnett J, Brown K, Marshall N, Brien KO (2012) Cultural dimensions of climate change impacts and adaptation. Nat Clim Change 3(2):112–117

Aggarwal PK, Mall RK (2002) Climate change and rice yields in diverse agro-environments of India. II Effect of uncertainties in scenarios and crop models on impact assessment. Clim Change 52(3):331–343

Alexander LV, Arblaster JM (2009) Assessing trends in observed and modelled climate extremes over Australia in relation to future projections. Int J Climatol 29(3):417–435

Ali H, Mishra V (2018) Increase in subdaily precipitation extremes in India under 1.5 and 2.0 °C warming worlds. Geophys Res Lett 45(14):6972–6982

Akhtar R, Palagiano C (2018) Climate change and air pollution: an introduction. In: Akhtar R, Palagiano C (eds) Climate change and air pollution. Springer, Cham. https://doi.org/10.1007/978-3-319-61346-8_1

Arora A, Arabameri A, Pandey M, Siddiqui MA, Shukla UK, Bui DT, Mishra VN, Bhardwaj A (2020) Optimization of state-of-the-art fuzzy-metaheuristic ANFIS based machine learning models for flood susceptibility prediction mapping in the middle ganga plain, India. Sci Total Environ 750:141565

Asadieh B, Krakauer NY (2017) Global change in streamflow extremes under climate change over the 21st century. Hydrol Earth Syst Sci 21(11):5863–5874

Asseng S, Cammarano D, Basso B, Chung U, Alderman PD, Sonder K et al (2017) Hot spots of wheat yield decline with rising temperatures. Glob Chang Biol 23(6):2464–2472

Åström DO, Forsberg B, Ebi KL, Rocklöv J (2013) Attributing mortality from extreme temperatures to climate change in Stockholm, Sweden. Nat Clim Chang 3(12):1050–1054

Auffhammer M, Ramanathan V, Vincent JR (2006) Integrated model shows that atmospheric brown clouds and greenhouse gases have reduced rice harvests in India. Proc Natl Acad Sci 103(52):19668–19672

Avnery S, Mauzerall DL, Liu J, Horowitz LW (2011) Global crop yield reductions due to surface ozone exposure: 2. Year 2030 potential crop production losses and economic damage under two scenarios of O3 pollution. Atmos Environ 45(13):2297–2309

Awasthi S, Pande VK (1997) Seasonal pattern of morbidities in preschool slum children in Lucknow, North India. Indian Pediatr 34:987–993

Azhar GS, Mavalankar D, Nori-Sarma A, Rajiva A, Dutta P, Jaiswal A et al (2014) Heat-related mortality in India: excess all-cause mortality associated with the 2010 Ahmedabad heat wave. PLoS One 9(3):e91831

Banerjee T, Shitole A, Mhawish A, Anand A, Ranjan R, Khan MF, Srithawirat T, Latif MT, Mall R (2021) Aerosol climatology over South and Southeast Asia: aerosol types, vertical profile and source fields. J Geophys Res Atmos 126:e2020JD033554. https://doi.org/10.1029/2020JD033554

Beniston M, Stephenson DB, Christensen OB, Ferro CAT, Frei C, Goyette S, Halsnaes K, Holt T, Jylhä K, Koffi B, Palutikof J, Schöll R, Semmler T, Woth K (2007) Future extreme events in European climate: an exploration of regional climate model projections. Clim Change 81 (SUPPL. 1):71–95

Bhatla R, Mandal B, Verma S, Ghosh S, Mall RK (2019) Performance of regional climate model in simulating monsoon onset over the Indian subcontinent. Pure Appl Geophys 176(1):409–420

Bhatla R, Sarkar D, Verma S, Sinha P, Ghosh S, Mall RK (2020a) Regional climate model performance and application of bias corrections in simulating summer monsoon maximum temperature for agro-climatic zones in India. Theor Appl Climatol 142(3):1595–1612

Bhatla R, Verma S, Ghosh S, Gupta A (2020b) Abrupt changes in mean temperature over India during 1901–2010. Journal of earth system science, 129(1):1–11

Bhatt D, Sonkar G, Mall RK (2019) Impact of climate variability on the rice yield in Uttar Pradesh: an agro-climatic zone-based study. Environ Process 6(1):135–153

Bhatt D, Kundu A, Mall R, Raju KP (2020) Dynamics of vegetation response to seasonal rainfall in the Gomati River basin (India) using earth observation data sets. J Sci Res 64(01):20–31

Bhatt D, Mall RK, Raju KP, Suryavanshi S (2022) Multivariate drought analysis for the temperature homogeneous regions of India: lessons from the Gomati River basin. Meteorological applications. RMetS 29(2):e2044. https://doi.org/10.1002/met.2044

Burney J, Ramanathan V (2014) Recent climate and air pollution impacts on Indian agriculture. Proc Natl Acad Sci 111(46):16319–16324

Byjesh K, Kumar SN, Aggarwal PK (2010) Simulating impacts, potential adaptation and vulnerability of maize to climate change in India. Mitig Adapt Strat Glob Chang 15(5):413–431

Campbell-Lendrum D, Manga L, Bagayoko M, Sommerfeld J (2015) Climate change and vector-borne diseases: what are the implications for public health research and policy. Philos Trans R Soc Lond B Biol Sci 370(1665):20130552. https://doi.org/10.1098/rstb.2013.0552

Cao J, Xu H, Xu Q, Chen B, Kan H (2012) Fine particulate matter constituents and cardiopulmonary mortality in a heavily polluted Chinese city. Environ Health Perspect 120(3):373–378

CCSP (2008) Weather and climate extremes in a changing climate. Regions of focus: North America, Hawaii, Caribbean, and US Pacific islands. A report by the US climate change science program and the subcommittee on global change research, 164

Chakraborty P, Chakraborty A, Ghosh D, Mandal J, Biswas S, Mukhopadhyay UK, Bhattacharya SG (2014) Effect of airborne Alternaria conidia, ozone exposure, PM 10 and weather on emergency visits for asthma in school-age children in Kolkata city, India. Aerobiologia 30(2): 137–148

Challinor AJ, Watson J, Lobell DB, Howden SM, Smith DR, Chhetri N (2014) A meta-analysis of crop yield under climate change and adaptation. Nat Clim Chang 4(4):287–291

Chanana-Nag N, Aggarwal PK (2020) Women in agriculture, and climate risks: hotspots for development. Clim Change 158(1):13–27

Chaubey PK, Srivastava PK, Gupta A, Mall RK (2021) Integrated assessment of extreme events and hydrological responses of indo-Nepal Gandak River basin. Environ Dev Sustain 23:8643–8668

Chaubey PK, Mall RK, Jaiswal R, Payra S (2022) Spatio-temporal changes in extreme rainfall events over different Indian river basins. Earth and space science. AGU 9:e2021EA001930. https://doi.org/10.1029/2021EA001930

Chauhan BS, Mahajan G, Randhawa RK, Singh H, Kang MS (2014) Global warming and its possible impact on agriculture in India. Adv Agron 123:65–121

Choudhary A, Dimri AP (2019) On bias correction of summer monsoon precipitation over India from CORDEX? SA simulations. Int J Climatol 39(3):1388–1403

Chowdhury S, Dey S, Smith KR (2018) Ambient PM 2.5 exposure and expected premature mortality to 2100 in India under climate change scenarios. Nat Commun 9(1):1–10

Cline WR (2007) Global warming and agriculture: impact estimates by country. Peterson Institute

Cooper RT (2019) Projection of future precipitation extremes across the Bangkok metropolitan region. Heliyon 5(5):e01678

CRED (2016) EM-DAT | the international disasters database website. In: Centre for Research on the epidemiology of disasters (CRED). Université Catholique de Louvain. https://www.emdat.be/

D'amato G (2002) Outdoor air pollution, climate and allergic respiratory diseases: evidence of a link. Clin Exp Allergy 32(10):1391–1393

Dai A, Zhao T, Chen J (2018) Climate change and drought: a precipitation and evaporation perspective. Curr Clim Change Report 4(3):301–312

Das S, Dey A, Pal A, Roy N (2015) Applications of artificial intelligence in machine learning: review and Prospect. Int J Comput Applic 115(9):31–41

DeBeer CM, Wheater HS, Carey SK, Chun KP (2016) Recent climatic, cryospheric, and hydrological changes over the interior of western Canada: a review and synthesis. Hydrol Earth Syst Sci 20(4):1573–1598

Dey S, Shukla UK, Mehrishi P, Mall RK (2021a) Appraisal of groundwater potentiality of multilayer alluvial aquifers of the Varuna river basin, India, using two concurrent methods of MCDM. Environ Dev Sustain 23:17558–17589

Dey S, Dey AK, Mall RK (2021b) Modeling long-term groundwater levels by exploring deep bidirectional long short-term memory using hydro-climatic data. Water Resour Manag 2021. https://doi.org/10.1007/s11269-021-02899-z

Dhiman RC, Pahwa S, Dhillon GPS, Dash AP (2010) Climate change and the threat of vector-borne diseases in India: are we prepared? Parasitol Res 106(4):763–773

Dhiman R, Vishnu Radhan R, Eldho TI, Inamdar A (2019) Flood risk and adaptation in Indian coastal cities: recent scenarios. Appl Water Sci 9(1):1–16

Duan W, Hanasaki N, Shiogama H, Chen Y, Zou S, Nover D, Zhou B, Wang Y (2019) Evaluation and future projection of Chinese precipitation extremes using large ensemble high-resolution climate simulations. J Climate 32(8):2169–2183

Endo H, Kitoh A, Mizuta R, Ishii M (2017) Future changes in precipitation extremes in East Asia and their uncertainty based on large ensemble simulations with a high-resolution AGCM. SOLAIAT 13:7–12

EPA (2020) Air Quality and Climate Change Research. Accessed from: https://www.epa.gov/air-research/air-quality-and-climate-change-research on 6th Aug 2021

Farrar DS, Awasthi S, Fadel SA, Kumar R, Sinha A, Fu SH, Wahl B, Morris SK, Jha P (2019) Seasonal variation and etiologic inferences of childhood pneumonia and diarrhea mortality in India. eLife 8:e46202. https://doi.org/10.7554/eLife.46202

Field CB, Levy AN, Mach KJ (2014) Climate change 2014: impacts. Adapt Vulnerabil 107:301–316

Fischer EM, Schär C (2010) Consistent geographical patterns of changes in high-impact European heatwaves. Nat Geosci 3(6):398–403

Galbraith H, Jones R, Park R, Clough J, Herrod-Julius S, Harrington B, Page G (2002) Global climate change and sea level rise: potential losses of intertidal habitat for shorebirds. Waterbirds 25(2):173–183

Ghosh S, Krishnan R (2020) Extreme storms. Assessment of climate change over the Indian region: a report of the Ministry of Earth Sciences (MoES), Government of India, 155

Ghosh S, Vittal H, Sharma T, Karmakar S, Kasiviswanathan KS, Dhanesh Y, Sudheer KP, Gunthe SS (2016) Indian summer monsoon rainfall: implications of contrasting trends in the spatial variability of means and extremes. PLoS One 11(7):e0158670

Gómez-Martín MB (2021) Tourism climatology: past, present, and future. Atmos 12(5):639–643

Gosling SN, Hondula DM, Bunker A, Ibarreta D, Liu J, Zhang X, Sauerborn R (2017) Adaptation to climate change: a comparative analysis of modeling methods for heat-related mortality. Environ Health Perspect 125(8):087008

Greene D, Hoffmann AL, Stark L (2019) Better, nicer, clearer, fairer: a critical assessment of the movement for ethical artificial intelligence and machine learning. Proceedings of the 52nd Hawaii international conference on system sciences

Gupta R, Somanathan E, Dey S (2017) Global warming and local air pollution have reduced wheat yields in India. Clim Change 140(3–4):593–604

Gusain A, Mohanty MP, Ghosh S, Chatterjee C, Karmakar S (2020) Capturing transformation of flood hazard over a large River Basin under changing climate using a top-down approach. Sci Total Environ 726:138600

Hagemann S, Chen C, Clark DB, Folwell S, Gosling SN, Haddeland I, Hanasaki N, Heinke J, Ludwig F, Voss F, Wiltshire AJ (2013) Climate change impact on available water resources obtained using multiple global climate and hydrology models. Earth Syst Dynam 4(1):129–144

Hajat S, Vardoulakis S, Heaviside C, Eggen B (2014) Climate change effects on human health: projections of temperature-related mortality for the UK during the 2020s, 2050s and 2080s. J Epidemiol Community Health 68(7):641–648

Han J, Liu S, Zhang J, Zhou L, Fang Q, Zhang J, Zhang Y (2017) The impact of temperature extremes on mortality: a time-series study in Jinan, China. BMJ Open 7(4):e014741

Hari V, Dharmasthala S, Koppa A, Karmakar S, Kumar R (2021) Climate hazards are threatening vulnerable migrants in Indian megacities. Nat Clim Chang:1–3

Hasegawa A, Gusyev M, Iwami Y (2016) Meteorological drought and flood assessment using the comparative SPI approach in Asia under climate change. J Disaster Res 11(6):1082–1090

Huang J, Zhang J, Zhang Z, Xu C, Wang B, Yao J (2011) Estimation of future precipitation change in the Yangtze River basin by using statistical downscaling method. Stoch Env Res Risk A 25(6):781–792

Huntingford C, Jeffers ES, Bonsall MB, Christensen HM, Lees T, Yang H (2019) Machine learning and artificial intelligence to aid climate change research and preparedness. Environ Res Lett 14(12)

IPCC (2001) Climate change 2001. Synthesis report (n.d.)

IPCC (2014) Working group II, fifth assessment report of the intergovernmental panel on climate change, summary for policymakers. Climate change 2014: impacts, adaptation, and vulnerability. Part a: global and sectoral aspects. Cambridge University Press, Cambridge/New York

IPCC (2021) In: Masson-Delmotte V, Zhai P, Pirani A, Connors SL, Péan C, Berger S, Caud N, Chen Y, Goldfarb L, Gomis MI, Huang M, Leitzell K, Lonnoy E, Matthews JBR, Maycock TK, Waterfield T, Yelekçi O, Yu R, Zhou B (eds) Climate change 2021: the physical science basis. Contribution of working group I to the sixth assessment report of the intergovernmental panel on climate change. Cambridge University Press

Jaiswal R, Mall RK, Singh N, Kumar TVL, Niyogi D (2022) Evaluation of Bias correction methods for regional climate models: Downscaled Rainfall analysis over diverse agro-climatic zones of India. Earth Space Sci:e2021EA001981. https://doi.org/10.1029/2021EA001981

Janssen NA, Hoek G, Simic-Lawson M, Fischer P, Van Bree L, Ten Brink H et al (2011) Black carbon is an additional indicator of the adverse health effects of airborne particles compared with PM10 and PM2. 5. Environ Health Perspect 119(12):1691–1699

Jehanzaib M, Sattar MN, Lee JH, Kim TW (2020) Investigating the effect of climate change on drought propagation from meteorological to hydrological drought using multi-model ensemble projections. Stoch Env Res Risk A 34(1):7–21

Jyoti J, Swapna P, Krishnan R, Naidu CV (2019) Pacific modulation of accelerated South Indian Ocean Sea-level rise during the early 21st century. Climate Dynam 53(7):4413–4432

Kahn ME, Mohaddes K, Ng RNC, Pesaran MH, Raissi M, Yang J-C (2019) Long-term macroeconomic effects of climate change: a cross-country analysis, Cambridge working papers in economics 1965, Faculty of Economics. University of Cambridge

Kilbourne EM (1999) The spectrum of illness during heat waves. Am J Prev Med 16(4):359–360

Knowlton K, Rotkin-Ellman M, King G, Margolis HG, Smith D, Solomon G, Trent R, English P (2009) The 2006 California heat wave: impacts on hospitalizations and emergency department visits. Environ Health Perspect 117(1):61–67. https://doi.org/10.1289/ehp.11594

Kumar KK, Parikh J (2001) Indian agriculture and climate sensitivity. Glob Environ Chang 11(2): 147–154

Kumar M, Raju MP, Singh RS, Banerjee T (2017) Impact of drought and normal monsoon scenarios on aerosol induced radiative forcing and atmospheric heating in Varanasi over middle indo-Gangetic plain. J Aerosol Sci 113:95–107

Kumar KJ, Chowdary KA, Usha HC, Kulkarni M, Manjunath VG (2018) Etiology of community-acquired pneumonia among children in India with special reference to atypical pathogens. Lung India: Official Organ of Indian Chest Society 35(2):116

Kumar V, Shukla T, Mehta M, Dobhal DP, Singh Bisht MP, Nautiyal S (2021) Glacier changes and associated climate drivers for the last three decades, Nanda Devi region, central Himalaya, India. Quat Int 575-576(May):213–226

Lavers DA, Allan RP, Villarini G, Lloyd-Hughes B, Brayshaw DJ, Wade AJ (2013) Future changes in atmospheric rivers and their implications for winter flooding in Britain. Environ Res Lett 8(3)

Lesk C, Rowhani P, Ramankutty N (2016) Influence of extreme weather disasters on global crop production. Nature 529(7584):84–87

Lewis SC, King AD (2017) Evolution of mean, variance and extremes in 21st-century temperatures. Weather Clim Extremes 15:1–10

Lozano R, Naghavi M, Foreman K, Lim S, Shibuya K, Aboyans V, Abraham J et al (2012) Global and regional mortality from 235 causes of death for 20 age groups in 1990 and 2010: a systematic analysis for the Global Burden of Disease Study 2010. Lancet 380(9859):2095–2128. https://doi.org/10.1016/S0140-6736(12)61728-0

Lindell MK (2013) Disaster studies. Curr Sociol 61(5–6):797–825

Luterbacher J, Dietrich D, Xoplaki E, Grosjean M, Wanner H (2004) European seasonal and annual temperature variability, trends, and extremes since 1500. Science 303(5663):1499–1503

Maghsood FF, Moradi H, Massah Bavani AR, Panahi M, Berndtsson R, Hashemi H (2019) Climate change impact on flood frequency and source area in northern Iran under CMIP5 scenarios. Water 11(2):273

Mahmood R, Jia S, Zhu W (2019) Analysis of climate variability, trends, and prediction in the most active parts of the Lake Chad Basin, Africa. Sci Rep 9(1):6317

Mall RK, Singh R, Gupta A, Srinivasan G, Rathore LS (2006) Impact of climate change on Indian agriculture: a review. Clim Change 78(2–4):445–478

Mall RK, Kumar R, Bhatla R (2011) Climate change and disaster in India. J South Asian Disaster Stud 4(1):27–76

Mall RK, Bhatt D, Sonkar G, Banerjee T (2014) Simulation modeling and climate change: issues and challenges. Environ Sci Pollut Res 21(19):11605–11608

Mall RK, Sonkar G, Bhatt D, Baxla AK, Singh KK (2016) Managing impact of extreme weather events in sugarcane in different agro climatic zones of Uttar Pradesh. Mausam 67(1):233–250

Mall RK, Gupta A, Sonkar G (2017) Effect of climate change on agricultural crops. In: Current developments in biotechnology and bioengineering. Elsevier, pp 23–46

Mall RK, Singh N, Singh KK, Sonkar G, Gupta A (2018) Evaluating the performance of RegCM4. 0 climate model for climate change impact assessment on wheat and rice crop in diverse agro-climatic zones of Uttar Pradesh. India Clim change 149(3):503–515

Mall RK, Srivastava RK, Banerjee T, Mishra OP, Bhatt D, Sonkar G (2019) Disaster risk reduction including climate change adaptation over South Asia: challenges and ways forward. Int J Disaster Risk Sci 10(1):14–27

Mall RK, Chaturvedi M, Singh N, Bhatla R, Singh RS, Gupta A, Niyogi D (2021) Evidence of asymmetric change in diurnal temperature range in recent decades over different agro? Climatic zones of India. Int J Climatol 41(4):2597–2610

Maurya S, Srivastava PK, Yaduvanshi A, Anand A, Petropoulos GP, Zhuo L, Mall RK (2021) Soil erosion in future scenario using CMIP5 models and earth observation datasets. J Hydrol 594: 125851

Mazdiyasni O, Agha Kouchak A, Davis SJ, Madadgar S, Mehran A, Ragno E et al (2017) Increasing probability of mortality during Indian heatwaves. Sci Adv 3(6):e1700066

Milojevic-Dupont N, Creutzig F (2020) Machine learning for geographically differentiated climate change mitigation in urban areas. Sustain Cities Soc:102526

Mishra B, Babel MS, Tripathi NK (2014) Analysis of climatic variability and snow cover in the Kaligandaki River basin, Himalaya, Nepal. Theor Appl Climatol 116(3):681–694

Mishra S, Singh R, Kumar R, Kalia A, Panigrahy SR (2017) Impact of climate change on pigeon pea. Econ Aff 62(3):455–457

Mishra V, Aadhar S, Mahto SS (2021) Anthropogenic warming and intraseasonal summer monsoon variability amplify the risk of future flash droughts in India. Npj Clim Atmos Sci 4(1):1

Mohapatra SC, Mohapatra P, Singh IJ, Gaur SD (1989) Epidemiology of gastro-intestinal and respiratory tract diseases in rural areas of Varanasi (India). Eur J Epidemiol 5(1):117–122. https://doi.org/10.1007/BF00145058

Moss R, Babiker M, Brinkman S, Calvo E, Carter T, Edmonds J, Elgizouli I, Emori S, Erda L, Hibbard K, Jones R, Kainuma M, Kelleher J, Lamarque JF, Manning M, Matthews B, Meehl J, Meyer L, Mitchell J et al (2008) Towards new scenarios for analysis of emissions, climate change, impacts, and response strategies?: IPCC expert meeting report?: 19–21 September 2007, Noordwijkerhout, the Netherlands. Intergovernmental Panel on Climate Change, Geneva

Mukherjee S, Mishra A, Trenberth KE (2018) Climate change and drought: a perspective on drought indices. Curr Clim Change Report 4(2):145–163

Murari V, Singh N, Ranjan R, Singh RS, Banerjee T (2020) Source apportionment and health risk assessment of airborne particulates over central Indo-Gangetic Plain. Chemosphere 257. https://doi.org/10.1016/j.chemosphere.2020.127145

Myhre G, Samset BH, Schulz M, Balkanski Y, Bauer S, Berntsen TK et al (2013) Radiative forcing of the direct aerosol effect from AeroCom phase II simulations. Atmos Chem Phys 13(4): 1853–1877

Nagalapalli S, Kundu A, Mall RK, Thattai D, Rangarajan S (2019) An appraisal of flood events using IMD, CRU, and CCSM4-derived meteorological data sets over the Vaigai river basin, Tamil Nadu (India). Sustain Water Resour Manag 5:1–14

Niyogi D, Subramanian S, Mohanty UC, Kishtawal CM, Ghosh S, Nair US, Ek M, Rajeevan M (2018) The impact of land cover and land use change on the Indian monsoon region hydroclimate. In: Land-Atmospheric Research Applications in South and Southeast Asia. Springer, Berlin, pp 553–575

Nori-Sarma A, Anderson GB, Rajiva A, Shah Azhar G, Gupta P, Pednekar MS et al (2019) The impact of heatwaves on mortality in Northwest India. Environ Res 176:108546

NRDC (2021) Global warming 101. Accessed from - https://www.nrdc.org/stories/global-warming-101#causes on 6th Aug 2021

Nyhan M (2015) Coping with air pollution in an age of urbanisation. Angle Journal. Retrieved from http://anglejournal.com/article/2015-06-protecting-urban-populations-from-air-pollution-in-an-age-of-global-urbanisation/

Orlowsky B, Seneviratne SI (2012) Global changes in extreme events: regional and seasonal dimension. Clim Change 110(3):669–696

Pandey V, Srivastava P (2019) Integration of microwave and optical/infrared derived datasets for a drought Hazard inventory in a sub-tropical region of India. Remote Sens (Basel) 11(4):439

Pandey V, Srivastava PK, Singh SK, Petropoulos GP, Mall RK (2021) Drought identification and trend analysis using long-term CHIRPS satellite precipitation product in Bundelkhand, India. Sustainability 13(3):1042

Parthasarathy B, Rupa Kumar K, Munot AA (1992) Forecast of rainy season foodgrain production based on monsoon rainfall. Indian J Agric Sci 62:1–8

Pathak H, Ladha JK, Aggarwal PK, Peng S, Das S, Singh Y et al (2003) Trends of climatic potential and on-farm yields of rice and wheat in the indo-Gangetic Plains. Field Crop Res 80(3):223–234

Pathak SK, Singh P, Singh MM, Sharma BL (2019) Impact of temperature and humidity on sugar recovery in Uttar Pradesh. Sugar Tech 21(1):176–181

Patel S, Mall RK, Jaiswal R, Singh R, Chand R (2022) Vulnerability assessment of wheat yield under warming climate in Northern India using multi-model projections. Int J Plant Prod 1–16. https://doi.org/10.1007/s42106-022-00208-1

Peng RD, Bobb JF, Tebaldi C, McDaniel L, Bell ML, Dominici F (2011) Toward a quantitative estimate of future heat wave mortality under global climate change. Environ Health Perspect 119(5):701–706

Philipsborn RP, Chan K (2018) Climate change and global child health. Pediatrics 141(6)

Rai P, Choudhary A, Dimri AP (2019) Future precipitation extremes over India from the CORDEX-South Asia experiments. Theor Appl Climatol 137(3):2961–2975

Rajak R, Chattopadhyay A (2020) Short and long term exposure to ambient air pollution and impact on health in India: a systematic review. Int J Environ Health Res 30(6):593–617

Ramanathan V (2007) Global dimming by air pollution and global warming by greenhouse gases: global and regional perspectives. In: Nucleation and atmospheric aerosols. Springer, Dordrecht, pp 473–483

Ramanathan V, Chung C, Kim D, Bettge T, Buja L, Kiehl JT et al (2005) Atmospheric brown clouds: impacts on south Asian climate and hydrological cycle. Proc Natl Acad Sci 102(15): 5326–5333

Ray DK, Gerber JS, Macdonald GK, West PC (2015) Climate variation explains a third of global crop yield variability. Nat Commun 6:1–9

Ray DK, West PC, Clark M, Gerber JS, Prishchepov AV, Chatterjee S (2019) Climate change has likely already affected global food production. PLoS One 14(5):0217148

Riedl MO (2019) Human-centred artificial intelligence and machine learning. Human Behav Emerg Technol 1(1):33–36

Rosenzweig C, Tubiello FN (2007) Adaptation and mitigation strategies in agriculture: an analysis of potential synergies. Mitig Adapt Strat Glob Chang 12(5):855–873

Roxy MK, Ghosh S, Pathak A, Athulya R, Mujumdar M, Murtugudde R, Terray P, Rajeevan M (2017) A threefold rise in widespread extreme rain events over Central India. Nat Commun 8(1): 708

Saha S, Gogoi P, Gayen A, Paul GC (2021) Constructing the machine learning techniques based spatial drought vulnerability index in Karnataka state of India. J Clean Prod 314:128073

Saini P, Sharma M (2020) Cause and age-specific premature mortality attributable to PM2. 5 exposure: an analysis for million-plus Indian cities. Sci Total Environ 710:135230

Sanchez-pi N, Marti L, Abreu A, Maass A (2021) Artificial intelligence, machine learning and modeling for understanding the oceans and climate change to cite this version?: HAL Id?: hal-03138712 Artificial Intelligence. Machine Learning and Modeling for Understanding the Oceans and Climate Change

Sarthi PP, Agrawal A, Rana A (2014) Possible future changes in cyclonic storms in the Bay of Bengal, India under warmer climate. Int J Climatol 35(7):1267–1277

Satterthwaite D, Archer D, Colenbrander S et al (2020) Building resilience to climate change in informal settlements. One Earth 2(2):143–156. https://doi.org/10.1016/j.oneear.2020.02.002

Scott D, Stefan G, Hall CM, Wiley J (2012) International tourism and climate. 3(June):213–232

Sedova B, Kalkuhl M (2020) Who are the climate migrants and where do they go? Evidence from rural India q. World Dev 129:104848

Semenza JC, Ebi KL (2019) Climate change impact on migration, travel, travel destinations and the tourism industry. J Travel Med 26(5):1–13

Seneviratne SI, Nicholls N, Easterling D et al (2012) Changes in climate extremes and their impacts on the natural physical environment. In: Managing the Risks of Extreme Events and Disasters to Advance Climate Change Adaptation: Special Report of the Intergovernmental Panel on Climate Change. Cambridge University Press, Cambridge/New York, 9781107025, pp 109–230

Sharifi A (2021) Co-benefits and synergies between urban climate change mitigation and adaptation measures: a literature review. Sci Total Environ 750(July):141642

Sharma S, Mujumdar P (2017) Increasing frequency and spatial extent of concurrent meteorological droughts and heatwaves in India. Sci Rep 7(1):1–9

Singh BR, Singh O (2012) Study of impacts of global warming on climate change: rise in sea level and disaster frequency. In: Global warming impacts and future perspective. IntechOpen. https://doi.org/10.5772/50464

Singh N, Murari V, Kumar M, Barman SC, Banerjee T (2017) Fine particulates over South Asia: review and meta-analysis of PM2. 5 source apportionment through receptor model. Environ Pollut 223:121–136

Singh N, Mall RK, SINGH K, GUPTA A, Sonkar G (2018) Evaluation of RegCM4 climate model for assessment of climate change impact on crop production. Mausam 69(3):387–398

Singh N, Mhawish A, Ghosh S, Banerjee T, Mall RK (2019) Attributing mortality from temperature extremes: a time series analysis in Varanasi, India. Sci Total Environ 665:453–464

Singh H, Yadav M, Kumar N, Kumar A, Kumar M (2020) Assessing adaptation and mitigation potential of roadside trees under the influence of vehicular emissions: a case study of Grevillea robusta and Mangifera indica planted in an urban city of India. PLoS One 15(1):1–20

Singh N, Singh S, Mall RK (2020a) Urban ecology and human health: implications of urban heat island, air pollution and climate change nexus. In: Urban ecology. Elsevier, pp 317–334

Singh H, Singh N, Mall RK (2020b) Japanese encephalitis and associated environmental risk factors in eastern Uttar Pradesh: a time series analysis from 2001 to 2016. Acta Trop 212: 105701

Singh S, Mall RK, Singh N (2021a) Changing spatio? Temporal trends of heatwave and severe heatwave events over India: an emerging health hazard. Int J Climatol 41:E1831–E1845

Singh S, Mall RK, Dadich J, Verma S, Singh JV, Gupta A (2021b) Evaluation of CORDEX-South Asia regional climate models for heatwave simulations over India. Atmos Res 248:105228

Singh N, Mhawish A, Banerjee T, Ghosh S, Singh RS, Mall RK (2021c) Association of aerosols, trace gases and black carbon with mortality in an urban pollution hotspot over central indo-Gangetic plain. Atmos Environ 246:118088

Singh N, Mall RK, Banerjee T, Gupta A (2021d) Associations between climate, and infectious diseases among children in Varanasi City, India: A prospective cohort study. Sci Total Environ 796:148769

Sonkar G, Mall RK, Banerjee T, Singh N, Kumar TL, Chand R (2019) Vulnerability of Indian wheat against rising temperature and aerosols. Environ Pollut 254:112946

Sonkar G, Singh N, Mall RK, Singh KK, Gupta A (2020) Simulating the impacts of climate change on sugarcane in diverse agro-climatic zones of northern India using CANEGRO-sugarcane model. Sugar Tech 22(3):460–472

Soora NK, Aggarwal PK, Saxena R, Rani S, Jain S, Chauhan N (2013) An assessment of regional vulnerability of rice to climate change in India. Clim Change 118(3):683–699

Srivastava AK, Rai MK (2012) Sugarcane production: impact of climate change and its mitigation. Biodiversitas 13:214–227. https://doi.org/10.13057/biodiv/d130408

Srivastava P, Dey S, Srivastava AK, Singh S, Tiwari S (2019) Suppression of aerosol-induced atmospheric warming by clouds in the indo-Gangetic Basin, northern India. Theor Appl Climatol 137(3):2731–2741

Srivastava PK, Singh P, Mall RK, Pradhan RK, Bray M, Gupta A (2020) Performance assessment of evapotranspiration estimated from different data sources over agricultural landscape in northern India. Theor Appl Climatol 140(1):145–156

Srivastava PK, Pradhan RK, Petropoulos GP, Pandey V, Gupta M, Yaduvanshi A, Mall RK (2021) Long-term trend analysis of precipitation and extreme events over Kosi River basin in India. Water 13(12):1695

Stein AL (2020) Artificial intelligence and climate change. Yale J Regul 37(3)

Stocker TF, Qin D, Plattner GK, Tignor M, Allen SK, Boschung J, Nauels A, Xia Y, Bex V, Midgley PM (2018) Climate change 2013: the physical science basis. Contribution of Working Group I to the Fifth Assessment Report of the Intergovernmental Panel on Climate Change

Swapna P, Jyoti J, Krishnan R, Sandeep N, Griffies SM (2017) Multidecadal weakening of Indian summer monsoon circulation induces an increasing northern Indian Ocean Sea level. Geophys Res Lett 44(20):10,560–10,572

Swapna P et al (2020) Sea-level rise. In: Krishnan R, Sanjay J, Gnanaseelan C, Mujumdar M, Kulkarni A, Chakraborty S (eds) Assessment of climate change over the Indian region. Springer, Singapore. https://doi.org/10.1007/978-981-15-4327-2_9

Tabari H (2020) Climate change impact on flood and extreme precipitation increases with water availability. Sci Rep 10(1):1–10

Thrall JH, Li X, Li Q, Cruz C, Do S, Dreyer K, Brink J (2018) Artificial intelligence and machine learning in radiology: opportunities, challenges, pitfalls, and criteria for success. J Am Coll Radiol 15(3):504–508

Tien Bui D, Shirzadi A, Chapi K, Shahabi H, Pradhan B, Pham B, Singh V, Chen W, Khosravi K, Bin Ahmad B, Lee S (2019) A hybrid computational intelligence approach to groundwater spring potential mapping. Water 11(10):2013

Tigchelaar M, Battisti DS, Naylor RL, Ray DK (2018) Future warming increases the probability of globally synchronized maize production shocks. Proc Natl Acad Sci 115(26):6644–6649

Tyagi S, Singh N, Sonkar G (2019) Sensitivity of evapotranspiration to climate change using DSSAT model in sub-humid climate region of eastern Uttar Pradesh. Model Earth Syst Environ 5(1):1–11

Uk RJN, Poh P, Hay J, Zealand N, Ragoonaden S, Arblaster J, Uk BB, Canada DF, Uk JH, Uk SK, Uk JL, Uk SR (2007) Coastal systems and low-lying areas coordinating lead authors. Lead Authors: Contributing Authors, pp 315–356

Ullah Z, Al-Turjman F, Mostarda L, Gagliardi R (2020) Applications of artificial intelligence and machine learning in smart cities. Comput Commun 154:313–323

UNICEF. Diarrhoeal disease. Available online: http://data.unicef.org/childhealth/diarrhoeal-disease.html. Accessed 12 May 2020

Vautard R, Gobiet A, Jacob D, Belda M, Colette A, Déqué M, Fernández J, García-Díez M, Goergen K, Güttler I (2013) The simulation of European heat waves from an ensemble of regional climate models within the EURO-CORDEX project. Climate Dynam 41(9–10): 2555–2575

Verma RR, Srivastava TK, Singh P (2019) Climate change impacts on rainfall and temperature in sugarcane growing upper Gangetic Plains of India. Theor Appl Climatol 135(1–2):279–292

Vinjamuri KS, Mhawish A, Banerjee T, Sorek-Hamer M, Broday DM, Mall RK, Latif MT (2020) Vertical distribution of smoke aerosols over upper indo-Gangetic plain. Environ Pollut 257: 113377

Viswanathan B, Kavi Kumar K (2015) Weather, agriculture and rural migration: evidence from state and district level migration in India. Environ Dev Econ 20(4):469–492. https://doi.org/10.1017/S1355770X1500008X

Vittal H, Karmakar S, Ghosh S (2013) Diametric changes in trends and patterns of extreme rainfall over India from pre? 1950 to post? 1950. Geophys Res Lett 40(12):3253–3258

Wasko C, Westra S, Nathan R, Orr HG, Villarini G, Villalobos Herrera R, Fowler HJ (2021) Incorporating climate change in flood estimation guidance. Phil Trans R Soc A 379(2195): 20190548

Watts N, Adger WN, Ayeb-Karlsson S, Bai Y, Byass P, Campbell-Lendrum D et al (2017) The lancet countdown: tracking progress on health and climate change. Lancet 389(10074): 1151–1164

WHO. World Health Organization (2007) Health relevance of particulate matter from various sources: report on a WHO workshop, Bonn, Germany 26–27 March 2007. WHO Regional Office for Europe, Copenhagen

WHO (2017) World Health Organisation Global health estimates: leading causes of death, cause specific-mortality 2009–2019. Accessed from https://www.who.int/data/gho/data/themes/mortality-and-global-health-estimates/ghe-leading-causes-of-death

Xu K, Xu B, Ju J, Wu C, Dai H, Hu BX (2019) Projection and uncertainty of precipitation extremes in the CMIP5 multimodel ensembles over nine major basins in China. Atmos Res 226:122–137

Yadav SS, Lal R (2018) Vulnerability of women to climate change in arid and semi-arid regions: the case of India and South Asia. J Arid Environ 149:4–17

Yadav MK, Singh RS, Singh KK, Mall RK, Patel CB, Singh MK (2015) Assessment of climate change impact on the productivity of different cereal crops in Varanasi, India. J Agrometeorol 17(2):179–184

Yadav MK, Singh RS, Singh KK, Mall RK, Patel C, Yadav SK, Singh MK (2016) Assessment of climate change impact on pulse, oilseed and vegetable crops at Varanasi, India. J Agrometeorol 18(1):13–21

Yadav SK, Yadav M, Patel C, Singh RS, Singh KK, Mall RK, Singh MK (2021) Assessment of climate change impact on different pigeon pea maturity groups in north Indian condition. J Agrometeorol 23(1):82–91

Zarekarizi M, Rana A, Moradkhani H (2018) Precipitation extremes and their relation to climatic indices in the Pacific Northwest USA. Climate Dynam 50(11–12):4519–4537

Zhang Y, Bi P, Hiller JE (2007) Climate change and disability-adjusted life years. J Environ Health 70(3):32–38

Zhang J, Liu Y, Cui LL, Liu SQ, Yin XX, Li HC (2017) Ambient air pollution, smog episodes and mortality in Jinan, China. Sci Rep 7(1):1–8

Zhang J, Li Y, Tao W, Liu J, Levinson R, Mohegh A, Ban-Weiss G (2019) Investigating the urban air quality effects of cool walls and cool roofs in Southern California. Environ Sci Technol 53(13):7532–7542. https://doi.org/10.1021/acs.est.9b00626

Chapter 3
Assessing the Historical and Future Relationship Between Climatic Factors and the Production of Different Crops over India

Bushra Praveen, Kanak Singh, Pritee Sharma, Shaghla Parveen, and Swapan Talukdar

Abstract This analysis explores the effects of climate variations over 15 different food and industrial crop productions in India using the econometric model and forecasts these crop outputs using the Artificial Neural Network (ANN) model up to 2030. The unit root test was used to inspect the stationarity time series datasets in agricultural crop productions and climatic variables. The Dynamic Ordinary Least Square (DOLS) and Fully Modified Dynamic Ordinary Least Square (FMOLS) models were used to test the accuracy of the outcomes. The results showed that land productivity declines if the annual mean temperature increases, demonstrating the sensitivity of Indian agricultural crop production to the temperature variance. Meanwhile, results also showed that if the temperature increases, rainfall would become more unstable, triggering a reduction of rice, wheat, tuar, cotton, and tea crop productions. Major crop productions, namely sesamum (til), gram, sugarcane, groundnut, jowar, and bajra would experience a positive impact and subsequently increase slightly up to 2030. The outcomes of the study can add insights into sustainable agricultural crop productions in India, and livelihood patterns, particularly among minor and marginal agricultural families.

Keywords Climate change · Agriculture productivity · Forecasting · Artificial neural network · Dynamic ordinary least square (DOLS)

B. Praveen (✉) · K. Singh · P. Sharma
School of Humanities and Social Sciences, Indian Institute of Technology Indore, Indore, India
e-mail: psharma@iiti.ac.in

S. Parveen
Department of Geography, Aligarh Muslim University, Aligarh, India

S. Talukdar
Department of Geography, University of Gour Banga, Malda, India

N. Khare (ed.), *Science, Policies and Conflicts of Climate Change*, Springer Climate,
https://doi.org/10.1007/978-3-031-16254-1_3

Introduction

Climate change is a hot issue affecting agricultural crop production that is sensitive to climatic factors and their fluctuations (Kumar and Gautam 2014). Several climatic factors such as rising temperature, highly unstable rainfall, recurrent drought, sea level rise and glacier melting have constantly reminded us that climate change has already impacted ecosystems as well as humans across the world. Although the fluctuation of climatic factors is natural, anthropogenic activities like the production of greenhouse gases, deforestation, expansion of urbanization, advanced industrial development, usage of new machinery and technology, advanced economic enlargement and transportation have accelerated their fluctuation in recent decades (Ahmad et al. 2011). Consequently, significant reduction in several crop's production and productivity, particularly in developing countries, has led to food insecurity, famine, malnutrition and other problems as their livelihoods mainly depend on agricultural crop production (Nath and Behera 2011). However, the climatic factors may be different in mid-latitude developed countries because of the application of advanced technology in agriculture, less dependency on agriculture, and awareness of agricultural science like crop rotation, mixed farming, etc. Therefore, investigating the repercussions of climate variation on agricultural crop production is paramount, especially for countries like India.

The outturns of climate variance on agricultural productions are being measured using three approaches: biophysical, hedonic, and panel data. Among the three approaches, the most important and popular method is the biophysical approach also known as the crop modeling technique (Adams et al. 2013; Aggarwal 2009; Kurukulasuriya et al. 2006; Lal 2000; Rao and Sinha 1994; Rosenzweig and Iglesias 1994; Reilly et al. 1994; Aggarwal and Sinha 1993; Kane et al. 1992). Apart from this, the hedonic or Ricardian approach has been applied to explore the long-term phenomenal variations of climate in the cropping system of production while considering adaptation (Mendelsohn 2014; Mendelsohn et al. 2010; (Mendelsohn and Dinar 2009; Deschênes and Greenstone 2007; Kurukulasuriya et al. 2006; Kumar and Parikh 2001). Furthermore, the panel data approach is also a popular method to quantify the reverberations of climate variation upon agricultural productivity. The change in climate may influence the economy in various ways. For example, high fluctuation of rainfall leads to irregularity and severity in floods, resulting in a huge number of crop losses. The interrelation between climate change and agricultural productivity is quantified through the aforementioned technique. The increase in surface temperature raises the mean sea level which affects the livelihood of inhabitants in the coastal regions of the world. IPCC projected that the temperature will increase by 1–6 °C by 2100. Consequently, the rise in the degree of heat intensity and irregular precipitation patterns will have a severe impact on crop productivity. Therefore, climate change has been severely affecting food production throughout the world including in India, which causes food insecurity (Meeting 2006).

There have been many types of research that report the negative consequence of climate variation over cropping patterns as well as crop production in India, but very few studies have been performed using empirical techniques regarding this issue. Kumar and Parikh (2001) have estimated the production of a few prominent crops like wheat, maize, barley, sorghum, and arhar. These crops were prone to adversity due to high climate sensitivity; therefore, it is also critical for the security of food in India. In addition to this, the production of commercial crops such as cotton, sesamum, and sugarcane have also declined since 1990 (Singh 2012) due to the elevated temperature. It is estimated by 2060 that the changes in climate would result in a decline in the production of potatoes and paddy and potentially endanger food safety for nearly one billion of the country's population. For food grain, any fluctuation in temperature below normal has a dismal and numerically noteworthy influence on linseed production (Singh 2012). Kumar et al. (2011) reported that irrigated areas of maize, mustard, wheat, rice, and sorghum in the seaside region, North Eastern Region, and Sahyadri region or the Western Ghats have been declining because of the negative effects of climate variations. Hundal (2007) stated that as the mean temperature spiraled by 1–3 ° C above the normal range, it resulted in a decrease in paddy production and wheat production, respectively, at 3% and 10% inside the state of Punjab. The unreliability in the pattern of precipitation occurrences has inimically slowed down the cropping of Jowar in Karnataka which led to food insecurity among farmers (Kaul and Ram 2009). According to Geethalakshmi et al. (2011), the production of paddy crops in Tamilnadu has declined by more than 41% due to the temperature reaching 40 °C. Furthermore, Saseendran et al. projected the paddy crop and temperature in the state of Kerala up to 2049 and found that temperature would cross 50 °C which would significantly reduce the production of paddy in the state. Saseendran et al. reported that each 1-degree temperature increase would reduce crop production by up to 6. Srivastava et al. argued that the production of monsoonal crops, i.e., sorghum may decline by 14% and 2%, respectively, by the year 2020 in regions of central India and southern central India in response to the phenomenal climatic variance. Empirical evidence indicates that an increase in atmospheric heat has limited consequences on the production of rice, maize, jawar (sorghum), bajra, and barley (Kalra et al. 2008; Geethalakshmi et al. 2011). The agricultural production outlook for gram and ragi is bleak due to an increase in the maximal degree of heat; however, the produced output of wheat and tuar have increased firmly due to the increase in the maximal intensity of heat (Kumar and Parikh 2001; Kaur and Hundal 2007). Kapur et al. (2009) estimated a potential 30% reduction in crop production in the middle of the twenty-first century as intensification occurred in surface warming with a change in the rate of precipitation, which could lead to a decline in arable land triggering crop production stresses in India. Evidence shows climate variability has negatively influenced agricultural crop production. Moreover, most of the empirical studies have either covered single crops or multiple crops with limited geographical coverage. Therefore, it is imperative to determine the overall effect of climate variability on major food grains along with a focus on profitable and sustainable crop production choices for ensuring food security in India (Hollaender 2010). Our study aims at testing the hypothesis which says, agricultural output in the country of India has a climate-sensitive nature, and any variation in rainfall precipitation and temperature patterns significantly influence the production of the food grain. The development of a proper policy guideline related to the irrigation system could possibly alleviate the detrimental consequences of climate sensitiveness on wheat, paddy, sorghum, tuar, and bajra, along with managing food safety in India (Kar and Kar 2008; Singh 2012; Ranuzzi and Srivastava 2012; Sing 2012).

The development of the agricultural sector executes a pivotal share in the overall socio-economic well-being of the people in India. Critically, the amount of agricultural share in Gross Domestic Product (GDP) as well as in employment has reduced, although the benefit of cropping systems is still highly significant and important for economic development (Mall et al. 2006). In India, extensive works were done in previous years assessing the nature and magnitude of crop yield fluctuations with a focus on climate change issues in the past decade. Nevertheless, those cited works concentrated on the reverse impact of climate changeability on some of the selected crops only. Considering limited coverage in earlier studies at the time, we have set two broad objectives here; the first is to analyze the repercussions of climate variability over different crops and measure the robustness of the methods using different statistical techniques. The second objective is to simulate and estimate the future climatic factors and the trends in different crop production using an Artificial Neural Network (ANN).

In this study, we applied a linear regression model to explore the trend of future climatic factors and the production of different crops. The application of artificial neural networks to forecast future production is a comparatively new contribution to the agricultural sector.

Method and Data Description

Study Area and Data Description

India is in South Asia situated from 8°4 to 37°6 north latitude and 68°70 to 9°25° east longitude north of the equator, circumambient by water bodies on three sides – the Arabian Sea on the West, the Bay of Bengal on the East, and the Indian Ocean on the South. The Indian Meteorological Department (IMD) divided the entire country into 34 meteorological subdivisions where the climate data were recorded (Fig. 3.1). In this analysis, we obtained meteorological subdivision wise data on temperatures and rainfall from the Indian Meteorological Department (IMD), a database of the Government of India from 1901 to 2015. Under this analysis, the continuous data was used for a period from 1976 to 2016, with no missing values.

In this study, to examine the effect of climate change on crop productivity, 16 major crops were taken into consideration, including bajra, rice, wheat, barley, arhar, ragi, maize, jowar, gram, and mustard plus every grain crop that are non-food such as linseed, sugarcane, groundnut, rapeseed, cotton, til and tea, covering over 75% of the total cropland. To estimate the climate change consequences on the cropping pattern as well as its production, we considered per unit area as a dependent

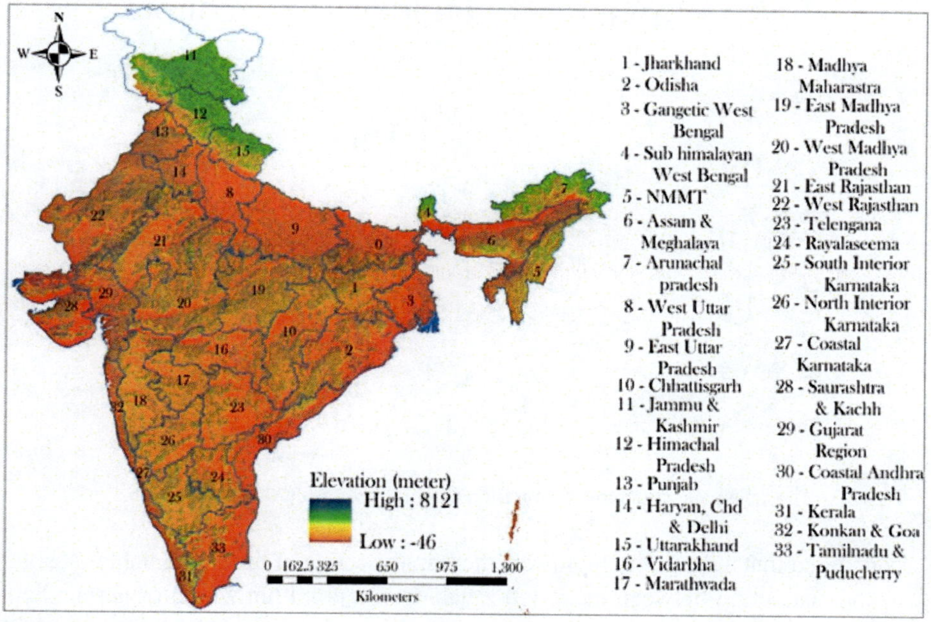

Fig. 3.1 Geographical location of the study area

variable by using the panel econometric analysis from 1967 to 2016. The Centre Monitoring Indian Economy (CMIE) provided the production dataset.

Descriptive Statistics of the Climatic Variables and Crops Production

The data analysis was performed using basic statistical techniques such as mean, standard deviation, kurtosis, and skewness to explore the real situation of these variables. In this study, descriptive statistical techniques have been applied to the climatic variables and agricultural productivity for the period 1967 to 2018.

Innovative Trend Analysis Method

As suggested by Sen, when two similar time series are plotted against each other, all points on the Cartesian coordinate system (Fig. 3.2) are scattered on the 1:1 line regardless of their distribution form, sample length, serial dependency, or trend form. Data points collected on the 1:1 (45°) line in the scatter plot indicates they were trendless (data without trend). If the data points fall above the upper triangular area, the line of 1:1, then it is said that the statistic undergoes a trend of upliftment. If the data points accumulated below the line of 1:1 in the triangular area, it can be

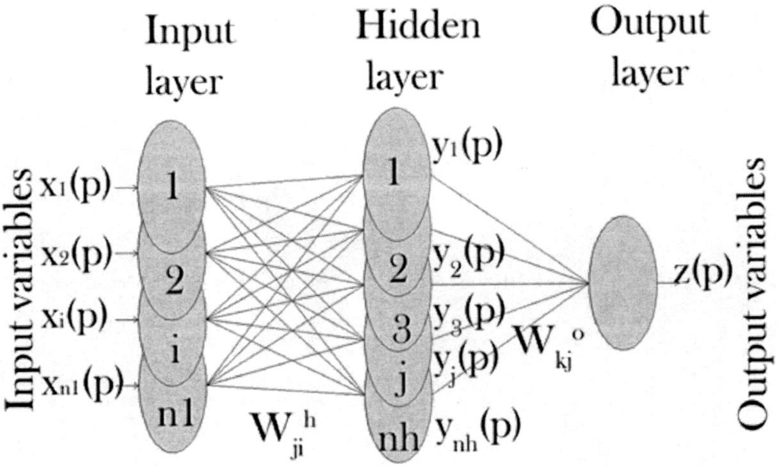

Fig. 3.2 The schematic structure of an artificial neural network (ANN)

concluded that a declining trend was there in the series of time. The value (absolute) of the distinction between the y and x values of a point horizontal or perpendicular distance from the one, suggests the amount of an enlarging or a reducing trend. The key benefits of revolutionary approaches can be understood with the fact that they are not based on any assumptions such as serial correlation, non-normality, sample size, and low, medium, and high data trends could be detected with less effort through the approach. Therefore, it is useful for measuring a trend, as well as the average climate change and agricultural production variations.

Over the years, a good number of researchers have frequently used ITA for identifying trends in hydrological and climatological parameters applied the Mann-Kendall (MK) test and ITA in Kizilirmak River, Turkey to analyzing the trends of parameters for water quality. The author has used the successful application of the ITA technique for trend assessment. Kisi implied the ITA method over six stations in Turkey to assess the trend system on monthly pan evaporation data and observed the negative and positive trends over the study area. Cui et al. used the ITA method, MK test, linear regression method, and Sen's slope estimator to analyze the trend in annual and seasonal rainfall and temperature of the Yangtze River Basin, China, and a significant increase in those climatic variables was reported. This analysis used the ITA method for the first time in India for agricultural research.

Multiple Regression

Artificial Neural Network (ANN)

The ANN is an abstract mathematical and black box model. It has been applied in countless fields, including decision making, pattern recognition, automatic controlling systems, robotics, and others. It can handle complex, non-linear, and unbalanced data sets. Therefore, it can imitate the functioning of the human brain and is even able to generalize and predict the output from a large number of complex inputs. For this reason, researchers across the world have widely used it to solve different problems in different fields. The ANN model can perform like an expert, which can detect complex predictive patterns, which is not apparent to the non-expert. It can act on the category, continuous and binary data without violating the assumption and characters of the data.

Several architectures of the neural network have widely been used. In this study, the feed-forward based multilayer perceptron (MLP) architecture was used. The standard MLP consists of three layers, such as an input, one or more hidden and output layers of non-linear activation nodes (Table 3.1). Each layer contains many neurons or nodes, which are connected with a certain weight to every node in the next layer. Their work is to transfer the information. Thus, the neural network has been formed. The MLP uses the backpropagation algorithm for training the network until the minimum errors are achieved between the anticipated and output values of the network. Thus, the ANN model generates the results.

Result and Discussion

Descriptive Statistics Analysis of Climatic Variables and Different Crop Production

The descriptive statistics are provided in Table 3.2, which summarizes the fundamental properties of all the variables under study. The mean annual rainfall and temperature were measured at 1388.56 mm and 24.87 °C, respectively, for climate variables. Negative kurtosis was found for all climatic and agricultural variables, except barley and ragi. The observed kurtosis values for all variables are not above −1 and 1 except for three variables, so the data is acceptable for further analysis. The rainfall, rice, jowar, and wheat have negative skewness and the rest of the variables have positive and moderate skewed distribution and can be called approximate symmetry data, except for two variables (Table 3.1).

Trend Analysis of the Climatic Variables and Crops Production

Figure 3.3 provides trend analysis results for climatic variables and agriculture production using the ITA method (Table 3.3). The rainfall and temperature showed that no trend was detected for rainfall, but the trend was slightly negative for the past 3 years. The high degree of the increasing trend was found in the case of temperature, which was very high for the past few years. The results of an innovative trend for cotton, tea, and rice showed a high degree of the positive trend while rapeseed has a

Table 3.1 Calculated parameters of the algorithms for different climates and crops used in the study

	Number of inputs (Lag)	Number of hidden layer	Seed	Activation function in hidden unit	Number of iterations	Learning algorithm	Learning rate	Momentum
Rainfall	10	8	15	Sigmoid	2000	Back propagation	0.1	0.2
Temperature	15	8	15	Sigmoid	2000	Back propagation	0.1	0.2
Cotton	10	8	15	Sigmoid	2000	Back propagation	0.1	0.2
Rice	15	8	15	Sigmoid	2000	Back propagation	0.1	0.2
Tea	10	8	15	Sigmoid	2000	Back propagation	0.09	0.2
Rapeseed & Mustard	10	8	10	Sigmoid	2000	Back propagation	0.1	0.2
Linseed	15	8	10	Sigmoid	1500	Back propagation	0.09	0.2
Maize	10	8	12	Sigmoid	2000	Back propagation	0.1	0.2
Arhar	15	8	20	Sigmoid	2000	Back propagation	0.09	0.2
Sesamum (Til)	15	8	5	Sigmoid	2200	Back propagation	0.1	0.23
Jowar	15	9	20	Sigmoid	42,000	Back propagation	0.19	0.4
Groundnut	10	8	20	Sigmoid	3200	Back propagation	0.19	0.3
Bajra	15	9	20	Sigmoid	50,000	Back propagation	0.19	0.42
Barley	15	8	20	Sigmoid	5000	Back propagation	0.19	0.22
Ragi	10	8	20	Sigmoid	30,000	Back propagation	0.19	0.22
Wheat	15	8	22	Sigmoid	38,000	Back propagation	0.18	0.22
Gram	15	7	20	Sigmoid	1000	Back propagation	0.19	0.22

Table 3.2 Descriptive analysis of the climatic variables and crop production

Climatic variables and production of various crops	Descriptive statistics					
	Mean	Standard Deviation	Kurtosis	Skewness	Minimum	Maximum
Rainfall	1388.56	108.11	−0.15	−0.21	1131.86	1611.18
Temperature	24.87	0.86	−1.29	0.63	23.74	26.46
Cotton	591.75	138.15	−0.77	0.37	347.70	893.00
Rice	70172.27	22622.23	−1.29	−0.02	30437.90	106645.50
Tea	728.47	231.10	−0.63	0.37	376.00	1208.70
Rapeseed & Mustard	315.71	133.17	−1.00	0.33	125.50	597.80
Linseed	6595.12	1437.33	−0.61	0.26	4091.60	9713.90
Maize	6636.81	2057.97	−0.37	0.40	3271.90	12109.30
Arhar	14050.71	3145.08	−0.93	0.28	8347.10	20368.10
Sesamum (Til)	1990.48	2648.94	0.62	1.53	31.80	8178.70
Jowar	54306.49	23826.01	−1.12	−0.01	11392.80	95849.80
Groundnut	6595.12	1437.33	−0.61	0.26	4091.60	9713.90
Bajra	227464.73	81291.28	−1.23	0.10	92826.10	362332.80
Barley	1855.98	546.55	1.04	1.24	1196.10	3503.60
Ragi	10912.51	5693.94	0.07	1.11	4893.60	24259.50
Wheat	54306.49	23826.01	−1.12	−0.01	11392.80	95849.80
Gram	5492.08	1346.31	0.95	0.97	3356.30	9526.30

very low degree of negative trend. Linseed, jowar, maize, groundnut, and bajra have a high magnitude of the increasing trend of their production. The increasing tendency has been shown for the past few years, which indicates that overall agricultural production has been increasing. In the case of til and barley, however, very low trends have been detected, which have become more apparent in the past few years. The increasing trend has been more prominent for the past few years for ragi and gram. The trend analysis showed that most edible food grains like rice, gram, and ragi have increased significantly, despite having low agricultural land and very low technological development. Therefore, due to the rising production of agriculture, the result of climate change cannot be detected from direct observation.

Relationship Between Climatic Factors and Crops Production

Forecasting of Climatic Parameters and Different Crops

We performed ANN methods for forecasting climatic parameters and production of different crops up to 2030. It is essential to check the performance of the model by predicting existing data before it is used for the forecast. It is discussed that if there is

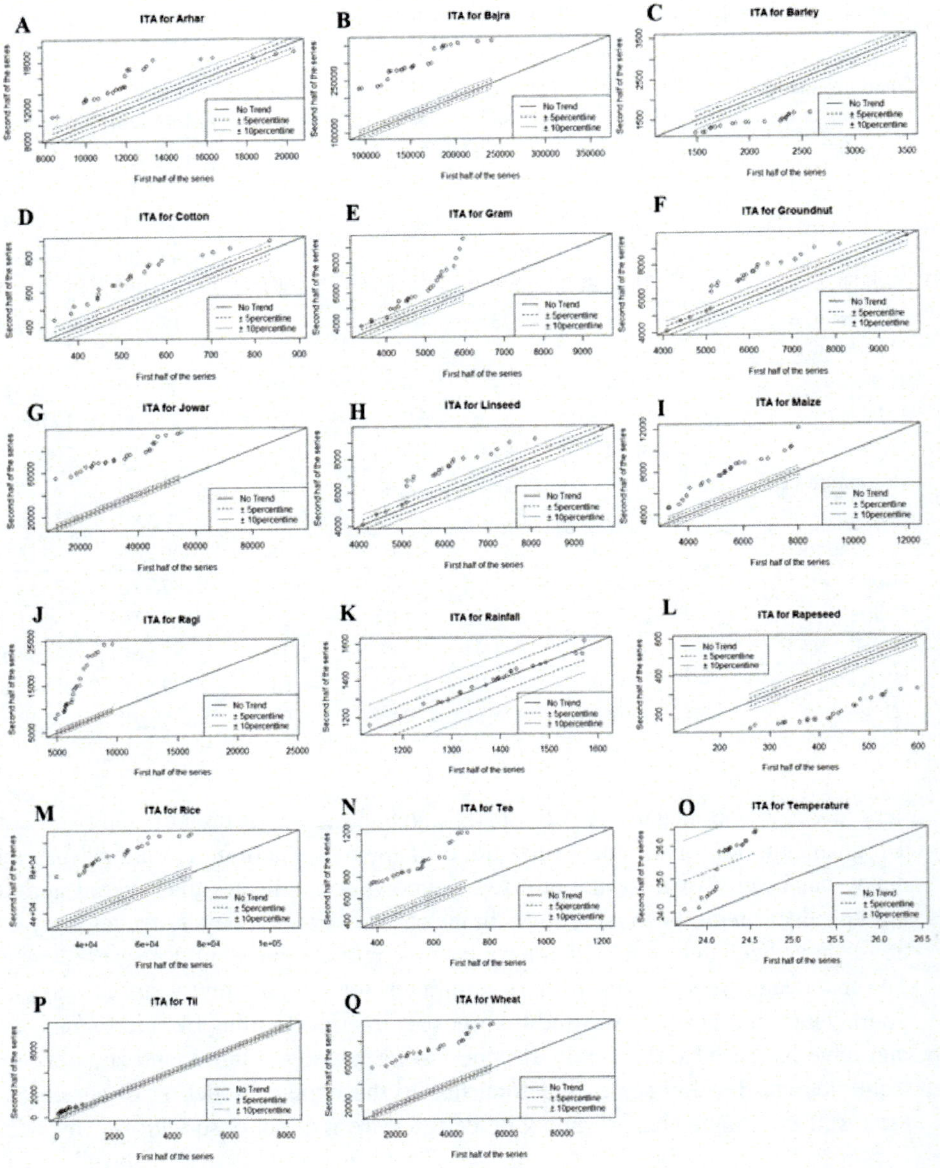

Fig. 3.3 Innovative trend analyses in climatic variables and agriculture production

a close adjacency between actual data and predicted data and error measures suggest that the performance of the model is good enough, then we can use it for further work. We predicted climatic parameters and different crops by optimizing the parameters of ANN and found that there was a very close adjacency between actual and predicted climatic parameters and different crops (Fig. 3.3). The coefficient of determination suggests that the performance of the ANN was measured by the error between actual and predicted values of climatic parameters and different crop production (Fig. 3.4) in the form of mean absolute error, mean absolute percentage

Table 3.3 Descriptive analysis of the climatic variables and crop production

Variables	Trend indicator	Slope
Rainfall	0.42025	0.07595
Temperature	0.05126	0.529
Cotton	5.61328	2.69052
Rice	1556.21	7.67065
Tea	15.2262	7.07355
Rapeseed	−8.5438	−5.0555
Linseed	45.7965	1.90101
Maize	97.7765	4.51448
Arhar	116	2.30145
Til	−84.411	−6.9289
Jowar	1633.41	12.0498
Groundnut	45.7965	1.90101
Bajra	5593.93	8.87696
Barley	−28.932	−3.2616
Ragi	335.932	12.509
Wheat	1633.41	12.0498
Gram	54.0624	2.80622

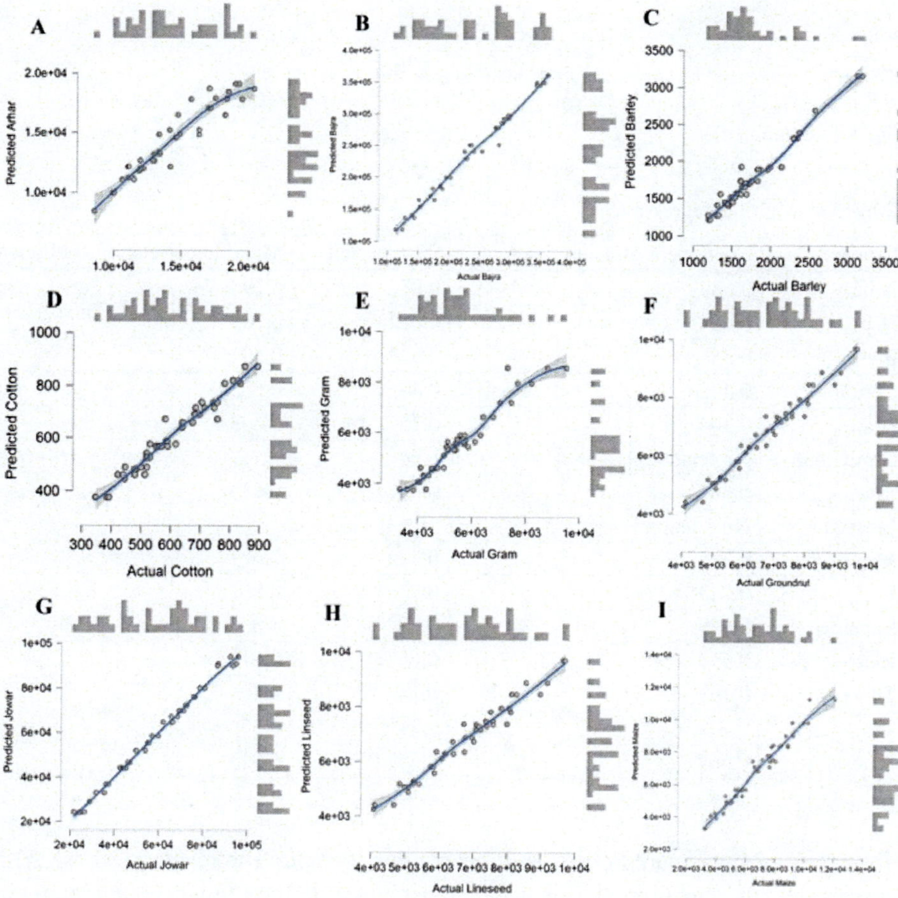

Fig. 3.4 Comparison between actual and predicted values of climatic parameters and different crops

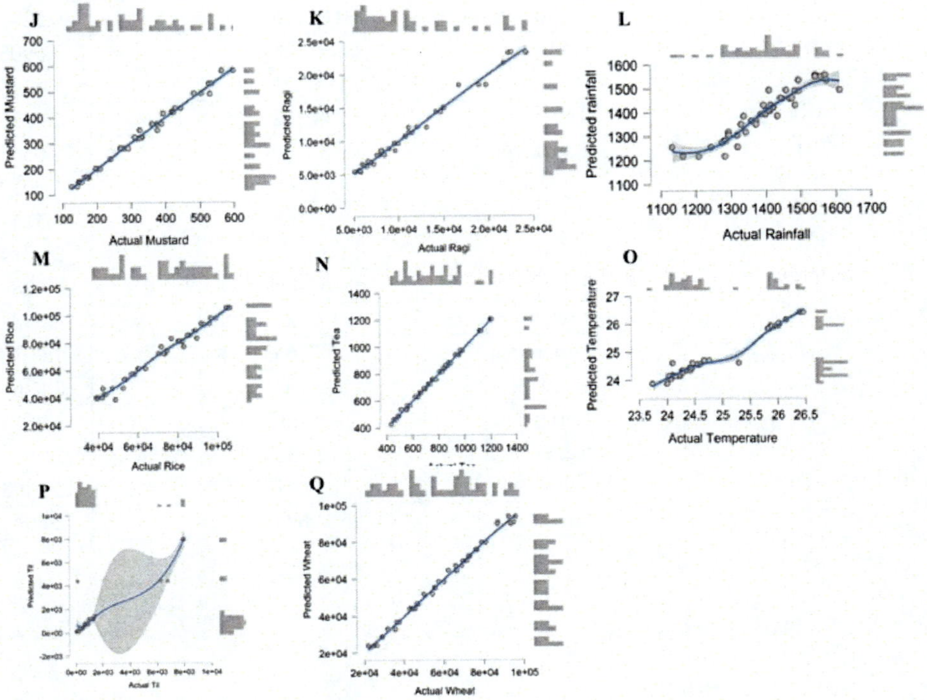

Fig. 3.4 (continued)

Table 3.4 Multiple regression

Crops	Coefficient of determination	R2-adjusted	Multiple correlation coefficient
Arhar	0.089	0.07	0.298
Bajra	0.62	0.612	0.788
Barley	0.356	0.343	0.597
Cotton	0.442	0.43	0.665
Gram	0.364	0.35	0.603
Groundnut	No relation		
Jowar	0.664	0.657	0.815
Linseed	No relation		
Maize	0.385	0.372	0.621
Ragi	0.743	0.738	0.862
Rapeseed	0.699	0.692	0.836
Rice	0.657	0.650	0.811
Til	0.161	0.143	0.401
Tea	0.722	0.716	0.85
Wheat	0.664	0.657	0.815

error, root means squared error, and mean squared error. The results of these errors suggest that the performance of the model for predicting climatic parameters and different crop productions is quite good (Table 3.4).

Figures 3.5 and 3.6 project that with increasing temperature, precipitation will slightly decrease up to 2030. On the other hand, production of some crops in future such as sesamum (til), gram, groundnut, jowar, bajra sugarcane is expected to rise, while a few crops, for example, wheat, arhar, cotton, rice and tea production, being climatic sensitive, would be affected the most adversely as temperature increases. We explored that fluctuation of rainfall and increasing trend of temperature have impacted different crops in different manners (Table 3.5). The increasing temperature and rainfall have benefitted arhar, as the favorable conditions during its vegetation period are a fairly moist and warm climate (Table 3.5). The analysis also

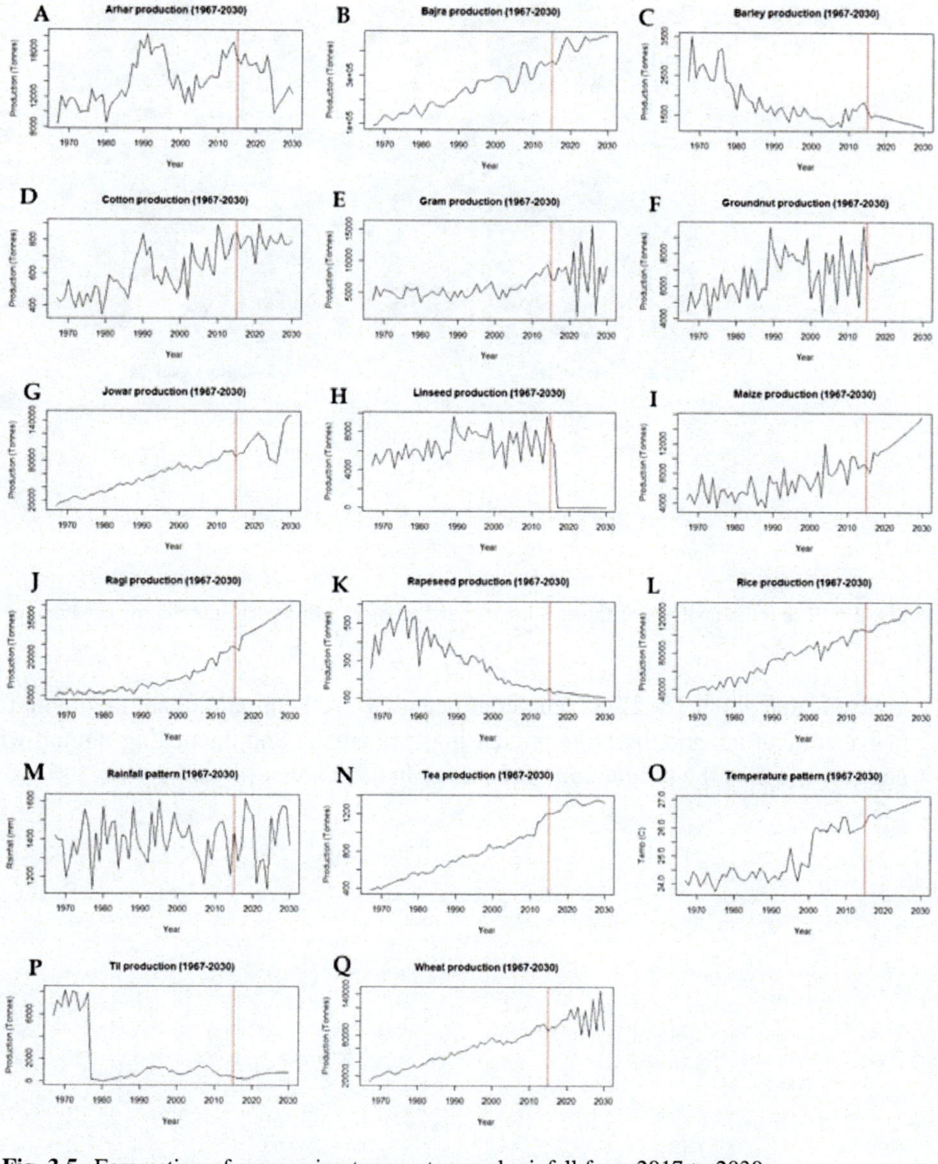

Fig. 3.5 Forecasting of crops using temperature and rainfall from 2017 to 2030

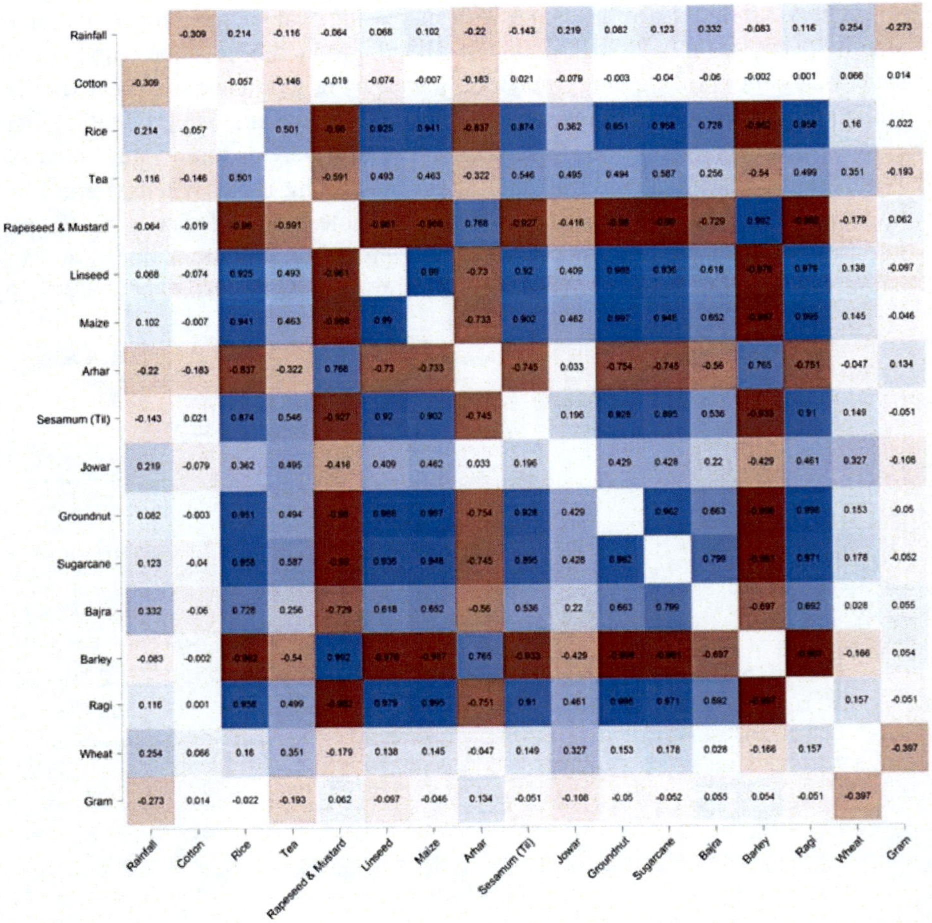

Fig. 3.6 The trend analysis of future climatic parameters and different future crops production

indicates that about 18–27 °C temperature and 50–100 cm rainfall are favorable for maize cultivation, and therefore, increasing temperature and fluctuating rainfall will negatively affect the production of maize in India (Tables 3.5 and 3.6).

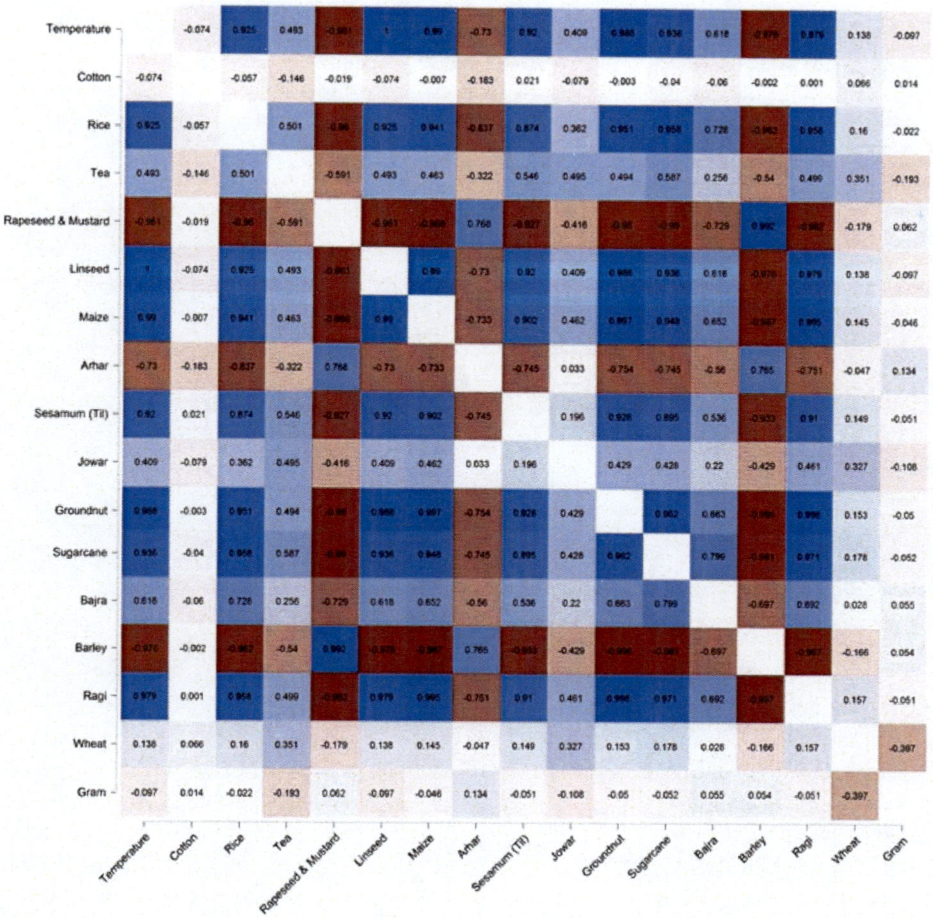

Fig. 3.6 (continued)

Table 3.5 Model performance evaluation using different error measures

Error measures	Rainfall	Temperature	Cotton	Rice	Tea	Rapeseed & Mustard	Linseed	Maize
Mean absolute error	2.68	0.082	5.47	155	14.01	5.85	23.1	42.63
Mean absolute percentage error	0.19	0.33	0.96	0.85	1.74	2.62	2.86	0.74
Root mean squared error	5.92	0.1	9.13	19	17.49	8.71	11.34	57.96
Mean squared error	35.15	0.011	83.34	380	306.21	76.03	95.36	3359

Error measures	Arhar	Sesamum (Til)	Jowar	Groundnut	Sugarcane	Bajra	Barley	Ragi	Wheat	Gram
Mean absolute error	54.32	54.59	85.07	3.76	63.12	84.21	14.58	68.24	39.58	10.08
Mean absolute percentage error	0.36	8.02	0.13	0.055	4.78	0.032	0.96	0.65	0.068	0.16
Root mean squared error	74.88	71.9	104.14	5.03	16.32	98.16	17.77	84.13	64.02	14.29
Mean squared error	5607.6	5183.42	1084.5	25.32	152.18	963.6	315.8	7078.65	4099.78	204.45

Relationship Between Future Climatic Factors and Crops Production

Table 3.6 Multiple regression

Crops	Coefficient of determination	R2-adjusted	Multiple correlation coefficient
Arhar	0.534	0.495	0.731
Bajra	0.382	0.331	0.618
Barley	0.952	0.949	0.976
Cotton	No relation		
Gram	No relation		
Groundnut	0.977	0.975	0.988
Jowar	No relation	0.657	0.815
Linseed	0.972	0.97	0.99
Maize	0.979	0.977	0.989
Ragi	0.959	0.955	0.979
Rapeseed	0.924	0.918	0.961
Rice	0.856	0.844	0.925
Til	0.846	0.833	0.919
Tea	No relation		
Wheat	No relation		

Discussion

Mall and Singh (2000) suggest that minor temperature fluctuations during the growing season of crops have affected annual wheat yield over the years. Pathak et al. (2003) assessed that intensifying minimum temperature and notable negative trends in solar radiations have decreased the crop yields of wheat and rice in India's Gangetic plains. During the rabi and kharif seasons after 1980, a minimal downward trend in solar radiation was also observed, which indicated a trend of warming. Since the production of the crop is largely associated with the incoming radiation of the sun, any downward trend would have a considerable reduction in agricultural productivity. Proliferation at the lowest temperatures increases the crop requirements for maintenance respiration and further decreases net yield and growth (Aggarwal 2003).

Krishna Kumar et al. (2004) assessed Indian crop–climate relationships with historical crop production data for major crops such as sorghum, rice, wheat, sugar cane, groundnut oilseed, pulses, and cereals. The aggregate of production in agriculture (except sugarcane and sorghum), the production of monsoon agriculture (except sorghum), and seasons of post-monsoon (excluding rice and sorghum) are notably associated with summer precipitation in India. The research also manifests that agricultural productivity has become stagnant with a decreasing course pattern in some of the regions. This mode of change in temperature perspicuously shows a reduction in production for rice–wheat cropping patterns.

It is arduous to determine varying climatic conditions for crop productivity in India due to the change in agricultural cropping structure and the adoption of technical know-how across the cropping systems. Limited research has been done in India to assimilate the immensity and phenomena of yield gains or crop failures at local levels due to CO_2 in the atmosphere and associated climatic change (Sinha and Swaminathan 1991; Abrol et al. 1991; Aggarwal and Sinha 1993; Rao and Sinha 1994; Aggarwal and Kalra 1994; Mathauda and Mavi 1994; Gangadhar Rao et al. 1995; Mohandass et al. 1995; Aggarwal and Mall 2002; Mall and Aggarwal 2002; Attri and Rathore 2003; Aggarwal 2003; Mall et al. 2004).

An increase in temperature by 2 °C will reduce the probable productivity of wheat in maximum regions, which varies across rain-fed and irrigated systems. The wheat performance in India influenced by climatic variance has been observed to be reduced due to the detrimental consequences of temperature on the crop during the filling of grain and maturity stage of the growth is stated by Gangadhar Rao and Sinha (1994). The outcomes here demonstrate that grain filling duration sensitivity to temperature plays a crucial role in defining the impacts of change in climate on the productivity of the crop.

Based on previous studies, the effects of change in climate on agriculture in India can be reported as undefined. The overall average effect can be good or adversely based on climatic scenarios (temperature increasing in 2 °C, 3 °C, 4 °C, increase in CO_2, and interaction of a rise in temperature and CO_2). Therefore, the aftermath of climate variation on agriculture fluctuates by crop, agronomic management stage, area, and season both qualitatively and quantitatively. The cropping method of rabi (winter season) from the central and southern regions of India is most likely to be riskier than the seasonal impacts. Nonetheless, the overall positive effect of climate change on Indian agriculture would be important until 2050 as suggested by most projections. By the year 2080, with an exponential rise in temperature, Indian agriculture will suffer the most. It can be said, in other words, that the production of food will not be in danger till 2050. However, by the year 2080, the cropping pattern and its system in India will be challenged and it is of utmost importance to confirm the security of food. In the Indian context, the likely effects of change in climate on agricultural crop production were shown optimistically (Mall et al. 2006).

Conclusions and Policy Implications

This study has analyzed the climatic sensitivity in Indian agricultural crop productions using a panel dataset from 1967 to 2016. Crop productions were designed according to temperature and precipitation datasets. The analysis shows mixed results in India as the country has diversity in terms of its weather conditions (or climate change factor). We found a mixed result in India with diverse climatic variability which also differs because of different geographic, territorial, and socio-economic facets. Therefore, agricultural crop production also differs throughout the country due to diverse agro-climatic conditions. The state of West Bengal for instance is comparatively better in rice production compared to other regions of the country. The findings of this study also echo similar outputs.

We have also executed the ANN model for a particular crop prediction. Comparing observed and projected values at the 95% confidence level, the ANN model provides an admissible result. Inevitably, the model enables us to understand forthcoming proposals, policy-making, and plan alleviations in agro-based cropping systems in the country. The outputs of the model are vital for determining crop calendars based on the local climatic conditions. Meanwhile, crop production is going through the ill effects of climatic shifting to irregular precipitation occurrences, which will lead to an adjustment in the cropping pattern.

Based on the empirical outcome, the aforementioned key suggestions may be delineated for handling issues of food security in India. First, the intensity of cropping patterns perhaps may increase the sequence of agro-based products that will inevitably increase food production in the country. Second, policymakers are required to increase additional irrigation conveniences to enhance crop production alongside food safety. Third, the availability of fertilizers and government expenditure in agriculture is essential to pacify ill impacts of climate variance (ADB 2012; Hollaender 2010). Fourth, the incorporation of substitutes in the consumption of biotic fertilizers may perhaps enhance the efficiency of agricultural products such as gram, wheat, rice, arhar, barley, ragi, jowar, and maize. Fifth and last, the predicted outcome indicates a few crops such as rice, wheat, tea, cotton, arhar, and jowar, some gram, sesamum (Til), groundnut, bajra, and sugarcane respond positively to the climatic change as an increase in temperature or rainfall increases their production. However, future studies are encouraged at the territorial and zonal levels so that they might assist in regulating sustainable cropping in the long run.

Acknowledgement The authors of the manuscript are grateful to the Indian Meteorological Department for providing the rainfall data of all subdivisions for undertaking this study. We are also thankful to the HSS Department of IIT Indore and the Department of Geography, the University of Gour Banga for providing laboratory space along with other support to bring about this piece of work.

Conflicts of Interest None

Funding The first author of the article would like to thank the University Grants Commission (UGC Ref. No.2942/(NET-DEC.2014), New Delhi, India, for providing financial support as Junior Research Fellowship (JRF) for conducting the research work.

References

Abrol YP, Bagga AK, Chakravorty NVK, Wattal PK (1991) In: Abrol YP et al (eds) Impact of rise in temperature on the productivity of wheat in India. Impact of global climate changes in photosynthesis and plant productivity, pp 787–789

Adams HD, Williams AP, Xu C, Rauscher SA, Jiang X, McDowell NG (2013) Empirical and process-based approaches to climate-induced forest mortality models. Front Plant Sci 4:438

Aggarwal PK (2003) Impact of climate change on Indian agriculture. J Plant Biol New Delhi 30: 189–198
Aggarwal PK (2009) Vulnerability of Indian agriculture to climate change: current state of knowledge. National Workshop – Review of Implementation of Work Programme Towards Indian Network of Climate Change Assessment
Aggarwal PK, Kalra N (1994) Simulating the effect of climatic factors, genotype and management on productivity of wheat in India. Division of Environmental Science, Indian Agricultural Research Institute, New Delhi
Aggarwal PK, Mall RK (2002) Climate change and rice yields in diverse agro-environments of India. II. Effect of uncertainties in scenarios and crop models on impact assessment. Clim Chang 52:331–343
Aggarwal PK, Sinha SK (1993) Effect of probable increase in carbon dioxide and temperature on wheat yields in India. 農業気象 48:811–814
Ahmad J, Alam D, Haseen MS (2011) Impact of climate change on agriculture and food security in India. Int J Agric Environ Biotechnol 4:129–137
Asian Development Bank (ADB) (2012) Food security and poverty in Asia and the Pacific: key challenge and policy issues. Mandaluyong City (Philippines)
Attri SD, Rathore LS (2003) Simulation of impact of projected climate change on wheat in India. Int J Climatol 23:693–705
Deschênes O, Greenstone M (2007) The economic impacts of climate change: evidence from agricultural output and random fluctuations in weather. Am Econ Rev 97:354–385
Gangadhar Rao D, Katyal JC, Sinha SK, Srinivas K (1995) Impacts of climate change on sorghum productivity in India: simulation study. Climate change and agriculture: analysis of potential international impacts, pp 325–337
Geethalakshmi V, Lakshmanan A, Rajalakshmi D, Jagannathan R, Sridhar G, Ramaraj AP, Bhuvaneswari K, Gurusamy L, Anbhazhagan R (2011) Climate change impact assessment and adaptation strategies to sustain rice production in Cauvery basin of Tamil Nadu. Curr Sci 101:342–347
Hollaender M (2010) Human right to adequate food: NGOs have to make the difference. CATALYST. Newsletter of Cyriac Elias Voluntary Association (CEVA) 8:5–6
Hundal SS (2007) Climatic variability and its impact on cereal productivity in Indian Punjab. Curr Sci 92:506–512
Kalra N, Chakraborty D, Sharma A, Rai HK, Jolly M, Chander S, Kumar PR, Bhadraray S, Barman D, Mittal RB (2008) Effect of increasing temperature on yield of some winter crops in northwest India. Curr Sci 94:82–88
Kane S, Reilly J, Tobey J (1992) An empirical study of the economic effects of climate change on world agriculture. Clim Chang 21:17–35
Kapur D, Khosla R, Mehta PB (2009) Climate change: India's options. Econ Polit Wkly 44:34–42
Kar J, Kar M (2008) Environment and changing agricultural practices: evidence from Orissa, India. Indus J Man Soc Sci 2:119–128
Kaul S, Ram G (2009) Impact of global warming on production of Jowar in India (special issue: sustainable agriculture in the context of climate change). Agric Situ India 66:253–256
Kaur P, Hundal SS (2007) Effect of temperature rise on growth and yield of wheat: a simulation study. J Res 44:6–8
Krishna Kumar K, Rupa Kumar K, Ashrit RG, Deshpande NR, Hansen JW (2004) Climate impacts on Indian agriculture. Int J Climatol 24:1375–1393
Kumar, R., Gautam, H.R., 2014. Climate change and its impact on agricultural productivity in India. Journal of Climatology & Weather Forecasting
Kumar KK, Parikh J (2001) Indian agriculture and climate sensitivity. Glob Environ Chang 11: 147–154
Kumar SN, Aggarwal PK, Rani S, Jain S, Saxena R, Chauhan N (2011) Impact of climate change on crop productivity in the Western Ghats, coastal and northeastern regions of India. Curr Sci 101: 332–341

Kurukulasuriya P, Mendelsohn R, Hassan R, Benhin J, Deressa T, Diop M, Eid HM, Fosu KY, Gbetibouo G, Jain S (2006) Will African agriculture survive climate change? World Bank Econ Rev 20:367–388

Lal M (2000) Climatic change-implications for India's water resources. J Soc Economic Develop 3: 57–87

Mall RK, Aggarwal PK (2002) Climate change and rice yields in diverse agro-environments of India. I. Evaluation of impact assessment models. Clim Chang 52:315–330

Mall RK, Singh KK (2000) Climate variability and wheat yield progress in Punjab using the CERES-wheat and WTGROWS models. Yayu Mandal 30:35–41

Mall RK, Lal M, Bhatia VS, Rathore LS, Singh R (2004) Mitigating climate change impact on soybean productivity in India: a simulation study. Agric For Meteorol 121:113–125

Mall RK, Singh R, Gupta A, Srinivasan G, Rathore LS (2006) Impact of climate change on Indian agriculture: a review. Clim Chang 78:445–478. https://doi.org/10.1007/s10584-005-9042-x

Mathauda SS, Mavi HS (1994) Impact of climate change in rice production in Punjab, India. In: Climate change and rice symposium. IRRI, Manila, pp 38–39

Meeting, J.F.E.C. on F.A (2006) Residue evaluation of certain veterinary drugs: joint FAO/WHO expert committee on food additives, 66th meeting 2006. Food & Agriculture Org

Mendelsohn R (2014) The impact of climate change on agriculture in Asia. J Integr Agric 13:660–665

Mendelsohn RO, Dinar A (2009) Climate change and agriculture: an economic analysis of global impacts, adaptation and distributional effects. Edward Elgar Publishing

Mendelsohn R, Arellano-Gonzalez J, Christensen P (2010) A Ricardian analysis of Mexican farms. Environ Dev Econ 15:153–171

Mohandass S, Kareem AA, Ranganathan TB, Jeyaraman S (1995) Rice production in India under current and future climates. Modeling the Impact of Climate Change on Rice Production in Asia, CAB International, UK, pp 165–181

Nath PK, Behera B (2011) A critical review of impact of and adaptation to climate change in developed and developing economies. Environ Dev Sustain 13:141–162

Pathak H, Ladha JK, Aggarwal PK, Peng S, Das S, Singh Y, Singh B, Kamra SK, Mishra B, Sastri A (2003) Trends of climatic potential and on-farm yields of rice and wheat in the Indo-Gangetic Plains. Field Crop Res 80:223–234

Ranuzzi A, Srivastava R (2012) Impact of climate change on agriculture and food security. ICRIER Policy series 16, p 33

Rao GD, Sinha SK (1994) Impact of climate change on simulated wheat production in India. In: Implications of climate change for international agriculture: crop modelling study 2. EPA, pp 1–10

Reilly J, Hohmann N, Kane S (1994) Climate change and agricultural trade: who benefits, who loses? Glob Environ Chang 4:24–36

Rosenzweig C, Iglesias A (1994) Implications of climate change for international agriculture. US Environmental Protection Agency, Washington, DC

Singh, A., 2012. Impact of sustainable agriculture on food production and challenges for food security in India

Sinha SK, Swaminathan MS (1991) Deforestation, climate change and sustainable nutrition security: a case study of India. In: Tropical forests and climate. Springer, pp 201–209

Part II
Geopolitics and Conflicts

Chapter 4
Geopolitics of Climate Change and Water Security in South Asia: Conflict and Cooperation

Dhanasree Jayaram and Gaurica Sethi

Abstract South Asia's climate vulnerabilities have direct and indirect implications for water security. While the effects of climate change on water access, safety, and affordability in the region are becoming increasingly adverse, the transboundary nature of both climate change and water security, especially on account of shared water resources, complicates responses. The shared river basins, such as the Indus, Ganges, Meghna, and Brahmaputra, have been marred by geopolitical tensions, which renders transboundary climate cooperation to deal with water-related problems difficult. Although the region has seen different forms of cooperation in the form of river water sharing treaties/arrangements as well as regional organizations' role in advancing cooperation, these efforts rarely take cognizance of the effects of climate change on water security. In this context, the chapter provides an analysis of the challenges posed by climate change to water security and geopolitics of transboundary conflicts and cooperation in South Asia, as well as explores opportunities for cooperation among the countries of the region.

Keywords Low-carbon technology · Early warning systems · Climate services · Transboundary framework · Water security

D. Jayaram (✉)
Manipal Academy of Higher Education, Manipal, India

Alexander von Humboldt Foundation – International Climate Protection Fellow, Berlin, Germany
e-mail: dhanasree.j@manipal.edu

G. Sethi
Department of Geopolitics and International Relations, Manipal Academy of Higher Education, Manipal, India

N. Khare (ed.), *Science, Policies and Conflicts of Climate Change*, Springer Climate,
https://doi.org/10.1007/978-3-031-16254-1_4

Introduction

The impacts of climate change such as more frequent and intense extreme weather events (cyclones, heat waves, etc.), droughts and crop failures, coastal erosion, climate change-induced migration, and major biodiversity shifts have geopolitical and geoeconomic implications for South Asia. Climate change coupled with the global health crisis in the form of the COVID-19 pandemic has led to adverse consequences for the region. With the complex and unprecedented nature of risks posed by climate change to the entire world, the level of urgency and immediacy attached to the phenomenon has increased – leading to heightened demand for cooperation at all levels, including within regions such as South Asia (Saeed and Fuentes-Hutfilter 2021). Hence, regional organizations are engaged in the coordination of joint initiatives, policy alignment, and the establishment of strategies to tackle climate change-related risks. Despite the existence of common problems and numerous efforts to implement synchronized policies at the regional level (Schoeters 2002), cooperation has largely been hampered by various obstacles, including political, economic, and technical ones (Das and Bandyopadhyay 2015).

Among the various issues, water security, in particular, is already being affected in adverse ways in the region. Water security in South Asia is a local, state, national, and regional issue, as a large proportion of the river basins, including the Indus, Ganges, and Brahmaputra, are transboundary. The impacts of climate change on water security have worsened over the years with monsoonal variations and drought-like conditions becoming more common and unpredictable. Water stress or scarcity has cascading impacts on food, health, livelihood, and human security. Moreover, transboundary disputes and conflicts over reducing water levels in the river basins could also increase, as water demand has increased in all the countries of the region in recent decades due to growing population, industrialization, agricultural demand, urbanization, etc. This poses a critical challenge to South Asian countries that need to be addressed through both the existing bilateral, regional, and multilateral mechanisms, apart from generating new platforms that could potentially enhance interactions and collaborations among them.

In this context, this chapter aims to assess climate change concerns with a focus on water security, particularly transboundary river water sharing. The chapter looks into both conflictual and cooperative aspects of the climate change–water security nexus in the region using examples of India's complex hydro-relations with its neighbours along transboundary rivers as well as that of efforts taken by regional organizations such as the South Asian Association for Regional Cooperation (SAARC), Bay of Bengal Initiative for Multi-Sectoral Technical and Economic Cooperation (BIMSTEC), and International Centre for Integrated Mountain Development (ICIMOD). It provides an overview of the challenges to regional and transboundary cooperation in South Asia, as well as explores opportunities for cooperation among the countries of the region.

Major Water Security-Related Climate Change Impacts in the South Asian Region

The elements of water security are: "(1) water access; (2) water safety; and (3) water affordability so that every person can lead a clean, healthy and productive life while ensuring that the natural environment is protected and enhanced" (Mishra et al. 2021). All these elements are hampered by climate change in different ways. Water has emerged as a scarce commodity in parts of the South Asian region. Besides physical scarcity, inaccessibility also leads to water insecurity. South Asia has seen a decrease of 70% in the availability of water resources since the 1950s (Roy 2010). The per capita availability is also on the decline in most countries, which has implications for food, livelihood, health, and other sectors of human security (Lacombe et al. 2019). Many river basins in the region are transboundary, and they are facing problems related to water scarcity. These include the Indus, Ganges, and Brahmaputra basins. These rivers are the lifelines of more than 1 billion people in the region. Climate change has major impacts on these rivers, with geopolitical implications. Although river water sharing treaties have been signed between the region's countries, these arrangements do not consider the climate factor. Moreover, the stability of these treaties and agreements is continuously challenged by geopolitical factors, such as contested borders, cross-border terrorism, and domestic politics. Hence, as climate change affects water availability, this may spiral into tensions between the riparian countries, if appropriate frameworks are not established to tackle this transboundary problem.

Before going further into the geopolitical implications of impacts of climate change in South Asia in terms of water security, the implications of water insecurity caused by climate change need to be discussed. Climate change is already affecting food and livelihood securities in the region due to the negative impacts of climate change on water availability. An increase in the temperatures and a reduction in mean annual precipitation during the monsoons in many parts of the region are linked with low agricultural yields. For example, studies indicate that there was a reduction in wheat yield by 5.2% during 1981–2009 in India, despite adaptation measures (Aryal et al. 2019). In another case, Pakistan which depends excessively on its agricultural sector for economic security is highly climate-vulnerable, rendering nearly 37% of its population susceptible to food insecurity, according to a study (Anwar 2020). Apart from variability in water availability, disasters such as floods also have wide-ranging impacts on crop yields. Moreover, the dependence of a large portion of the population on sectors such as agriculture that depend to a large extent on water causes livelihood insecurities with far-reaching impacts such as poverty, migration, and gender inequalities.

Besides triggering the visible variability in the South West monsoon, climate change also threatens water security in South Asia by affecting the glaciers in the Hindu Kush Himalaya (HKH) region. Although the glaciers of the Tibetan Plateau are often focused upon, other countries such as Afghanistan which houses nearly 4000 glaciers are also witnessing the adverse effects of climate change. The glaciers are a major contributor to river flows in the entire region; while some river basins such as the Indus Basin are more dependent on glacial meltwater, others are less so. In any case, the variability in glacial meltwater due to glacial retreat caused by rising temperatures has a direct impact on water security. As per a study, glacial meltwater contributes to nearly "9% of wheat production, 15% of rice, 28% of cotton, and 17% of sugarcane annually" in the Indus Basin (Lutz et al. 2016). Variabilities in the meltwater can lead to both floods and droughts in the region.

There are other ways in which climate change affects water security. For instance, there are many water-related slow and rapid onset disasters that adversely affect water security with wide-ranging implications for health and other sectors. Disasters such as extreme precipitation events or floods that are exacerbated by climate change are known to pollute drinking water sources and/or disrupt water services (Grigg 2016). Slow onset phenomena such as sea level rise, on the other hand, could lead to intrusion of seawater into fresh water, thereby affecting water quality and rendering the water unusable for drinking, agricultural, and other purposes (Mills et al. 2021). These changes have direct and indirect implications for health and livelihood securities, and they also affect population mobility patterns. For example, subsistence farmers who are constantly affected by these disasters are forced to migrate from rural to urban areas for better livelihood opportunities. At times, they also migrate to other countries, as seen in the case of the seasonal migration of people from Nepal to India (Gautam 2017). Similarly, water insecurity caused by climate change also affects human health adversely by triggering water-borne diseases such as cholera, diarrhoea, and typhoid (Sen et al. 2017).

Climate Change, Water Security, and Conflict in South Asia

The nexus between climate change, water security, and conflict are crucial in South Asia. The region's countries have a set of interests as far as resources are concerned that are tied to their historical, geographic, socio-economic, political, and cultural backgrounds. Often, these interests overlap, merge, or compete with each other – leading to different dynamics of resource sharing relations, particularly river water sharing. The Indus, Ganges, and Brahmaputra are three major river basins shared by many of the region's countries, indicating a complex interdependence between them. Even though the Hindu Kush Himalayan region that feeds most of the rivers of South Asia is one of the world's largest freshwater reservoirs, many scholars believe that water scarcity may likely lead to conflictual situations between the region's countries. Since the waters of these rivers are critical for hydroelectricity production, industrial expansion, agricultural stability, urbanization, and other developments, any form of water scarcity would be seen as a huge security dilemma (Tripathi 2011). Depletion of surface water levels in several parts of the region in light of these developmental activities has led to over-exploitation of groundwater, which has also emerged as a major concern. These problems are compounded by the influence of climate change, which is altering the river flows during both dry and monsoon seasons. In such a scenario, the requirement for stable water levels may cause bilateral water-sharing agreements to be reviewed in the future to incorporate climate change concerns within the agreements (Dutta 2021). In this section, India's hydro-diplomacy with its neighbours provides an overview of the existing disputes over water sharing and the potential impacts of climate change on these relations.

The Qinghai–Tibetan Plateau in China is the source of water for major river basins in South Asia. China, being an upper riparian country, has a strategic advantage over the lower riparian countries as it could control the flow of water through its engineering projects, especially large-scale dams that it is famous for constructing all over the country. Despite signing several Memorandums of Understanding (MoUs) with the lower riparian countries, China is not very transparent about its dam constructions on the transboundary rivers. Even though China has agreed to share hydrological data about transboundary rivers with India, there have been occasions when China did not share the data with India, citing technical reasons. For instance, this happened during the Doklam border standoff between Indian and Chinese armies. If one takes the case of the River Brahmaputra that originates in Tibet, both India (the north-eastern region) and Bangladesh are dependent on its waters. China has laid out proposals to build several dams on the river that are sources of concern for the downstream countries (Krishnan 2021). Not only can it use these dams to create floods or flood-like situations in the downstream countries, but also pollution and other factors could reduce water quality and affect the populations. As has been made clear by China already, it could also use this as a strategic asset during conflictual situations.

In the case of India–Bangladesh river water sharing relations, longstanding disagreements exist over River Ganges and River Teesta. The two countries share 54 rivers. Although the two countries have signed the Ganges Water Sharing Treaty, there have been several disputes raised by Bangladesh over sharing of water by India with Bangladesh during the dry season as well as the low availability of flow at the Farakka Barrage, which is the centre-point of the Indo-Bangladesh dispute. In fact, with climate change and other factors, the treaty has failed to assess the future projected flow appropriately (Rahman et al. 2019). River Teesta is another shared river that joins the Brahmaputra in Bangladesh, on which India and Bangladesh have so far not been able to reach an agreement due to domestic political concerns within India, particularly in West Bengal. Bangladesh has sought an agreement that follows the principle of equitable distribution of the waters. This river is vital for Sikkim and West Bengal in India, and the Rangpur region of Bangladesh, especially for the agricultural sector (Gambhir 2021). Furthermore, China has also entered the fray by expressing interest in financing Bangladesh Government's Teesta River Comprehensive Management and Restoration Project (Bhattacharjee 2020).

Nepal and India have signed water cooperation agreements on major rivers such as the Kosi, Gandaki, Karnali, and Mahakali, primarily for hydropower and irrigation projects, including dams or barrages. Most of the projects are still in the pipeline, with a major exception being the Kosi barrage that has been held responsible for floods in Nepal. Many analysts in Nepal also accuse India of infringing upon Nepal's sovereignty by constructing the dam as well as wilful ignorance about the challenges faced by Nepal's population due to the barrage. The embankments have not been effective in reducing sedimentation and flooding (Bagale 2020). Since the signing of the 1994 Kosi Agreement, the disagreements on water rights-related issues have largely remained unresolved. Rivers with smaller catchment areas have been overlooked, leading to a non-holistic view of water sharing between the two countries. River Mahakali forms a natural boundary between India and Nepal (demarcated during the British colonial era) and this has also long been a source of contention between the two countries. Although they arrived as the Mahakali Treaty in 1996, the provisions of the treaty are yet to be implemented. Owing to the disagreements over interpretations of water entitlements, perceived Indian unilateralism in decision-making, and cost-benefit sharing mechanisms, many storage dam projects have been stalled and Indo-Nepal water cooperation has reached a deadlock (Bagale and Adhikari 2020). Hence, the scope for further cooperation on issues such as climate change is contingent on the success of the existing arrangements.

When it comes to India and Bhutan, there have so far not been conflicts. The two countries have been engaging in hydropower cooperation for over five decades. The cooperation began with the construction of small-scale hydroelectric projects such as Tala, Chukha, and Kurichu. Bhutan has a 30,000-megawatt hydroelectric power producing capability (Shree 2020). Both countries signed a 35-year 'Power Purchase Agreement' in 2006, allowing Bhutan and India to generate and import 1020 MW of hydropower, respectively. Although the two countries have shared friendly relations, there have been small-scale opposition to India-funded dam projects in Bhutan due to their environmental impacts. Furthermore, in 2020, several Assamese farmers claimed that Bhutan restricted the flow of River Kalanadi's water to their agricultural fields, but Bhutan rejected this claim and clarified that the river water supply to Assam was disrupted due to heavy rains (Yashwant 2020).

Perhaps the most conflictual relationship that exists in South Asia, as far as water security is concerned, is the one between India and Pakistan. India–Pakistan river water sharing relations have been marred by disputes despite the signing of the 1960 Indus Waters Treaty. Although some scholars have pointed towards the possibility of a water war between the two countries, they have so far used the treaty as a confidence building measure and adhered to the treaty obligations even during wars, small-scale conflicts, and border skirmishes. The division of the basin between the two countries follows the model of land partition, which reduces the scope of joint management of the basin despite it being the need of the hour with worsening climate change. The Indus is a geopolitical trump card for India, as it is the upper riparian and all the rivers, including the ones allocated to Pakistan flow through India. Some literature also points to the linkages drawn between the river water sharing issue and the Kashmir issue, as the headwaters of major rivers such as the River Indus flow through Jammu and Kashmir. Most of the disputes over the treaty are addressed through the dispute resolution mechanisms, but lately with the growing number of cases of Pakistan-sponsored cross-border terrorism and domestic rhetoric to not stop the flow of the rivers in response to these attacks (such as Uri, Pathankot, Pulwama, etc.), the calls for scrapping the treaty have grown stronger in India. Similarly, in Pakistan too, the demand for reviewing the treaty in light of

Indian dams on the western rivers (allocated to Pakistan) – River Jhelum and River Chenab – has gained momentum. While certain disputes such as the one on India's Baglihar dam were resolved by the World Bank (that mediated the treaty negotiations), others such as the one on India's Kishanganga hydroelectric project were taken by Pakistan to the Permanent Court of Arbitration (PCA). In both these examples, the verdicts were favourable to India (Jayaram 2020a, b). Yet, for Pakistan, which is entirely dependent on the Indus basin for its water supply, the fear that India could restrict the flow of water to its territory is a longstanding one and will not cease to exist anytime soon (Qamar et al. 2019). At the same time, for Indian states that are dependent on the western rivers, primarily Jammu and Kashmir, the ability to use these waters is linked with the capacity to spur developmental and agricultural activities.

Regional Cooperation on Climate Change in South Asia

As already discussed in the earlier section, several treaties have been signed by the region's countries on transboundary rivers. However, these treaties largely fail to address the newly emerged and future challenges, including climate change. Nevertheless, climate change has emerged as an area of cooperation in regional organizations as well as at the bilateral levels. This section primarily looks into the role of regional organizations in promoting climate change cooperation.

The South Asian Association for Regional Cooperation (SAARC), which includes India, Bangladesh, Maldives, Nepal, Pakistan, Sri Lanka, Bhutan, and Afghanistan, was founded in 1985 to advance economic and social welfare in the region. The South Asian countries began to focus on environmental issues in the early 1980s, as evident from the establishment of the South Asia Cooperative Environment Scheme in 1982. This forms a bedrock of major agreements reached by the SAARC member states thereafter, including the SAARC Plan of Action on Climate Change. SAARC recognized the need for a regional approach to addressing climate change-related concerns through this Action Plan as well as the 2010 Thimphu statement on climate change. As a result, various regional projects were initiated, including the setting up of regional centres that recognize various aspects of climate change impacts. For instance, the SAARC Environment Centre was merged with the SAARC Energy Centre (SEC) to protect and efficiently utilize natural resources by adopting sustainable natural resource management practices through cooperation and research among member states. Similarly, the SAARC Disaster Risk Management Centre (SDMC) was tasked to advance effective disaster management in South Asia through policy advocacy and capacity-building (Islam and Kieu 2020).

The 2010 Thimphu statement sought to reinforce the timely implementation of SAARC's previous declarations and action plans on climate change. It also laid the foundation for the establishment of an intergovernmental expert group on climate change to augment policy making for regional cooperation. Furthermore, the statement focussed on afforestation and reforestation campaigns in the region; commissioning of a joint study on climate risks in consonance with the social, economic, and environmental challenges; promotion of low-carbon technology and renewable energy, etc. Most importantly, it also sowed the seeds of the SAARC Monsoon Initiative Programme (to study the effects of climate change on the monsoon patterns), SAARC Marine Initiative (shared water bodies), and SAARC Intergovernmental Mountain Initiative (shared mountain ecosystems, especially the glaciers). However, SAARC's efforts have mostly concentrated on declarations, endorsements, and agreements, indicating a lack of political will among the member nations to implement them due to geopolitical rivalries (Islam and Kieu 2020).

Another regional organization, the Bay of Bengal Initiative for Multi-Sectoral Technological and Economic Cooperation (BIMSTEC) also deals with climate change as a priority area of concern. The South Asian countries belonging to BIMSTEC are Bangladesh, Bhutan, India, Nepal, and Sri Lanka. On the one hand, the Bay of Bengal region is extremely vulnerable to extreme weather events such as cyclones that have become more intense in recent years, which is attributed to climate change. On the other hand, three major river basins – Ganga, Meghna, and the Brahmaputra – lie in this region, facing the brunt of climate change impacts in the form of variabilities in monsoon, extreme precipitation events, glacial meltwater, etc. Sub-regional groupings such as BBIN (Bangladesh, Bhutan, India, and Nepal) have also been focussing on cooperation in river water sharing through initiatives such as hydropower generation and water-borne transportation of goods. Proposals to establish an arrangement that aims at promoting basin-wide water resources management have also been floated, but they have not taken off yet (Karim 2020). BIMSTEC countries have shown solidarity with the Paris Agreement in unison and agreed to cooperate to tackle the effects of climate change. It has also set up the BIMSTEC Centre on Weather and Climate, of which the first meeting of the Governing Board and Scientific Advisory Council was held in India in 2018. However, BIMSTEC's initiatives have not gained momentum either, despite being seen as a constructive alternative to the SAARC. While BIMSTEC has made some strides in areas such as disaster management, the implementation of various initiatives has been delayed largely due to a reactive approach to climate change (Roy 2017).

The International Centre for Integrated Mountain Development (ICIMOD) is yet another intergovernmental organization that has targeted programmes for climate change and water security in the Hindu Kush Himalaya (HKH) region. The organization is composed of eight countries – Afghanistan, Bangladesh, Bhutan, China, India, Myanmar, Nepal, and Pakistan. The ICIMOD works towards building adaptive and resilience strategies that could be implemented to protect mountain ecosystems and communities. Considering that the HKH region is not only one of the most climate-vulnerable regions in the world but also remains backwards in terms of socio-economic development makes it imperative for regional organizations such as the ICIMOD to focus on sustainable development by mobilizing multidisciplinary research on the region's vulnerabilities and requirements (Macchi et al. 2009).

The ICIMOD has a River Basins and Cryosphere (RBC) Regional Programme under which it carries out research on water and cryosphere science; promotes gender-sensitive climate change, water, and disaster risk reduction initiatives; supports transboundary cooperation on water security and disaster management, etc. ICIMOD's approach is considered to include 'managing water resources, prioritizing the sustainable use of water and land resources to enhance livelihoods, understanding the role of the cryosphere and scenarios for future water availability, improving the health of environments, and reducing water-related disaster risks in river basins.'[1] It focusses on a range of activities, including 'real-time flood outlook, community-based flood warning, drought monitoring and so on,' and 'emphasises the importance of developing resilient systems to address concerns regarding rapid urbanisation, water scarcity, migration and food insecurity, and the associated security challenges' (Jayaram 2020b). However, the ICIMOD's effectiveness is also hampered by geopolitical conundrums, including the recent border conflicts between India and China in Doklam and Eastern Ladakh. These tensions essentially impede the prospects of joint research on climate change between them. In addition, Chinese inroads into South Asian countries such as Nepal, even in terms of natural resource management through infrastructure (dams) development, leads to competition with India and renders regional cooperation secondary to national and geopolitical interests (Pillai 2021).

The Way Forward: Prospects for Cooperation

Based on the climate change-related water security vulnerabilities in the region, there is ample scope for transboundary cooperation, despite geopolitical entanglements. The protection and sustenance of the HKH region should be a priority for the countries of the region that are dependent on the rivers that originate in this region. This is critical for water, food, livelihood, and energy security. Countries such as Nepal and Bhutan are excessively dependent on hydropower generation from rivers of the Ganga–Meghna–Brahmaputra basin for their energy security and economic development. Hence, sustainable and efficient use of water resources should be the way forward for natural resource management and climate change adaptation. Any transboundary efforts towards joint river basin management need to be built upon nation-level water management policies and practices so that more effective knowledge and data sharing mechanisms and institutional arrangements can be developed under transboundary frameworks. This would also help gauge each country's resource (water, energy, etc.) requirements, thereby also helping build a better understanding between them in terms of costs and benefits of cooperative resource

[1] More information about ICIMOD's "River Basins and Cryosphere (RBC) Regional Programme" is available at: https://www.icimod.org/initiative/about-river-basins-and-cryosphere (retrieved on January 17, 2022).

management strategies. In this context, strengthening and sharing technical solutions such as monitoring systems, early warning systems, and climate services could pave the way for deepened regional cooperation.

As much as focussing on bottom-up approaches, there is a need for strong diplomatic and political will at the top for developing trust and designing such regional and transboundary frameworks in a region like South Asia. This is particularly relevant in the case of transboundary river water sharing treaties that need to be climate-proofed. For instance, due to flow variability in the rivers on account of climate change, all water sharing arrangements should keep allocation mechanisms more flexible. Similarly, since many disasters are also transboundary, South Asian countries could design mechanisms and incorporate special provisions on climate change adaptation and disaster risk reduction within the treaties as areas of cooperation. This could potentially engender initiatives to build adaptive capacities of the most vulnerable communities, develop locally led programmes on resource management and climate change adaptation, and reduce conflicts over resource sharing within and across borders. There needs to be additional focus on groundwater levels in addition to surface water management, as the former tends to be ignored in transboundary arrangements. Variabilities in surface water availability have led to over-extraction of groundwater in several parts of the region that needs to be addressed urgently. Furthermore, in the HKH region, springs are drying up faster or are becoming seasonal due to impacts of climate change as well as infrastructure development, which portends a major threat to water security in the region (Bhattacharya 2019). Similarly, transboundary river pollution is yet another issue that brings attention to the need for including aspects of water quality too within such arrangements. In most cases, data sharing includes hydrological (flow) information only. Hence, establishing institutions for knowledge creation and joint scientific assessments, and ultimately for joint river basin management and climate change adaptation strategies should be a priority for South Asian countries (Jayaram 2019).

Cooperation among South Asian countries is not easy to achieve due to several hurdles discussed in this chapter, including geopolitical rivalries and the failure of the regional organizations in scaling up cooperation in various sectors. It is imperative that the existing regional organizations, which have already initiated processes to develop cooperative climate action mechanisms, are strengthened to find and implement climate solutions cost-effectively since the region's countries are either developing or least developed. One could also argue that the way forward for South Asian countries in addressing climate change and water security could emphasize a concerted and coordinated approach towards the goals enshrined under the Paris Agreement (climate change), the Sendai Framework for Disaster Risk Reduction 2015–2030 and the Sustainable Development Goals (SDGs). Individually, these countries have instituted policies to implement these international frameworks nationally, but there is scope for scaling them up to the regional level as well as building synergies between them. When it comes to climate change, the South Asian countries can achieve more by joining hands with each other rather than acting on climate change alone.

Acknowledgements The first author would like to thank Aishwarya R. J., Postgraduate Student in the Department of Geopolitics and International Relations, Manipal Academy of Higher Education (MAHE), Karnataka, India, for providing support in copy-editing this chapter.

References

Anwar R (2020, November 4) The impact of climate change on Pakistan's food security. Centre for Strategic and Contemporary Research. https://cscr.pk/explore/themes/energy-environment/the-impact-of-climate-change-on-pakistans-food-security/

Aryal JP, Sapkota TB, Khurana R, Khatri-Chhetri A, Rahut DB, Jat ML (2019) Climate change and agriculture in South Asia: adaptation options in smallholder production systems. Environ Dev Sustain 22(6):5045–5075. https://doi.org/10.1007/s10668-019-00414-4

Bagale DR (2020) Nepal–India water cooperation: consequences of mutuality or hegemony? Water Policy 22(6):1098–1108. https://doi.org/10.2166/WP.2020.135

Bagale DR, Adhikari KD (2020) Mahakali treaty: delay in implementation and resulting impacts from Nepal's perspective. Water Policy 22(4):658–669. https://doi.org/10.2166/WP.2020.141

Bhattacharya B (2019, March 29) Water expert Aditi Mukherji: India's drying Himalayan springs is a severe crisis. Live Mint. https://lifestyle.livemint.com/news/talking-point/water-expert-aditi-mukherji-india-s-drying-himalayan-springs-is-a-severe-crisis-111641632484109.html

Bhattacharjee K (2020, August 22) What does Chinese interest in the Teesta mean for India? The Hindu. https://www.thehindu.com/news/international/the-hindu-explains-what-does-chinese-interest-in-the-teesta-mean-for-india/article32420205.ece

Das K, Bandyopadhyay RK (2015) Climate change adaptation in the framework of regional cooperation in South Asia. Law Review 9(1):40–54. https://www.jstor.org/stable/43859672

Dutta V (2021) Water conflicts and resistance: issues and challenges in South Asia. Routledge India

Gambhir M (2021, May 31) Teesta dispute and India-Bangladesh relations. Centre for Land Warfare Studies. https://www.claws.in/teesta-dispute-and-india-bangladesh-relations/

Gautam Y (2017) Seasonal migration and livelihood resilience in the face of climate change in Nepal. Mt Res Dev 37(4):436. https://doi.org/10.1659/mrd-journal-d-17-00035.1

Grigg NS (2016) Water security, disasters, and risk assessment. Integrated Water Resource Management, pp 375–393. https://doi.org/10.1057/978-1-137-57615-6_20

Islam MS, Kieu E (2020) Tackling regional climate change impacts and food security issues: a critical analysis across ASEAN, PIF, and SAARC. Sustainability 12(3):883. https://doi.org/10.3390/su12030883

Jayaram D (2019, November 12) Climate-Fragility Risk Brief: South Asia. Adelphi. https://climate-security-expert-network.org/sites/climate-security-expert-network.com/files/documents/csen_climate_fragility_risk_brief_-_south_asia_0.pdf

Jayaram D (2020a) India–Pakistan river water sharing: prospects for environmental peacebuilding. In: Bandyopadhyay S, Magsi H, Sen S, Ponce Dentinho T (eds) Water Management in South Asia. Springer, Cham, pp 175–202. https://doi.org/10.1007/978-3-030-35237-0_11

Jayaram D (2020b, April 1) How to strengthen climate services and climate diplomacy in South Asia. Climate Diplomacy. https://climate-diplomacy.org/magazine/cooperation/how-strengthen-climate-services-and-climate-diplomacy-south-asia

Karim T (2020) BIMSTEC and its future: connecting south and southeast Asia. National Security 3(1):12–17. https://www.vifindia.org/sites/default/files/national-security-vol-3-issue-1-essay-TKarim.pdf

Krishnan A (2021, March 8) China gives the green light for the first downstream dams on the Brahmaputra. The Hindu. https://www.thehindu.com/news/international/china-gives-green-light-for-first-downstream-dams-on-brahmaputra/article34014912.ece

Lacombe G, Chinnasamy P, Nicol A (2019) Climate risks and solutions: adaptation frameworks for water resources planning, development and management in south Asia: review of climate change science, knowledge and impacts on water resources in south Asia. International Water Management Institute. https://reliefweb.int/sites/reliefweb.int/files/resources/sawi-paper-2.pdf

Lutz AF, Immerzeel WW, Kraaijenbrink PDA, Shrestha AB, Bierkens MFP (2016) Climate change impacts on the upper Indus hydrology: sources, shifts and extremes. PLoS One 11(11): e0165630. https://doi.org/10.1371/journal.pone.0165630

Macchi M, Kollmair M., Sharma E, Dekens J (2009) Strategic framework on climate change adaptation. International Centre for Integrated Mountain Development. http://www.icimod.org/resources/575

Mills L, Janeiro J, Martins F (2021) Effects of sea level rise on salinity and tidal flooding patterns in the Guadiana Estuary. J Water Clim Change 12(7):2933–2947. https://doi.org/10.2166/wcc.2021.202

Mishra BK, Kumar P, Saraswat C, Chakraborty S, Gautam A (2021) Water security in a changing environment: concept, challenges and solutions. Water 13:490. https://doi.org/10.3390/w13040490

Pillai AV (2021, November 18) Reviving South Asian cooperation, climate first. The Third Pole. https://www.thethirdpole.net/en/regional-cooperation/climate-south-asian-cooperation-saarc/

Qamar MU, Azmat M, Claps P (2019) Pitfalls in transboundary Indus Water Treaty: a perspective to prevent unattended threats to the global security. Npj Clean Water 2(1). https://doi.org/10.1038/s41545-019-0046-x

Rahman KS, Islam Z, Navera UK, Ludwig F (2019) A critical review of the Ganges water sharing arrangement. Water Policy 21(2):259–276. https://doi.org/10.2166/wp.2019.164

Roy S (2010, December 1). Climate change and water sharing in South Asia: conflict or cooperation? The Asia Foundation. https://asiafoundation.org/2010/12/01/climate-change-and-water-sharing-in-south-asia-conflict-or-cooperation/

Roy A (2017, October 18) BIMSTEC and climate change: setting a common agenda. Observer Research Foundation. https://www.orfonline.org/research/bimstec-climate-change-setting-common-agenda/

Saeed F, Fuentes-Hutfilter U (2021, January 27) Renewed hope for tackling climate change – could it boost south asia cooperation? Climate Analytics. https://climateanalytics.org/blog/2021/renewed-hope-for-tackling-climate-change-could-it-boost-south-asia-cooperation/

Schoeters K (2002) Regional cooperation in South-Asia on climate change: issues and concerns. Dev Altern 12(10) https://www.devalt.org/newsletter/oct02/of_4.htm

Sen B, Dhimal M, Latheef AT, Ghosh U (2017) Climate change: health effects and response in South Asia. *BMJ*:j5117. https://doi.org/10.1136/bmj.j5117

Shree R (2020) Mapping the water disputes in India: nature, issues and emerging trends. In: Ranjan A (ed) Water issues in himalayan South Asia, pp 103–123. https://doi.org/10.1007/978-981-32-9614-5_5

Tripathi NK (2011) Scarcity dilemma as security dilemma: geopolitics of water governance in South Asia. Econ Polit Wkly 46(7):67–72. https://www.jstor.org/stable/27918149

Yashwant S (2020, June 26) Assam farmers allege that Bhutan is blocking the flow of the Kalanadi River to its paddy fields. Scroll. https://scroll.in/article/965781/did-bhutan-really-obstruct-flow-of-water-to-assam-or-was-it-just-a-misunderstanding

Part III
Water (Water/Sea-Level Rise/Precipitation)

Chapter 5
Vulnerability Analysis of Sea Level Rise (SLR) on a Micro-tidal Coast: A Case Study of Nagapattinam Coast

R S Kankara, Padmini Gunasekaran, and Dhanalakshmi Silamban

Abstract The IPCC sixth assessment report shows that the world will probably reach or exceed 1.5 degrees C (2.7 degrees F) of warming within just the next two decades and the oceans will rise between 10 and 30 inches (26–77 cm) by 2100 (Globally). The report highlights that the sea levels in South Asia could rise by 0.4 m in the low emissions scenario, about 0.5 m in the current pledges scenario, and 0.7 m in the high emissions scenario. Sea levels rise (SLR) has been rapidly increasing, with devastating effects on coastal habitats farther inland, it can cause destructive erosion, wetland flooding, aquifer and agricultural soil contamination with salt, and lost habitat for fish, birds, and plants. Studies by Indian scientists reveal that the trends of sea level rise are estimated to be 1.3 mm/year along the Indian coasts during the past 40–50 years (Unnikrishnan et al., Curr Sci 90(3), 2006). Also, impacts of storm surges, tsunamis, coastal floods, and coastal erosion in the low-lying coastal area cause gradual loss of coastal land. In the present study, the impact of sea level rise on the sandy coast of the Nagapattinam sector is assessed to understand the devastation incurred in both physical and socioeconomic aspects. In the analysis, it is further noted that the exposed vulnerable sectors to sea level rise are increasing with changing climate. The quantitative SLR impact assessment method suggested in the study forms the basic framework for coastal planners in disaster mitigation and building a climate-resilient community.

Keywords Sea level rise · Coastal slope · Regional mean sea level rise · Geomorphology · Shoreline Retreat · Climate Change

Introduction

Climate change is a global or regional change in weather patterns that lasts for a long period, and it has become a global concern over the past few decades. Climate change refers to the change in the environmental conditions of the earth. It is caused

R. S. Kankara (✉) · P. Gunasekaran · D. Silamban
National Centre for Coastal Research (NCCR), Ministry of Earth Sciences, Government of India, NIOT-Campus, Chennai, India
e-mail: rs.kankara@gov.in; kankara@nccr.gov.in

N. Khare (ed.), *Science, Policies and Conflicts of Climate Change*, Springer Climate,
https://doi.org/10.1007/978-3-031-16254-1_5

by many internal and external factors, i.e., by the increased levels of atmospheric carbon dioxide, which is produced by the use of fossil fuels, such as exhaust from cars and trucks, burning trash, and factory emissions. These are also contributing to global warming, which is a gradual increase in the overall temperature of the Earth's atmosphere. These climatic changes affect life on the Earth in various ways and have various impacts on the ecosystem and ecology. Due to these changes, many species of plants and animals have gone extinct. Another is the loss of sea ice and melting of ice sheets, which is caused by the increased temperatures. The melting ice sheets are destroying the wildlife population and increasing the sea level.

Sea level rise is not uniform across the globe. Coastal communities are affected by their local sea level rise, which reflects global sea level rise, changes in local land elevation, tides, and winds. Changes in sea level occur for many reasons in different scenarios. The global average sea level rise, local sea level rise – sometimes called "relative sea level rise" – happens at different rates in different places. Tide gauges record "relative sea level" change due to vertical land movements like natural isostatic movements, sedimentation, tectonic processes, and even anthropogenic activities. Differences in atmospheric pressure, winds, ocean currents, and density of seawater cause spatial and temporal variations in sea level concerning the Geoid. Over these timescales, the most important climate-related factors are likely to be the thermal expansion of the oceans and melting of land ice (but not floating ice shelves or sea ice). According to the 6th IPCC report on the Indian Ocean–South Pacific region, a new tide-gauge based reconstruction finds a Regional Mean Sea Level rise (RSLR) change of 1.33 [0.80–1.86] mm year^{-1} over 1900–2018 (Frederikse et al. 2020) compared to a Global Mean Sea Level (GMSL) change of around 1.7 mm year^{-1}. RSLR rates based on satellite altimetry for the period 1993–2018 increased to 3.65 [3.23–4.08] mm year^{-1} (Frederikse et al. 2020), compared to a GMSL change of 3.25 mm year^{-1}.

Indian Ocean sea level shows large regional variability in all temporal scales, interannual to decadal and multi-decadal scales. Regionally, sea level variations can deviate considerably from the global mean due to various geophysical processes. These include changes in ocean circulations, which partially can be attributed to natural, internal modes of variability in the complex Earth's climate system and anthropogenic influence. The Indian Ocean sea level trends since the 1960s exhibit a basin-wide pattern, with sea level falling in the southwest tropical basin and rising elsewhere. While sea level rises at a faster rate in some oceanic regions, such as in the north Indian Ocean, sea level has shown a fall in the thermocline ridge region south of the equator in the Indian Ocean. Since the extreme sea level is projected to increase with an increase in mean sea level and climate extremes, an extensive network of tide gauges with co-located GPS systems is needed along the Indian coastline, as our coastline is among the most vulnerable and densely populated regions of the globe.

Erosion of beaches and coastal lands, salinization of soils and farmlands, and saltwater intrusion into aquifers. Natural systems may not be able to keep up with the rapid rates of sea level rise. The salinity of freshwater rivers and wetlands may increase. Tidal wetlands and mangroves, which have some ability to migrate in response to rising seas, may find they have nowhere to migrate to because of human development that hems them in. When sea levels rise, it is not just those below the new high tide line that have to worry. Rising seas mean a higher potential for flood events of all kinds. Tidal flooding, storm surge, and other forms of flooding in low-lying coastal areas become even more likely; what the IPCC calls "extreme sea level" events. In the absence of adaptation, more intense and frequent extreme sea level events, together with trends in coastal development will increase expected annual flood damages by 2–3 orders of magnitude by 2100.

Study Area

Nagapattinam lies at 10.77°N 79.83°E. The town is bounded by the Bay of Bengal in the east, Uppanar River in the south, Thiruvarur district in the west, Thanjavur in the North West and Karaikal & Puducherry in the north. The town lies at sea level. The study area constitutes the coastal stretch of the deltaic region of the famous east-flowing river Cauvery (Fig. 5.1). Fluvial-marine interferences and deltaic plains supporting the agricultural sector influence the region. Paleo strandlines are throughout the regions south of the coast with dunes at some places. Five prominent E-W trending sand ridges alternating with tidal flats and mudflats referred to as cheniers are seen. These strandline deposits mainly comprise sand and salty clay. Beach along the coast occurs as narrow zones. Monsoon is often observed with cyclonic storms and flooding of the rivers and canals. The study area has a history of extreme events, for example, 1992 flooding, 1993 cyclone, and 2004 Indian Ocean tsunami. Tourism is one of the biggest industries along the coast, which flourishes with domestic and

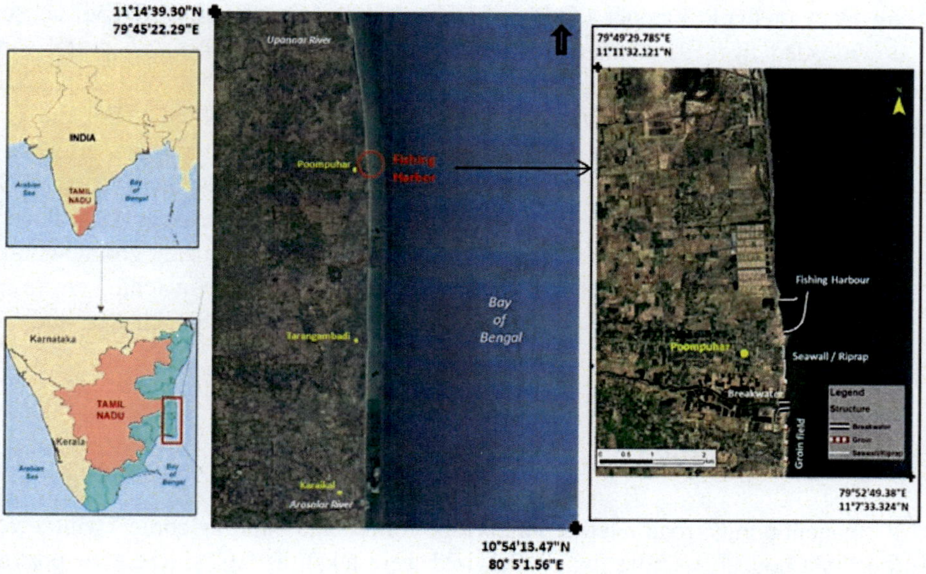

Fig. 5.1 Study area map

international visitors. However, frequent occurrences of tropical cyclones, floods, and other coastal erosion and tide have constantly affected the life and livelihood of people. Sea level rise is yet another threat faced by the coast. Climate change-induced sea level change will increase the impact of storm surges and coastal erosion on the coast. The study region's most crucial need is to assess the impact of sea level rise and identify regions vulnerable to SLR for effective planning and mitigations measures.

Methodology

In the present study, the erosion trend along the coast is assessed since 1990. Later, the shoreline change rate determined is used in the modified Brunn rule to map the inundation extents to assess the impact of sea level rise. Using the sea level change rate, a coastal exposure index attempts to identify the regions vulnerable to SLR, as shown in Fig. 5.2. The sections below elaborated on the data used and the methodology followed in the study.

Shoreline Change Rate Determination

Data Used

The 30 years of shoreline details since 1990 were obtained from satellite images of multiple sensors, as shown in Table 5.1. Satellite images of the years 1990 and 2000 are from thematic mapper (TM) and enhanced thematic mapper (ETM) sensors downloaded from the United States geological survey (USGS) web portal earth explorer-https://earthexplorer.usgs.gov/. Images for 2006, 2008, and 2012–2019 were procured from the National Remote Sensing Centre (NRSC). For the year 2020, sentinel 2 images downloaded from the earth explorer site were used. Data is chosen such that it is cloud-free and preferably of the same season in all years. NRSC provides level 1, i.e., both geometrically and radiometrically corrected data. Though the data are geometrically corrected, rectification is performed using ground control points (GCP) collected from the field to overcome possible edge matching errors and shifts in images, which highly affect the accuracy of the shoreline change study.

Image Rectification

Permanent points (road intersection, bridge corner, and building corner) which were locatable both in satellite image and field were taken as GCPs, and their position (longitude & latitude) was collected using a handheld Trimble geoexplorer global

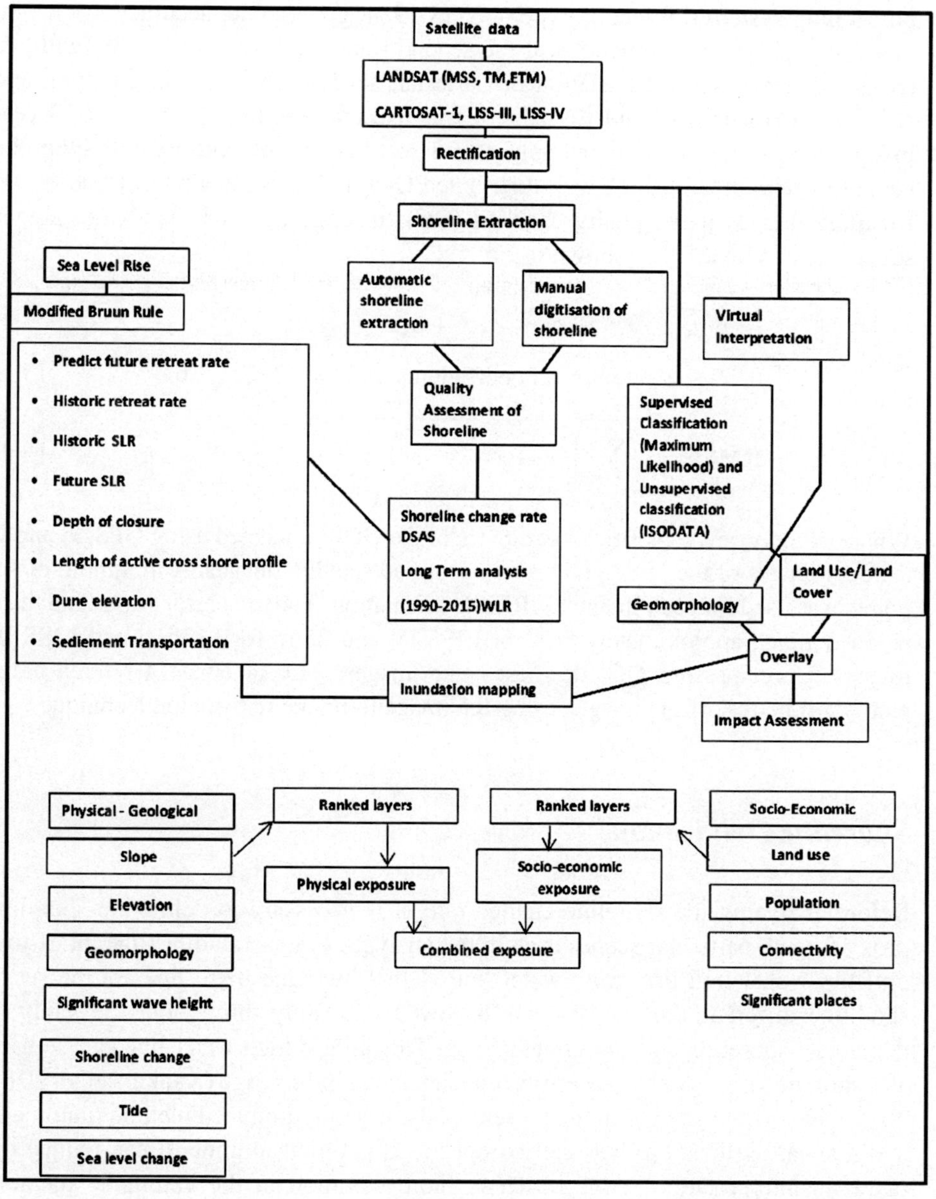

Fig. 5.2 Methodology workflow

Table 5.1 Details of satellite images used in the shoreline study

Satellite data sensor	Year	Resolution (in m)
TM (Landsat – 4,5)	1990	30
ETM+ (Landsat – 7)	2000	30
Pan (Cartosat – 1)	2006	2.5
LISS III (Resourcesat – 1)	2008	23.5
LISS IV (Resourcesat – 2)	2012–2019	5.8
Sentinnel 2	2020	10

positioning system (Projection type – WGS 1984, geographic, accuracy <5 m). In this study, polynomial functions of the second order, as shown in Eq. 5.1 with six terms for correction to translation in Cartesian axes, rotation, scaling, torsion and convexity (Mohammed and Eisa 2013) was used in rectification. First, 2013 year images were considered base images and rectified by second-order polynomial equations with at least 8–12 well-distributed GCP in Erdas imagine 2014 software. Rectified images were quality checked (QC) for edge matching, and root means square errors (RMSE), as shown in Eq. 5.2.

$$p(x, y) = a_1 + a_2x + a_3y + a_4x^2 + a_5\,xy + a_6y^2 \text{ with six coefficients } a_k, k = 1,2,3, \ldots..6. \quad (5.1)$$

$$\text{RMSE} = \sqrt{\sum \left[(x_s - x_r)^2 + (x_s - x_r)^2\right]^2}; r = 1,2\ldots.n \quad (5.2)$$

Where x_s and y_s are geospatial coordinates of the GCPs collected using GPS, x_r and y_r are coordinates of the same point on the rectified satellite images. A minimum of six points was used for each image for RMSE calculation. The total error was maintained below 1 pixel, approximately 5 m for LISS-IV and 25 m for Landsat and LISS-III images. Upon passing QC, the base year images were taken as a reference for rectifying the rest of the images using the image to image registration technique.

Shoreline Delineation

Before analyzing the shoreline change rate, it is necessary to define the shoreline proxy for any particular scenes used in the analysis. In general, dune line, bluff-top/ cliff top, vegetation line, high water line, debris line, and berm line are the set of shoreline differentiators. Anyone is followed uniformly throughout the study to demarcate shoreline and shoreline change. The highest high water line can be used as a shoreline proxy to reduce errors associated with tidal range (Kankara et al. 2014, 2015; Selvan et al. 2016). In the present study, a semi-automated method (automatic and manual digitization) was carried out to extract the shoreline. It can reduce the pixel misinterpretation error, which is more common in the automatic method. Digitized shorelines were validated using shoreline data collected using a handheld GPS of 5 m accuracy at the time of the field visit.

Approach to Estimating Shoreline Changes

The shoreline change analysis was carried out using a digital shoreline analysis system (DSAS) and added to the ArcGIS environment (Thieler et al. 2009). A baseline (onshore) was created parallel to the shoreline to generate transects intersecting the shoreline at every 20-m interval using DSAS. Shoreline change rate was estimated by statistical methods, viz., EPR and WLR. A short-term analysis that quantifies decadal and quadrennial changes was carried out using the endpoint rate (EPR) method. EPR at most supports two shorelines, with the rate of shoreline change being estimated in m/years by dividing shoreline movement along a transect and the time duration is taken for the change (Eq. 5.3).

$$\text{EPR} = \text{Distance between } S_y \text{ and } S_o / \text{Time duration between } S_y \text{ and } S_o \tag{5.3}$$

Where s_y is the youngest shoreline, and s_o is the oldest shoreline.

Long-term shoreline change rates were calculated using the WLR method. The WLR method is similar to the linear regression (LR) method, which involves fitting a least-squares regression line to all shoreline points for a particular transect so that the sum of residuals is minimized. The slope of such a regression line is the required shoreline change rate. The main difference between LR and WLR is the weightage value assigned to overcome errors that could incur due to the reliability of shoreline data. While assigning weightage, higher resolution data sets are given greater emphasis or weightage toward determining the best-fit line compared to unreliable data sets. The weight is defined as a function of the variance in the uncertainty of the measurement (Eq. 5.4).

$$w = \frac{1}{e^2} \tag{5.4}$$

Where W is weightage and e is shoreline uncertainty value.

Uncertainty in shoreline position (e): Several sources of errors affect the accuracy of shoreline positions. In the present study, two positional errors, viz., Season (E_s) and tide (E_{td}), along with three measurement errors, viz., digitization (E_d), rectification (E_r), and satellite resolution (E_p), are considered while determining the uncertainty value. Uncertainty value is the root mean square of shoreline positional errors mentioned before (Eq. 5.5).

$$e = \sqrt{\left[\mathrm{E_s}^2 + \mathrm{E}_{td}^2 + \mathrm{E_d}^2 + \mathrm{E_p}^2 + \mathrm{E_r}^2\right]} \tag{5.5}$$

Accordingly, along the study area, uncertainty values for shoreline obtained from Landsat, Cartosat, LISS III, LISS IV, and sentinel are determined to be 23 m, 5.8 m, 20 m and 6.1 m and 6.5 m, respectively. Shoreline change rates arrived at in DSAS using the WLR method for long term change analysis were classified into seven classes, viz., high erosion (< -5 my^{-1}), moderate erosion (-5 my^{-1} to -3 my^{-1}), low erosion (-3 my^{-1} to -0.5 my^{-1}), stable (-0.5 my^{-1} to 0.5 my^{-1}), low accretion (0.5 my^{-1} to 3 my^{-1}), moderate accretion (3 my^{-1} to 5 my^{-1}), and high accretion (>5 my^{-1}). In short-term analysis, EPR values in the range of 0.5 to $-$0.5 my^{-1} were characterized as stable coast. The shoreline changes $> +0.5$ my^{-1} were considered accretion, while changes < -0.5 my^{-1} were categorized as erosion.

Shoreline Retreat with Response to Sea Level Rise

The first step toward planning adaptation policies of SLR would be the projection of SLR at the local level at different time scales and at its different scenarios as devised by the Intergovernmental Panel on Climate Change (IPCC). A range of values can be used to develop an appreciation of the potential impact or determine thresholds in the magnitude of impacts, vulnerabilities, and adaptation options (Parry et al. 2007). In the sixth Assessment Report of IPCC, the scientists have defined a set of four scenarios, denoted as Representative Concentration Pathways (RCPs). They are identified by their approximate total radiative force in the year 2100. It includes four RCPs with one mitigation scenario leading to a very low force level (RCP 2.6), two stabilization scenarios (RCP 4.5 and RCP 6.0), and one scenario with very high greenhouse gas emissions (RCP 8.5), thus the RCPs can represent a range of twenty-first century climate policies. Climate models play a major role in projecting SLR under different scenarios and at different time slices. However, RCPs are based on a combination of integrated assessment models, simple climate models, atmospheric chemistry and global carbon cycle models under the framework of the Coupled Model Inter-comparison Project Phase 5 (CMIP5) of the World Climate Research Programme (Green Facts 2015). To provide SLR projection at different RCPs (i.e., RCP 2.6, RCP 4.5, RCP 6.0, and RCP 8.5) of IPCC AR6 on a different time scale (i.e., 2025, 2050, 2075, and 2100) at the local level using climate models.

Shoreline change rate and Projection of climate change-induced SLR at different RCPs (2.6, 4.5, 6, and 8.5) of IPCC AR6 on the different time scales (2025, 2050, 2075, and 2100) for the coasts of Nagapattinam are used for calculating Shoreline retreat. The Bruun Rule of shoreline response to sea level rise was formulated for shorelines where all sediment is assumed to remain within the active profile and applies to low-lying shores with sediment covered shore platforms (Bruun 1962). The Bruun Rule appears to provide shoreline positions that underestimate retreat by more than an order of magnitude (Nicholls and Stive 2004; Ranasinghe et al. 2007), even for coastal settings under which it is assumed to work best, dune and barrier coasts with no sediment transportation in the alongshore direction (Chandramohan and Nayak 1992). Obstructive assumptions were overwhelmed by adding parameters like dune elevation (B) and proportion of dune sediment, which is sufficiently large and remains within the active profile (P) (Weggel 1979; Hands 1983). This is defined as the percentage of sand and gravel within the retreating materials, since the material of this size is difficult to mobilize and, unlike silt-clay materials, remains within the active profile rather than being lost offshore and where sediment is released, a Modified Bruun Rule (Dean 1991) appears more suitable for predicting shoreline response (Bray and Hooke 1997). Shoreline retreat rates derived from eight shorelines (1990–2015) response models were tested to know the rates of historical shoreline retreat. Historical map evidence provided a baseline retreat rate between 1990 and 2015, with which accelerating sea level rise could be predicted. The model was applied in full spatial mode and was utilized to evaluate future shoreline change. Using ArcGIS tools, the inundation map was prepared, and the exposure assessment was carried out. The formula of the Modified Bruun Rule is

$$R_2 = R_1 + \frac{(S_2 - S_1)L}{P(B + h)} \tag{5.6}$$

Where,

R_2 – Predict future shoreline retreat rate (ma^{-1}),
R_1 – Historic shoreline retreat rate (ma^{-1}),
S_1 – Historic sea level rise (mma^{-1}),
S_2 – Future sea level rise (mma^{-1}),
L – Length of active cross shore profile (m),
h – Depth of closure (m),
P – Proportion of sediment that remains within the active profile (m^3/year) and
B – Elevation (m).

Identifying Exposed Areas of SLR

Susceptibility is the extent to which a geophysical, biological, and socioeconomic setup is affected by a hazard (sea level rise). So exposure to an event can be quantitatively assessed as a physical, social, and economic liability. While physical exposure quantifies the effect on the physical environment, social, and economic exposure elucidate the loss of life, property, infrastructure, and livelihood. For a better plan for resilience, it is crucial to integrate both physical and socioeconomic aspects. In this study, the physical exposure is assessed using seven physical–geological parameters, viz., coastal slope, Elevation, geomorphology, shoreline change, sea level rise, significant wave height (SWH), and tidal range. The coastal slope and elevation indicate the areas susceptible to inundation during the sea level rise and the associated land loss and gain are a direct function of the coastal slope. Geomorphology is yet another crucial variable in exposing the extent to which the coast is susceptible to destruction during an event. Waves and tides are the continuous coastal process that influences shoreline change. Rising sea levels due to climate change pose a great threat to coastal areas all around the world for land submergence and shoreline retreat. The entire variables are ranked on a scale of 1–5 as per the threshold collected through a literature review as shown in Table 5.2. Physical exposure is finally calculated as the arithmetic mean of the ranks of each variable in Eq. 5.7.

$$\text{Physical exposure} = \frac{\left(R_{\text{slope}} + R_{\text{Elevation}} + R_{\text{Geomorphology}} + R_{\text{SWH}} + R_{\text{Tide}} + R_{\text{Sea level change}} + R_{\text{Shoreline change}}\right)}{7} \tag{5.7}$$

In the socioeconomic exposure index, four socioeconomic variables are considered for analysis, viz., population, land use, connectivity, and significant places. Population and land use pattern details the loss of life and livelihood which could incur during a hazard. Connectivity refers to the rail and road networks present in the

Table 5.2 Physical variables used in physical exposure assessment

Parameter	Very low	low	Moderate	High	Very high	Data source	Period/scenario
Rank	1	2	3	4	5		
Physical – Geological							
Slope (in degree)	>45	20–45	10–20	6–10	< 6	C – MAP	NA
Elevation (in m)	>9	6–9	3–6	0–3	< 0	Aster DEM (30 m)	NA
Geomorphology	Rocky coast	Medium cliff, Indented coast	Estuary, Lagoon	Mangrove, Mudflat	Sandy beach	Landsat OLI	2018
Significant wave height (in m)	<0.55	0.55–0.85	0.85–1.05	1.05–1.25	>1.25	TCA Wave Atlas, 2014	At 25 m depth
Tide (in m)	<1	1–2	2–4	4–6	>6	Jain et al. (2010)	NA
Shoreline change (in m/year)	>5	3–5	−3 to 3	−3 to −5	< −5	Analysis of Shorelines in DSAS	1990, 2000, 2006, 2008, 2012–2020
Sea level change (in mm/year)	<5	5–10	10–15	15–20	>20	Ramachandran et al. (2017)	NA

Table 5.3 Socioeconomic variables used in the assessment

Parameter	Very low	Low	High	Very high	Data source	Period/ scenario
Rank	1	2	3	4		
Socioeconomic						
Land use	Barren land	Vegetated land or Open space	Agriculture or fallow land	Urban area, Salt pan	Sentinel 2	2020
Population	<50,000	>50,000 & <1 lakh	>1 lakh and <2 lakh	>2 lakh	District wise Census report	2011
Connectivity	2 km buffer	1 km buffer	500 m buffer	250 m buffer	GIS data	–
Significant places	NA	Absent	Present	NA	GIS data	–

area, which is essential in emergency conditions for easy mobilization of relief. Significant places include tourist spots and eco-sensitive areas like mangroves, corals, etc. Socioeconomic exposure is deduced using these four variables by ranking them on a scale of 1–4, as shown in Table 5.3, and the social exposure is calculated as the arithmetic means of their ranks, as in Eq. 5.8.

$$\text{Socio}-\text{economic exposure} \\ = \left(R_{\text{Land use}} + R_{\text{Population}} + R_{\text{Connectivity}} + R_{\text{Significant places}}\right)/4 \quad (5.8)$$

The physical exposure, socioeconomic exposure index (SEI), and coastal exposure index (combined impact) of the coast is determined as detailed in the workflow shown in Fig. 5.2. After physical exposure and SEI are estimated, combined impact is taken as their average, as shown in Eq. 5.9. Thus, the combined impact scores obtained in the process are grouped into five classes, viz., very low (<=0.75), low (>0.75 and <=1.55), medium (>1.55 and <=2.35), high (>2.25 and<=3.15), and very high (>3.15) liability. The combined impact class of very high is the most vulnerable to SLR.

$$\text{Combined Impact} = (\text{Physical exposure} + \text{socioeconomic exposure})/2 \quad (5.9)$$

Results & Discussion

Shoreline Change Rate

Based on shoreline change analysis, the long-term (1990–2020) WLR method was used and the results showed an erosion state with a maximum rate of –14.57 m/year and a minimum rate of –0.51 m/year with a mean rate of –2.36 m/year. The accretion state was observed to have a maximum rate of 8.65 m/year and a minimum rate of −0.6 m/year with a mean rate of 1.29 m/year. The overall long-term analysis has been analyzed as an erosion state. The EPR method was used for calculating both short-term and annual analysis. In the short term (2012–2020), the erosion factor was observed to have a maximum rate value of −62.73 m/year and minimum rate value of −0.01 m/year with a mean rate of −4.07 m/year, and based on the accretion state, the maximum rate value was 31.24 m/year and minimum rate value was 0.04 m/year with a mean rate of 6.06 m/year. The erosion state was observed with a maximum value of −112.93 m/year with a minimum rate of −0.09 m/year a mean rate of −12.4 m/year. The accretion rate was observed to be 621.3 m/year with a minimum rate value of 0.03 m/year with a mean rate of 24.7 m/year in annual (2019–2020) shoreline change analysis. The overall long-term analysis indicates that 78% of the coast is in a state of erosion (Fig. 5.3).

The Bruun Rule is observed to have certain limitations; therefore, the modified Bruun Rule has been used to calculate the inundation in the study area. The geography of a study area was observed to have an average elevation of 1 m. Using the modified Bruun Rule, by substituting length, depth, elevation, shoreline rate, historic and future SLR, sediment transportation, the inundation rate has been calculated. The shoreline retreat of the Nagapattinam coast with different RCPs (2.6, 4.5, 6, and 8.5) of IPCC AR6 on a different time scale (2025, 2050, 2075, and 2100)

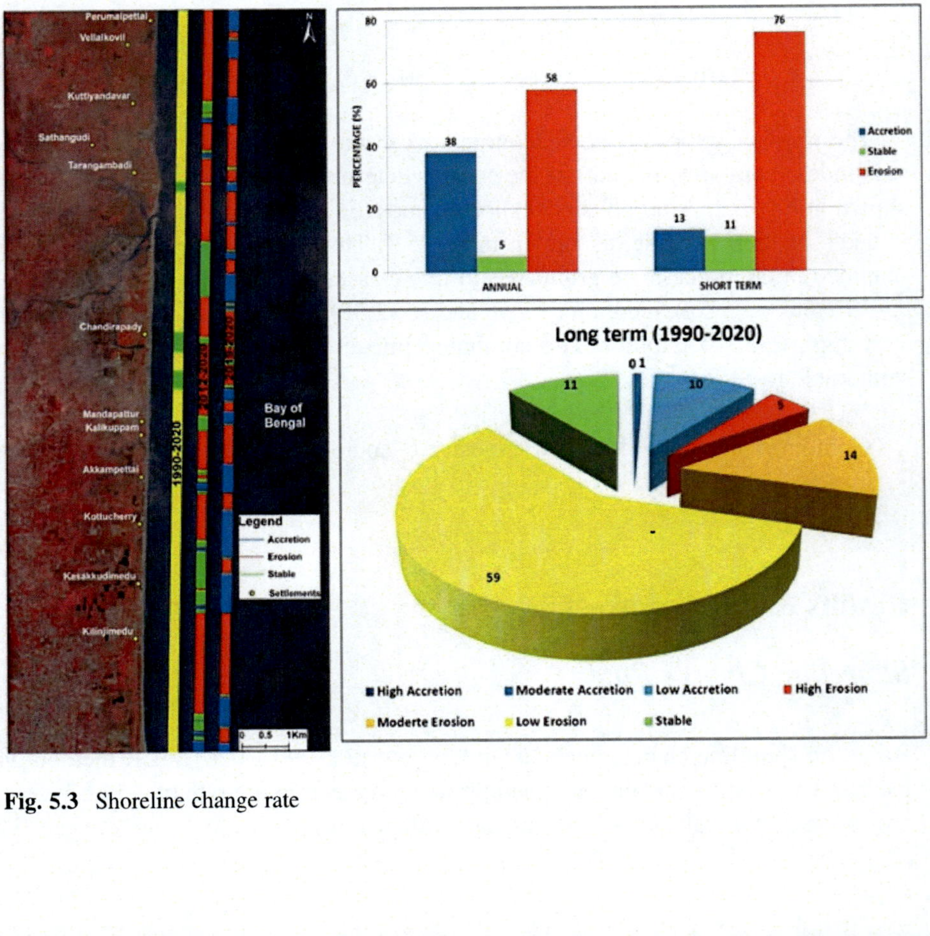

Fig. 5.3 Shoreline change rate

has been considered in this study. Projected SLR for Nagapattinam range from 17.97 cm (2050) to 36.99 cm (2100) for RCP 2.6, the inundation was observed to be 10.04 m to 190.8 m; 19.82 cm (2050) to 50.02 cm (2100) for RCP 4.5, inundation is 10.36 m to 222.7 m; 18.08 cm (2050) to 51.93 cm (2100) for RCP 6, inundation is 10.06 m to 227.37 m; 22.51 cm (2050) to 78.17 cm (2100) for RCP 8.5, inundation is 10.82 m to 291.6 m. The maximum inundation of the coast is observed to be approximately 300 m (2100) explained in Fig. 5.4. Ramasamy et al. (2017) stated that the constant shifting of the port city appears to be logical under the proven phenomenon of sea level rise. The observed MSL trend of the Nagapattinam tide gauge station is given as a key input to the SimCLIM SLR scenario generator to project that estimates the SLR with four different scenarios, and the scenario where RCP 2.6 agrees with our present shoreline has been considered for the coast to predict the area which is under exposure (Fig. 5.4).

Impact of SLR on Exposed Areas

In the present study, understanding the impact of SLR on the physical–geological environment and socioeconomic setup of the coast is prima facie for meticulous mitigation planning. The impact of SLR is determined in terms of exposure area, which is a function of both physical and socioeconomic aspects in the study area. Physical exposure is calculated using seven variables, viz., elevation, coastal slope, significant wave height (SWH), geomorphology, tide, shoreline change rate, and sea level change rate. Variables like elevation and coastal slope play a crucial role in

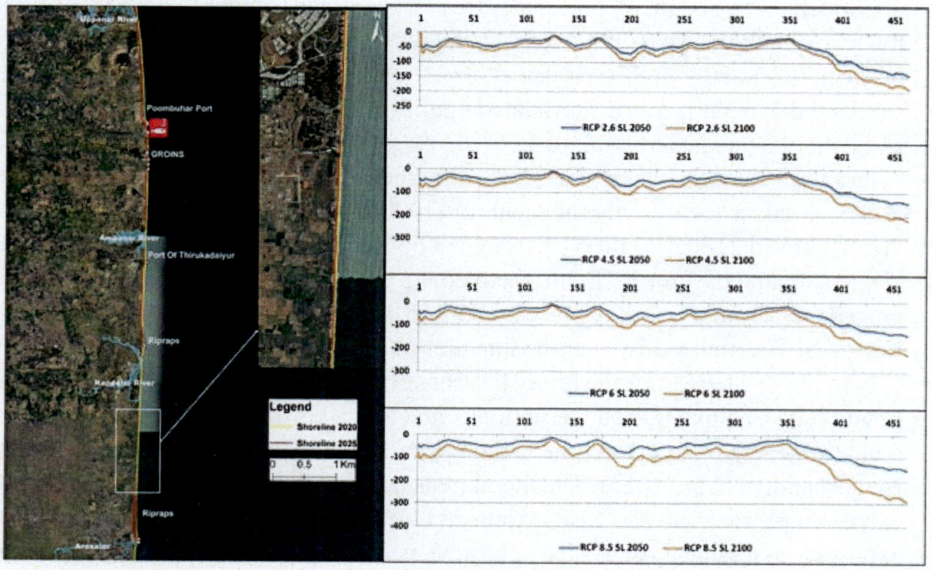

Fig. 5.4 Projection of climate change-induced sea-level rise

delineating the inundation extent and damage occurrences (Rao et al. 2009; Kumar et al. 2010). The SWH and tide are continuous physical processes seen on the coast. Geomorphology highlights the susceptible regions which are likely prone to destruction during an event. Though shoreline change occurs naturally, in places with structures like the fishing harbor, breakwater and groins as found in Poompuhar and Tharangambadi, the shoreline trend alters and increases the coast's exposure state (Selvan et al. 2016).

From the Aster DEM (digital elevation model), it can be observed that the study region's elevation has a maximum of up to 4 m at some places like near the newly constructed fishing harbor with an exposure rank of either 1 or 2. The elevation along the rest of the sector varied with a mean elevation of 1.2 m and was assigned ranks ranging from 3 to 5. The coastal slope is less than 20°, and the coast has a gentle slope of <6°. According to the ranking criterion, as shown in Table 5.2, coastal sectors with a slope < 6° were assigned a very high exposure rank of 5. Regions with slopes between 6° and 10° were assigned rank 4, and 10°–20° were assigned rank 3, respectively. The overall long-term analysis indicates that 78% of the coast is in a state of erosion. The areas prone to erosion are considered highly vulnerable compared to accreting areas. Therefore, they are assigned a high exposure rank of 5 & 4, while accreting areas are assigned the least rank of 1 & 2, respectively.

The coastal features from aeolian, fluvial, and marine influences are seen. The river Cauvery flows is a deltaic plain with sandy beaches along Poompuhar, Tharangambadi, and Karaikal. Exposure ranking is given according to the criterion as discussed in Table 5.2. The sandy beaches of Poompuhar and Tharangambadi were given a high exposure rank of 5. Significant wave height (SWH) was taken from TCA of the Indian Coast, in 2014. From the wave atlas, it can be seen that along the study area, SWH is in the order of 0.5–1 m. Hence, an exposure rank of 3 (Moderate) is assigned. The study area is in the micro-tidal range and does not fluctuate much in a year. The average tidal range of the region from 2011 to 2015 is between 0.3 and 0.9 m Priya Rajan et al. (2019) and Jain et al. (2010) also observed similar tidal ranges. Since the tidal range is well below 1 m, the study sector is assigned a low exposure rank of 1.

For the sea level change variable, SLR at the regional level in Unnikrishnan et al. (2006) is taken for the present scenario (1990–2020) and the futuristic scenarios, viz., 2050 and 2100 uses the SLR obtained by Ramachandran et al. (2017) along the Nagapattinam coast for RCP 2.6; the medium scenario is taken as the reference and ranked. Accordingly, for the present scenario, sea level change rank is given as very low as SLR estimated by Unnikrishnan et al. (2006) is a regional average rise of 1.29 mmy^{-1}, whereas for the futuristic scenarios (2050 and 2100) the observed MSL trend of the Chennai tide gauge station is given as a key input to the SimCLIM SLR scenario generator to project the estimates of SLR for the coast of Tamil Nadu by Ramachandran et al. (2017). The literature is taken as the reference in assigning high to very high ranks in the SLR exposure calculation. For the physical–geological exposure assessment, the nine physical–geological variables are given equal consideration and the physical exposure score is taken as the arithmetic means of the sum of the ranks of variables.

Socioeconomic variables are significant in determining the socioeconomic exposure of the coast as they enhance the accuracy of the assessment, which often involves the effect of hazards to loss incurred in human life and livelihood. Integration of different socioeconomic variables such population, land use, connectivity (rail and road network), and significant places are used in the present study. On the Poompuhar, Tarangambadi, and Karaikal coast, settlements are seen very near to the coast within a buffer range of 500 m; hence those regions were assigned a very high exposure rank of 5 in the land-use variable. Barren lands seen near the north of fishing harbor are assigned a very low rank of 1 as damage incurred in these places due to SLR will least affect the humans and their livelihood. The coast of Poompuhar, Tarangambadi, and Karaikal being tourist spots are flanked by both national and international travelers. Hence these locations along the coast are given significant places status and beaches in the sector are assigned a rank of 4 in the significant places variable.

As per the census of India 2011, the population density of the Poompuhar sector is more than 200,000 and was assigned a high exposure rank of 4 as the impact caused by SLR could affect many lives. Similarly, as per the population density mentioned in Table 5.3, respective exposure ranks were assigned in the study area. The rail and road networks present in an area help in identifying the economic development of a region. The destruction caused to these infrastructures can affect the mobility of people and rescue operations. The exposure rank for the railroad connectivity variable is assigned by classifying buffers of 250 m, 500 m, 1000 m, and 2000 m from the coast. Sections of the road/rail in the 250 m buffer near the coast were assigned a high exposure rank of 4 (high), while the sections in the 2000 m buffer and above were assigned the rank of 1 (very low). In the Poompuhar, Tarangambadi, and Karaikal sectors, certain pockets of the coast, especially near the tourist places where road density is increased, are assigned very high to high exposure rank of either 4 or 5. Socioeconomic exposure is taken as the arithmetic mean of the sum of the ranks of variables as discussed above.

Ranking the physical–geological and socioeconomic variables and integrating them into impact exposure assessment helps in understanding their rapid change spatially. Finally, the combined impact assessment of SLR on the coast is determined by the arithmetic mean of physical exposure and social exposure, obtained through the corresponding arithmetic mean of the sum of the ranks of physical and socioeconomic variables, respectively.

Impact of SLR on Exposed Area Present Scenario

For the present scenario of 1990–2020, the exposure scores computed ranged from a minimum of 2 to a maximum of 4.5 with a standard deviation of 0.13 and a mean of

2.4. The combined impact scores obtained in the process are grouped into five impact classes, viz., very low (<=0.75), low (>0.75 and<=1.55), medium (>1.55 and <=2.35), high (>2.25 and <=3.15), and very high (>3.15). The exposure class of very high is the most vulnerable sector to SLR. Accordingly in the study area, it can be observed that about 66% and 14% of the coast has high and very high exposure to the impacts of SLR, respectively. Certain pockets near the southern Poompuhar and the north are seen to exhibit a moderate impact on SLR. Based on the exposure score and the statistics it can be noted that the study region is highly vulnerable at the present.

Impact of SLR on Exposed Area Futuristic Scenario (2050 & 2100)

Assessing the impact of SLR in the futuristic scenarios of 2050 and 2100 in RCP 2.6 force under medium conditions helps in understanding the exposure areas of SLR in the possible realistic future climatic projections. In 2050, exposure scores are seen to increase in comparison with the present scenario. The 2050 scores ranged from a minimum of 2.2 to a maximum of 4.6 with an average of 2.5. Approximately 67% and 18% of the coast are seen to exhibit high and very high impacts on exposed areas to SLR, respectively. This is a significant increase in comparison to the present day scenario, as shown in Fig. 5.5. The impact of SLR on the physical–geological environment and socioeconomic setup of the coast is further seen to increase in 2100; wherein about 72% and 22% of the coast is seen to exhibit high and very high vulnerable sectors to sea level rise, respectively.

Conclusions

In this study, a methodology is conceptualized and demonstrated to quantitatively assess the impact of climate change-driven sea level rise. The example of the sandy coast of the Nagapattinam sector gives an initial warning that the coastal ecosystem, infrastructure, populations, and shore coast need to be preserved. Using different time frames and scenarios for the projected SLR gives source information for coastal policy planning and decision-makers to take SLR adaption. Since the Bruun rule is not full-fledged, the modified Bruun rule has been compared to demarcate the coastal disaster. The horizontal inundation overlaid on geomorphic landforms and land use/land cover features help in identifying the exposure areas during sea level rise. This study is useful for stakeholders, policymakers, coastal managers, scientists, and coastal livelihoods. This study provides an interactive means to identify the exposure zone. The output maps can be used to visualize the affected areas spatially.

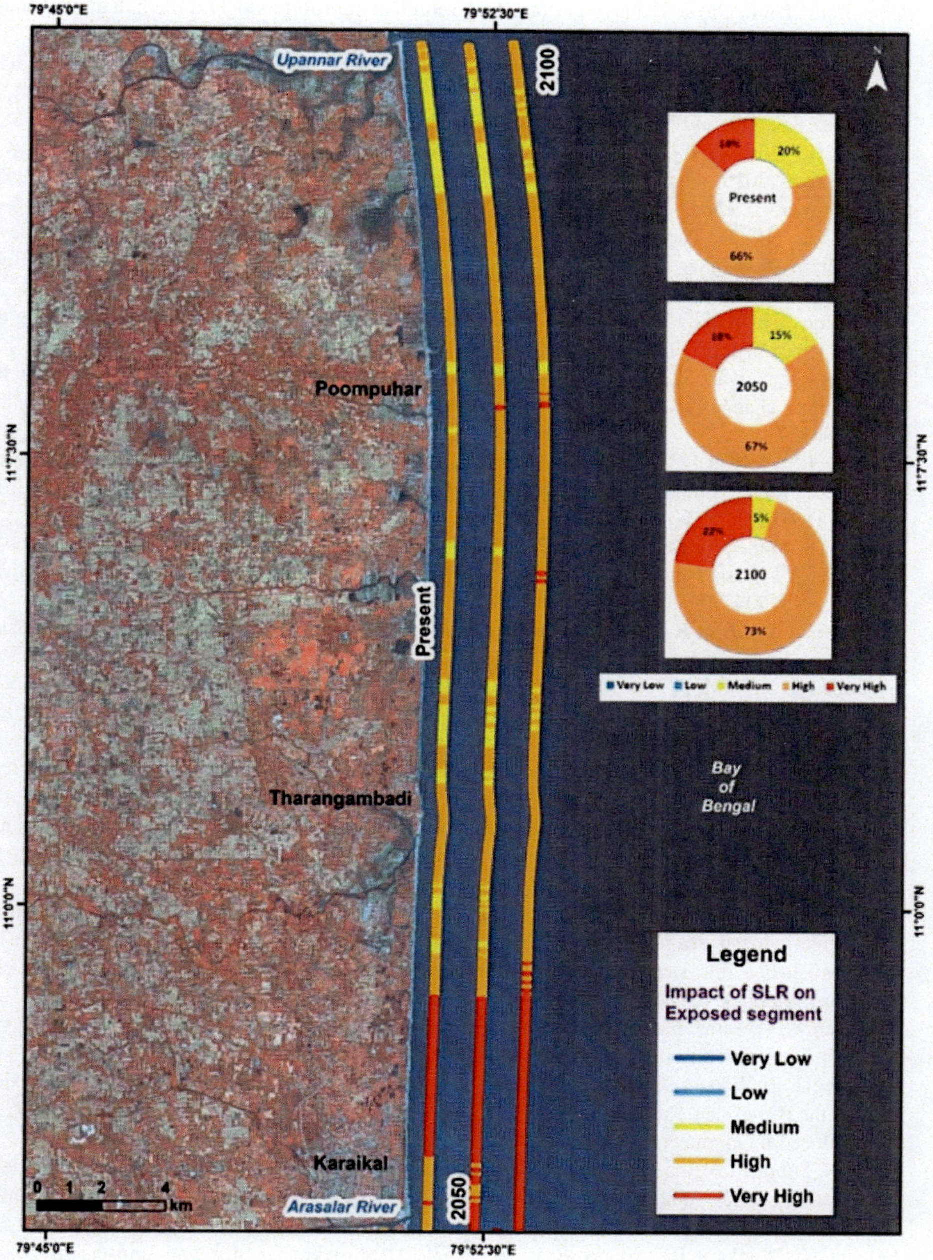

Fig. 5.5 Exposed area to SLR in 2050 and 2100

References

Bray JM, Hooke JM (1997) Prediction of soft-cliff retreat with accelerating sea-level rise. J Coast Res 13(2):453–467

Bruun P (1962) Sea level rise as a cause of shore erosion. J Waterw Harb Div, Proc Am Soc Civil Eng 88:117–130

Chandramohan P, Nayak BU (1992) Longshore sediment transport model for the Indian west coast. J Coast Res 8(4):775–787

Dean RG (1991) Equilibrium beach profiles: characteristics and applications. J Coast Res 7(1): 53–84

Frederikse T, Landerer FW, Caron L et al (2020) The causes of sea-level rise since 1900. https://doi.org/10.1038/s41586-020-2591-3

Green Facts (2015) Climate change IPCC 2013 update. http://www.greenfacts.org/en/climate-change-ar5-science-basis/l-3/5-climatesystemfuture.html

Hands EB (1983) The Great Lakes as a test model for profile responses to sea-level changes. In: Komar PD (ed) Handbook of coastal processes and erosion. CRC Press, Boca Raton, Florida, pp 176–189

Jain I, Rao AD, Jitendra V, Dube SK (2010) Computation of expected total water levels along the East Coast of India. J Coast Res 264:681–687. https://doi.org/10.2112/08-1170.1

Kankara RS, Selvan SC, Rajan B, Arockiaraj S (2014) An adaptive approach to monitor the shoreline changes in ICZM framework: a case study of Chennai coast. Ind J Mar Sci 43(7): 1266–1271

Kankara RS, Selvan SC, Markose VJ et al (2015) Estimation of long and short term shoreline changes along Andhra Pradesh coast using remote sensing and GIS techniques. Procedia Eng 116:855–862

Kumar TS, Mahendra RS, Nayak S, Radhakrishnan K, Sahu KC, Kumar TS et al (2010) Coastal vulnerability assessment for Orissa State, East Coast of India. J Coast Res 263:523–534. https://doi.org/10.2112/09-1186.1

Mohammed NZ, Eisa E (2013) The effect of polynomial order on georeferencing remote sensing images. Int J Eng Innov Technol 2:5–8

Nicholls RJ, Stive MJF (2004) Society and sea-level rise requires modeling. Sci E Lett 303:5665

Parry ML, Canziani OF, Palutikof JP, Van der Linden PJ, Hanson CE (2007) Climate change 2007: impacts, adaptation and vulnerability. Contribution of Working Group II to the Fourth Assessment Report of the Intergovernmental Panel on Climate Change. Cambridge University Press, Cambridge, 976p

Priya Rajan SM, Nellayaputhenpeedika M, Tiwari SP, Vengadasalam R (2019) Mapping and analysis of the physical vulnerability of coastal Tamil Nadu. Hum Ecol Risk Assess 0:1–17. https://doi.org/10.1080/10807039.2019.1602752

Ramachandran A, Khan AS, Palanivelu K, Prasannavenkatesh R (2017) Projection of climate change-induced sea-level rise for the coasts of Tamil Nadu and Puducherry, India using SimCLIM: a first step towards planning adaptation policies. J Coast Conserv 21(23):1–12. https://doi.org/10.1007/s11852-017-0532-6

Ramasamy SM, Saravanavel J, Kumanan CJ, Gunasekaran S (2017) Coordinates and chronology of the ancient port city of Poompuhar, South India. Curr Sci 112(6):25

Ranasinghe R, Watson P, Lord D, Hanslow D, Cowell P (2007) Sea level rise, coastal recession, and the Bruun rule. Proceedings of coasts and ports '07, Melbourne, VIC, Australia. On CD rom produced by engineers Australia

Rao KN, Subraelu P, Rao TV, Malini BH, Ratheesh R, Bhattacharya S, Rajawat AS (2009) Sea-level rise and coastal vulnerability: an assessment of Andhra Pradesh coast, India through remote sensing and GIS. J Coast Conserv 2008:195–207. https://doi.org/10.1007/s11852-009-0042-2

Selvan SC, Kankara RS, Markose VJ et al (2016) Shoreline change and impacts of coastal protection structures on Puducherry, SE coast of India. Nat Hazards 83:293–308

Thieler ER, Himmelstoss EA, Zichichi JL, Ergul A (2009) The Digital Shoreline Analysis System (DSAS) version 4.0-an ArcGIS extension for calculating shoreline change

Unnikrishnan AS, Kumar KR, Fernandes SE et al (2006) Sea level changes along the Indian coast: observations and projections. Curr Sci 90(3)

Weggel JR (1979) A method for estimating long term erosion rates from a long term rise in water level. Coast Engineering Tech Aid (CETA) 79-2. U.S. Army Coastal Engineering Research Center, Fort Belvoir, VA

Chapter 6
Lacustrine Records of Anthropogenic Change and Precipitation in the Monsoonal Core Zone of Central India since the Last Century

Samaya S. Humane, Sumedh K. Humane, and Snehal Juare

Abstract A sediment core from the Navegaon Bandh Lake (NBL) was studied for diatoms and sediment geochemistry to investigate the anthropogenic impact and climatic variability over the past century. The declined geochemical concentrations show three major impacts on the NBL coinciding with the drought events in the foremost parts of India around 1918, 1975, 1965 and 1991. The rise in the geochemical content of the NBL core was evident in the post dam construction period (i.e., ~1917–1919) with the two major peaks around 1975 and 1998. These major peaks are indicators of the higher rate of soil erosion and improved precipitation in the catchment of the NBL. Anthropogenic activities have caused five major changes in diatom assemblages and ecology in the NBL around 1946, 1956, 1972, 1978 and 2006. The increased agricultural activities in the catchment of the NBL are evident since ~1906–1966 as the fertilizers (K, P) are mainly derived other than soil particles of the catchment. Similarly, deforestation in the region was evident during ~1967–1975 coinciding with the period of the highest soil erosion. Another major phase of soil erosion caused by anthropogenic activities was noticed during ~1992–1998. The hypolimnetic oxygen of the NBL has declined for the past ~80 years indicating increased eutrophication in the lake till the present. The nutrient input in response to climatic conditions and anthropogenic activities has played a vital role in the diatom shift of the NBL. *Aulacoseira granulata* was predominant during ~1906–1914, ~1920–1922, ~1941–1943, ~1948–1956, ~1972–1976 and ~1973–1982 in the NBL signifying the increased soil erosion and flux, persistent wet period and increased nutrient levels. While the major period of the profusion of *Rhopaloidea musculus* in the NBL core was during ~1919–1921, ~1933–1940, ~1956–1972 and ~1978–2004, respectively, indicating a mesotrophic, Meso

S. S. Humane · S. K. Humane (✉)
Department of Geology, Rashtrasant Tukadoji Maharaj Nagpur University, Law College Square, Nagpur, Maharashtra, India

S. Juare
Department of Geology, Yashwantrao Chawhan Art, Commerce and Science College, Lakhandur, Maharashtra, India

N. Khare (ed.), *Science, Policies and Conflicts of Climate Change*, Springer Climate,
https://doi.org/10.1007/978-3-031-16254-1_6

euhyaline and alkalibiontic condition of the lake linked with the long dry period/less rainfall prevailed in the NBL watershed. The planktonic form *Discostella stelligera* was common during ~2007–2012 and indicates the existence of alkaline lake waters with moderate nutrients coinciding with the comparatively dry period/less rainfall and changing water level and deforestation in the catchment of the NBL. Overall, the investigation of the past evidence shows the importance of the study of fossil diatoms and geochemistry to understand the anthropogenic and precipitation changes along with the trophic status of the lake, which may be used for the management of the tropical wetlands in India.

Keywords Diatoms · Geochemistry · Soil erosion · Tropical Lake sediments · Eutrophication · Central India

Introduction

The sediments deposited at the bottom of a lake provide a variety of information that can be used to understand the past conditions of a lake, its watershed, and climate (Meyers and Teranes 2001). Lakes are mostly fed by some rivers or streams and hence the lake consists of autochthonous and allochthonous sediments (Smol 2008). A stream serves as the major source of allochthonous sediments. The allochthonous sediments carry with them the dissolved chemicals and particulate inorganic and organic matter (OM). The autochthonous sediments on the other hand have with them a significant assemblage of remains of macrophytes, phytoplankton, zooplankton, bacteria, microorganism and aquatic invertebrates, which thrive in that environment. Hence, the source of autochthonous material includes biological activity and chemical precipitation within the lake (Smol 2008). The geology of the lake's watershed, climate and land use, including anthropogenic activities, directly affect the quality and quantity of material that enters a lake ecosystem (Cohen 2003). The past changes in the pH, salinity, nutrient status, climatic changes and lake level fluctuations can be inferred by studying the sediment geochemistry and diatoms from the core extracted from the lakes/reservoirs (McFadden et al. 2005; Mullins et al. 2011). The paleolimnological studies were also done using diatoms owing to their sensitivity to environmental changes (Batterbee et al. 1999; Bennion et al. 1995; Liu et al. 2012; Schroeder et al. 2016).

The Indian economy is mainly based on agricultural production which depends upon favourable monsoonal conditions. The central Indian region has been experiencing unreliable climatic conditions with low precipitations for the past few decades affecting the farmers depending upon agricultural produce. The increasing agricultural practices in the region have also been impacting the water quality of the area. Thus, the present chapter provides a good insight into the shift in the past climatic patterns mainly the hydrologic conditions in the parts of the central Indian region determined based on the study of sediment geochemistry and diatoms from the core sediments of the Navegoan Bandh Lake (NBL), Central India. The NBL is one of the major lakes of the Gondia district present in the eastern part of the

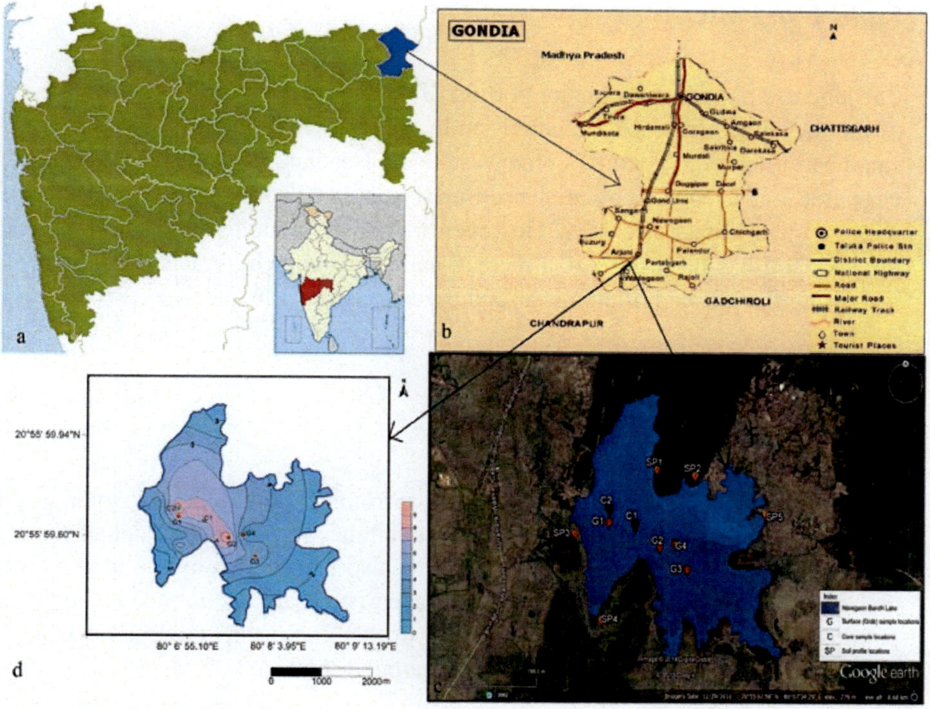

Fig. 6.1 (**a**) Map of Maharashtra State, India, showing the location of Gondia District, (**b**) Map of Gondia District showing the location of Navegaon Bandh Lake (NBL), (**c**) Satellite Image of NBL and (**d**) Bathymetric map of the NBL

Maharashtra state, India (Fig. 6.1a–c). The NBL (latitude 20°53′ to 20°56′N and longitude 80°06′ to 80°09′E) has a circumference of ~27 km and a water surface of ~20 km^2. The average depth of the lake is about 21 m with a catchment area of about 90 km^2 (Fig. 6.1d). Various factors such as adjoining lithology, soils, agricultural practices, animal excrements, aquatic plants and vegetal matter influence the water quality of the Navegaon Bandh Lake.

Climate and Rainfall

The Gondia district has tropical climatic conditions and experiences very hot summers and cold winters with temperature variations from ~7 °C to 48 °C, respectively. The summer starts in March and continues till mid-June, while the winter begins in October end and continues till February with December being the coldest month. The district receives rainfall from southwest summer monsoonal winds mainly from June to September. The maximum rainfall generally occurs during July and August with an average annual rainfall of 1197 mm (Gondia District Gazetteer 2013).

Land Use Pattern

The land use pattern of the major portion of the NBL watershed shows the presence of the deciduous forest, agricultural land (mainly used for kharip crop and a very small part used for rabi crops), some wastelands associated with the dense scrubs and very little land is covered by rural build-up (Fig. 6.1c). Fishing takes place in the lake with no other human activities. However, a zoological garden was developed near the lake and has been used as a recreational centre which attracts tourists thereby increasing the plastic pollutants around the lake.

Geological Setting

The Navegaon Bandh Lake is geologically surrounded by the granitic gneiss of the Tirodi gneissic complex, quartzite-gritty quartzite, amphibolites and hornblende schist of the Amgaon gneissic complex. It is also surrounded by the meta gabbros and quartz veins of the Mesoproterozoic age. There are occurrences of phyllites, pelitic schist and basic tuffaceous andesite of the Khairagarh group around it. The pockets of laterites are also distinctly seen around the lake (DRM 2000).

Materials and Methods

Core Sampling, Processing and Chronology

In January 2013, a sediment core (Latitude: N 20° 54′ 55.8″ and Longitude: E 80° 7′ 3.4″) of 78 cm length was recovered in the polycarbonate barrel from the deepest part of the NBL using a gravity corer. The total weight of the retrieved sediment core was calculated in the laboratory. Further, the sediment core was vertically cut into two halves and photographed in addition to the description of the core profile. One of the halves of the core was sub sectioned at an interval of 1 cm for the bottom sections and an interval of 0.5 cm for the top 20 cm. It was used for diatom processing and other analyses. The other half of the core was retained as an archive in the laboratory. The chronology for the present study of the NBL follows the already published paper (Humane et al. 2016). The results of the Constant Rate Supply (CRS) model for the sediments of each depth of the NBL core were further used to calculate the bulk sediment accumulation rates of the geochemical elements (g/m^2/year) by the related amount (mg/g) of each component in the bulk sediment (Fig. 6.2; Garrison and Laliberte 2010). The uncertainty of the age prevails in the older sediments deeper than ~50 cm of the core. Therefore, the ages of the NBL sediment core up to 48 cm were considered for the calculation of the accumulation rate of geochemical elements at various depths and discussed in this chapter (Fig. 6.3). Diatom analysis was

Fig. 6.2 Sediment accumulation rate ($g\ m^{-2}\ year^{-1}$) of NBL since 1861, using constant rate supply (CRS) model

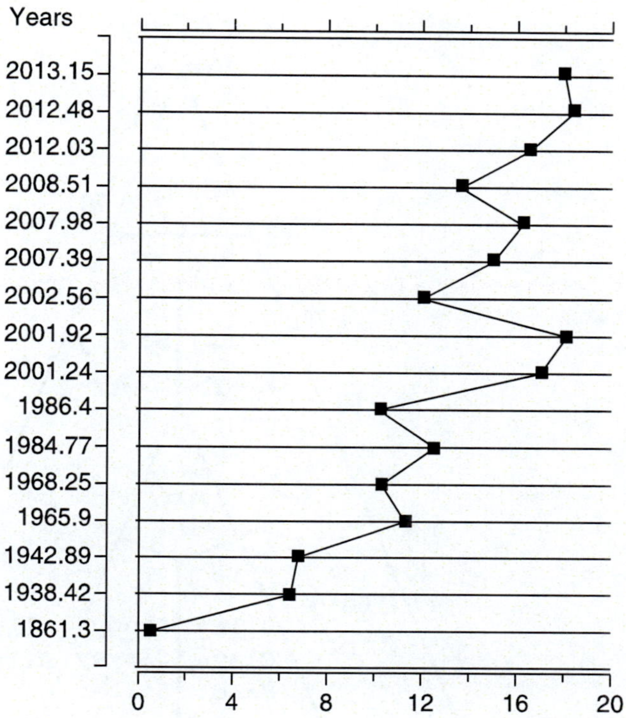

done by processing some uncrushed core samples. The accumulation rate of geochemical elements for a sediment core is shown in the vertical diagram prepared by the computer program C2, version 1.5 (Juggins 2007).

Maceration for Diatom Study

About 1 gm of a core sample obtained through conning and quartering at 1 cm intervals was used for the maceration to prepare diatom slides following the standard methods (Batterbee et al. 2001). The prepared slides were studied for diatom identification and enumeration under a Leica microscope (DM-350) at 630× and 1000× (oil immersion) along with the microphotography of each species. Minimum 300 valves were measured wherever possible in the slides. The occurrences of individual species were converted into a percentage. The identification of diatom species was done with the help of important literature (John 2012; Karthick et al. 2013; Liu et al. 2017; Metzeltin et al. 2005; Metzeltin and Lange-Bertalot 2007).

The recognition of the diatom groups was made based on their division in the complete length of the core and the constrained incremental sum of squares (CONISS) by the computer program TILIA (Grimm 1991).

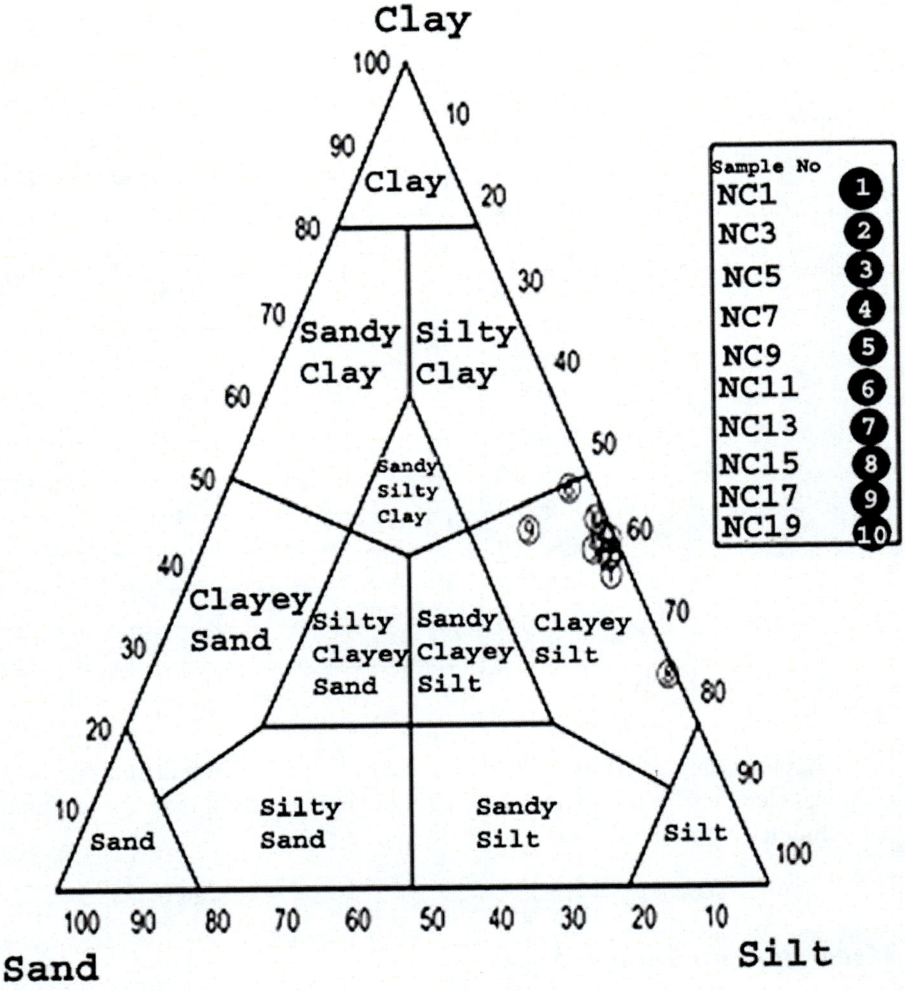

Fig. 6.3 Distribution of sand, silt and clay in the sediment cores of NBL

Results

Description of NBL Core Section

The colour of the entire core varies from light brown to dark brown with a smooth transition. The overall grain size of the core lies in between clayey silt, with intermediate variation from slight sandy-silt to silty clay (Fig. 6.4). The higher organic matter content in the core was demarcated by dark bands representing increased soil erosion coinciding with the wet periods. A few fragments of rotten plant roots/stems are also observed along with a few fish scales and shell fragments. The annual laminations are quite distinct in the core, although not very prominent in the photograph.

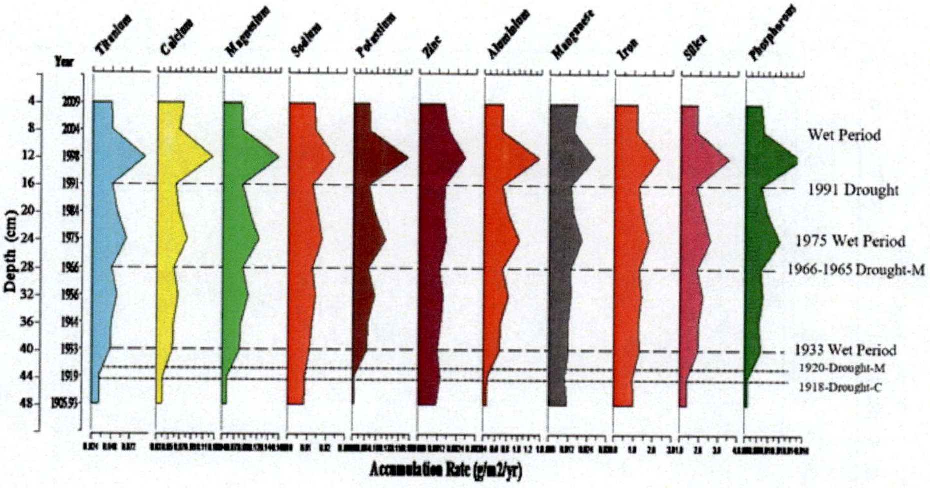

Fig. 6.4 Vertical profile of accumulation rate of the geochemical element in a sediment core of NBL (M – moderate; C – Calamitous)

Geochemical Analysis

The trophic status and changing precipitation patterns were investigated using major and trace elemental content of the NBL core. The past changes in the precipitation trends and the trophic condition of the NBL have been inferred employing the varying accumulation of geochemical elements. The impact of anthropogenic activities other than soil erosion on the trophic status of the NBL is also interpreted (Fig. 6.5).

Chemical Weathering Intensity (CWI)

The deforestation and the enhanced agricultural practices in the catchment have affected the NBL by augmenting the weathering and erosion of soils of the region. In this context, the chemical weathering intensity (CWI) in the catchment of this lake is calculated using the formula $\{(CaO+MgO+Na_2O)/Al_2O_3\}$ to know the strength of weathering and soil erosion (Sun et al. 2009; Fig. 6.6).

Diatom Stratigraphy

The stratigraphic data shows the abundant diatom species in the NBL (Fig. 6.7). The four diatom assemblage zones (DAZ)/units and ten sub-zones were identified for major species abundance and the constrained cluster analysis for the NBL core using

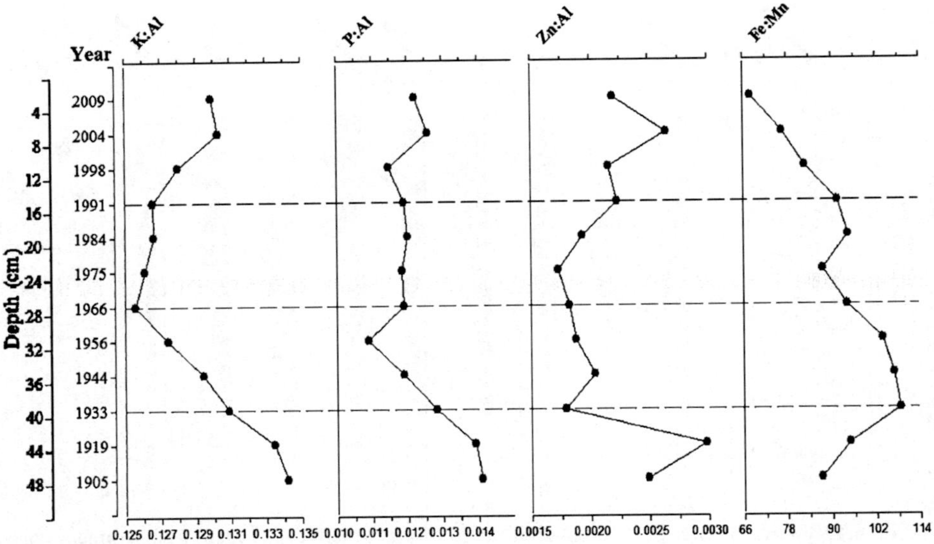

Fig. 6.5 Vertical profile of sediment core of NBL showing K: Al, P: Al, Zn: Al and Fe: Mn

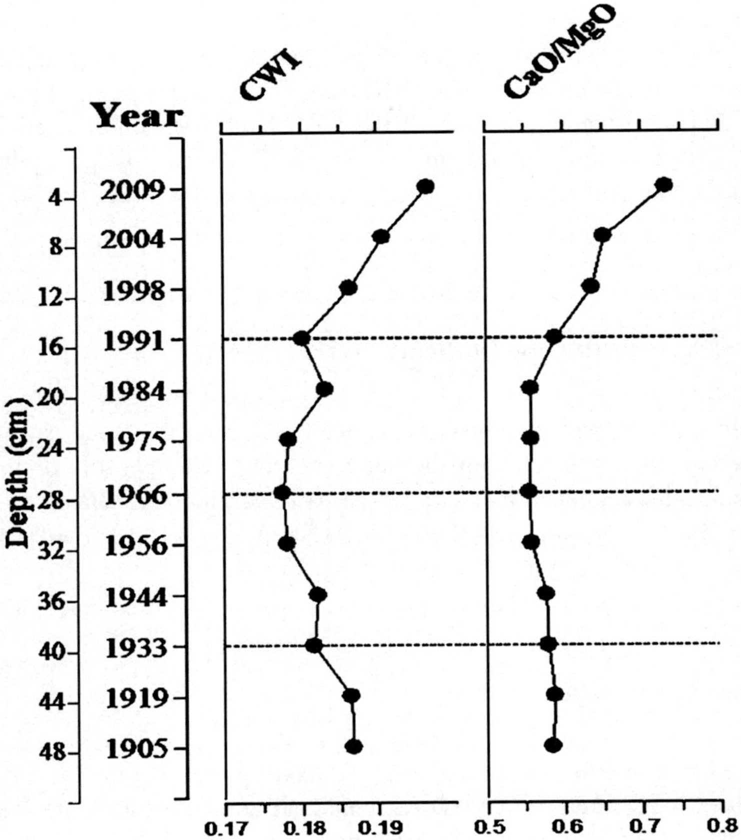

Fig. 6.6 Chemical weathering intensity (CWI) and CaO/MgO ratio of sediment core of NBL

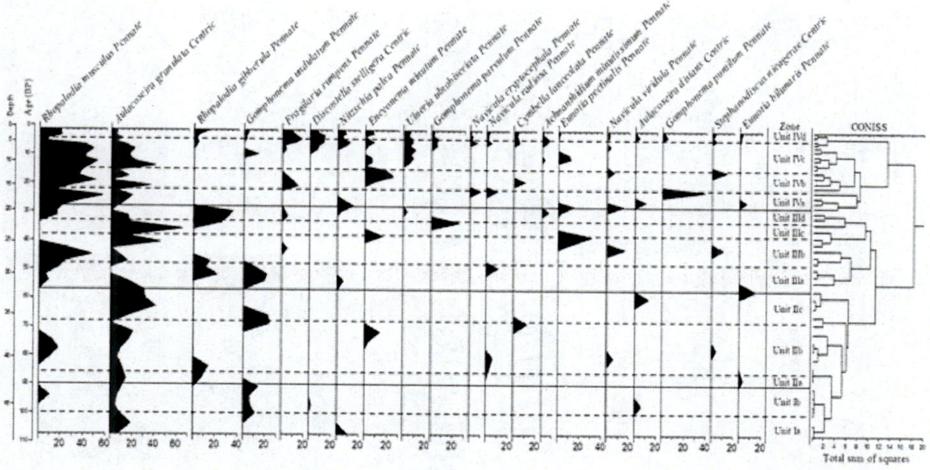

Fig. 6.7 Diatom succession in NBL between (1906 and 2012 A. D.). Only the major taxa (Species with ≥1% in at least one sample) were shown. The right is the constrained incremental sum of squares (CONISS)

the program TILIA (Grimm 1991). A total of 62 diatom taxa were observed in the NBL core of which 20 attained a maximum abundance of >1% in at least one sample. Diatom assemblages were categorized as 1) Planktonic diatoms, such as *Aulacoseira granulata* (Ehr.) Simonsen (4–65%), *Discostella stelligera* (Cleve & Grun.) Houk & Klee (2–14%), *Aulacoseira distans* (Ehr.) Simonsen (1–12%), *Stephanodiscus niagarae* Ehrenberg (0.8–13%) and 2) with benthic diatoms such as *Rhopalodia musculus* Müllar (6–57%), *Rhopalodia gibberula* Müllar (3–35%), *Gomphonema undulatum* Hustedt (3–25%), *Fragilaria rumpens* (Kutzing) Carlson (3–18%), *Nitzschia palea* (3–10%), *Encyonema minutum* (Hilse) Mann (4–26%), *Gomphonema parvulum* Kutzing (2–25%), *Navicula cryptocephala* Kutzing (0.6–10%) and *Eunotia bilunaris* (Ehr.) Schaarschmidt (1–15%).

The planktonic diatoms were dominated by *A. granulata* (Fig. 6.8a, b) and *D. stelligera* (Fig. 6.8k), whereas benthic diatoms were represented by *R. musculus* (Fig. 6.8e, i), *R. gibberula* (Fig. 6.8j), *Gomphonema undulatum* (Fig. 6.8g), *Encyonema minutum* (Fig. 6.8h), *Amphora ovalis* (Fig. 6.8c), *Diploneis ovalis* (Hilse) Cleve (Fig. 6.8f, g) and *Cocconeis placentula* Ehrenberg (Fig. 6.8d). *A. granulata* was abundant during ~1906–1980 and progressively decreased on the top of the core (~1983 and above), while *R. musculus* dominated the core with the decline in *A. granulata*. The other benthic diatoms present in the NBL core are *Ulneria ulnabiseriata* Liu et al., *C. placentula*, *D. ovalis*, *A. ovalis* Kutzing, *Epithemia adnata* (Kutzing) Brebisson, *Cymbella affines* Kutzing, *Cymbella lanceolata* (Agardh) Kirchner and *Rhopalodia gibba* Müllar. There are four Diatom Assemblage Zones (DAZ)/units identified in the NBL core (Fig. 6.7).

Unit I (48–43 cm; ~1906–1922) was categorized into two sub-zones. This zone is characterized by the dominance of planktonic diatoms in the entire core. Unit I-a (48–46 cm; ~1906–1918) shows an abundance of the planktonic species *A. granulata*

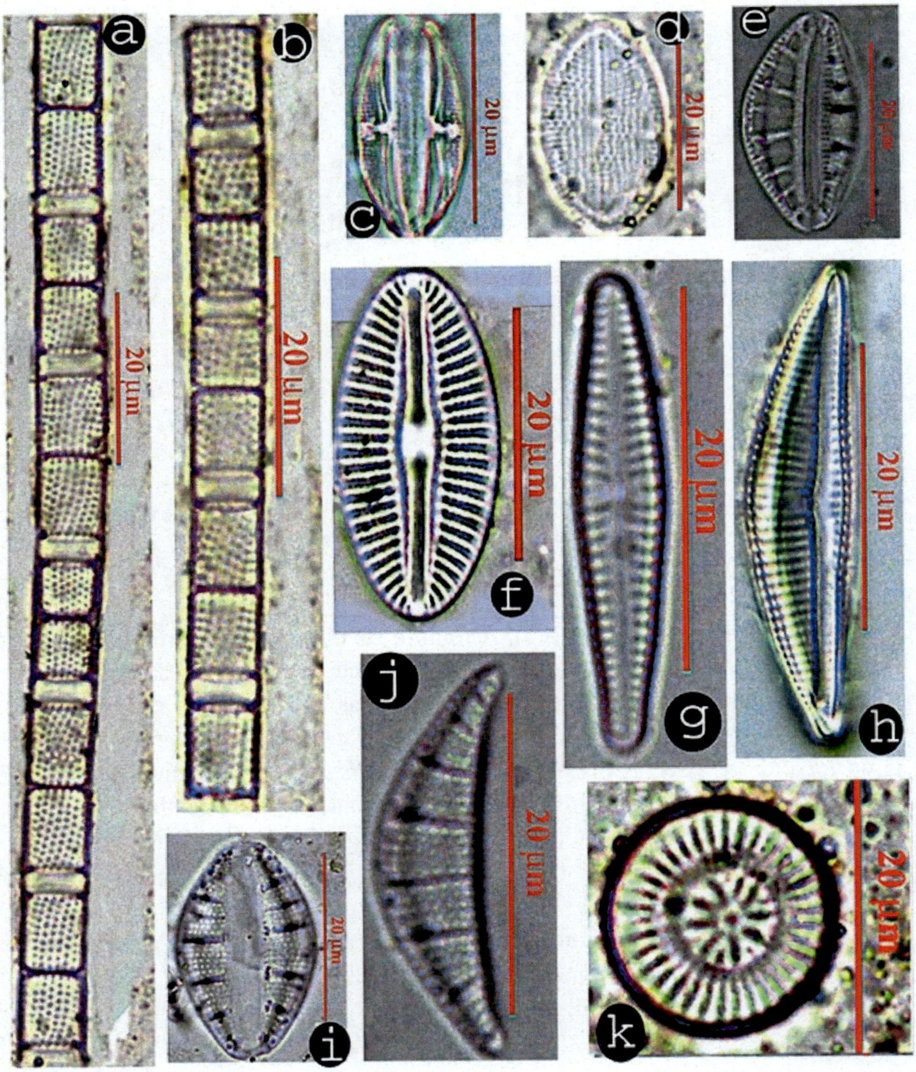

Fig. 6.8 (**a**) *Aulacoseira granulata*; (**b**) *Aulacoseira granulata*; (**c**) *Amphora ovalis*; (**d**) *Cocconeis placentula*; (**e**) *Rhopaloidea musculus*; (**f**) *Diploneis ovalis*; (**g**) *Gomphonema undulatum*; (**h**) *Encyonema minutum*; (**i**) *Rhopaloidea musculus*; (**j**) *Rhopaloidea gibberula*; (**k**) *Discostella stelligera*

(~13% on average) and benthic species (~10%) present in this zone. The ratio of Planktonic and benthic taxa (P/B) had an average of 2.63. Unit I-2 (45–43 cm; ~1919–1922) was represented by an abundance of benthic diatom species *R. musculus* (~10% on average) and *G. undulatum* (~8% on average). The planktonic diatoms present are *A. granulata* (~6%), *A. distans* (~7%) and *Discostella stelligera* (~2%). The ratio of Planktonic and benthic taxa (P/B) had an average of 0.82. Unit II (42–32 cm; ~1921–1956) is classified into three subzones. Unit II-a (42–40 cm; ~1921–1933) represents the abundance of benthic diatom taxa

G. undulatum (~14%), *R. gibberula* (~ 7%) and *Eunotia bilunaris* (~4%). The only planktonic diatom species present in this zone is *A. granulata* (~12%). The ratio of Planktonic and benthic taxa (P/B) had an average of 1. Unit II-b (39–36 cm; ~1932–1946) shows the dominance of benthic diatom taxa *R. gibberula* (~14%), *R. musculus* (~13%) and *Encyonema minutum* (~9%). The planktonic diatom species present in this zone are *A. granulata* (~13%) and *Stephanodiscus niagarae* (~ 4%). The ratio of Planktonic and benthic taxa (P/B) had an average of 0.76. Unit II-c (35–32 cm; ~1945–1956) has an abundance of planktonic diatom taxa *A. granulata* (~ 30%) and *A. distans* (~12%). The benthic diatom species present in this zone is *G. undulatum* (~23%). The ratio of Planktonic and benthic taxa (P/B) had an average of 1.71. Unit III (31–20 cm; ~1955–1983) is represented by the four subzones. Unit III-a (31–29 cm; ~1955–1966) shows an abundance of benthic diatom taxa such as *G. undulatum* (~17%), *R. gibberula* (~ 17%) and *R. musculus* (~11%). The other benthic diatom taxa are *Nitschia palea* and *Eunotia bilunaris*. The planktonic diatom species present in this zone is *A. granulata* (~23%). The ratio of Planktonic and benthic taxa (P/B) had an average of 0.29. Unit III-b (28–25 cm; ~1967–1972) has an abundance of the benthic diatom taxa *R. gibberula* (~ 15%), *R. musculus* (~33%) and *Fragilaria rumpens* (~5%). The common planktonic diatoms present in this zone are *A. granulata* (~23%) and *Stephanodiscus niagarae* (~11%). The ratio of Planktonic and benthic taxa (P/B) had an average of 0.52. Unit III-c (24–22 cm; ~1973–1978) shows the presence of the abundant planktonic diatom taxa such as *A. granulata* (~40%). The benthic diatom species present in this zone is *E. minutum* (~16%). The ratio of Planktonic and benthic taxa (P/B) had an average of 1.71. Unit III-d (21–19 cm; ~1979–1983) shows the dominance of the benthic diatom taxa *R. musculus* (~ 16%) and *R. gibberula* (31%). The other associated benthic diatom taxa are *F. rumpens, N. palea* and *G. parvulum.* The planktonic diatom species present in this zone is *A. granuata* (~16%). The ratio of Planktonic and benthic taxa (P/B) had an average of 0.18. Unit IV (18-0 cm; ~1982–2012) is divided into four subzones. This zone shows the maximum concentration of benthic diatom species. Unit IV-a (18–16 cm; ~1982–1991) indicates an increase in *R. musculus* (~38% on average) with a decrease in *A. granulata* (~ 13% on average). The other prominent benthic diatom species are *N. palea* (~11% on average) and *E. binularia* (~7% on average) and the other planktonic species commonly present is *A. distans*. The Planktonic and benthic ratio of taxa (P/B) had an average of 0.28. Unit IV-b (15–12 cm; ~1990–1998) was characterized by dominant benthic species *R. musculus* (~33% on average) and planktonic diatom species *A. granulata* (~19% on average). The second dominant benthic diatom species was *Encyonema minutum* (~18% on average) followed by *Fragilaria rumpens* (~10% on average) and *Navicula cryptocephala* (~10% on average). The planktonic and benthic (P/B) ratio is 0.21. Unit IV-c (11–6 cm; ~1999–2006) prominently shows an abundance of the benthic species *R. musculus* (~40% on average) and planktonic species *A. granulata* (~18% on average). The other benthic diatom taxa present in this zone are *R. gibberula, G. undulatum, F. rumpens, N. palea, E. minutum, G. parvulum* and *E. binularis*. The other planktonic diatom species present in this zone are *D. stelligera* and *S. niagarae*. The ratio of P/B is 0.34. Unit IV-d (5–0 cm;

~2007–2012) indicates a decline in *R. musculus* (~17% on average) and *A. granulata* (~6% on average). Other planktonic species *D. stelligera* (~11% on average), *A. distans* (~2% on average) and *S. niagarae* (~3% on average). The other benthic forms present in this zone are *R. gibberula* (~7% on average), *G. undulatum* (~3% on average), *F. rumpens* (~12% on average) and *N. palea* (~6% on average). The ratio of planktonic and benthic taxa (P/B) had an average of 0.27.

Discussion

The sediment core of the NBL, Central India is studied to investigate soil erosion, trophic state and past rainfall since the last century using some proxies. The watershed of the NBL is present in the tropical region and has had a nearly stable temperature since the last century. Hence, the amount of rainfall in the catchment area of the lake has been the controlling factor for the soil erosion, accumulation of geochemical elements and diatom assemblage. Therefore, various proxies such as (1) Geochemical elements and their normalized ratio with erosional indicators, (2) Chemical weathering intensity (CWI) and (3) Diatoms are studied from the sediment core of the NBL of Central India.

Soil Erosion and Trophic State

The intensity of the impact on the lakes of a different watershed can be analysed using geochemical variables (Table 6.1; Garrison and Laliberte 2007, 2010). The chemical elements titanium (Ti) and aluminium (Al) are derived from detrital aluminosilicates and thereby the fluctuations in their profiles will suggest the transformation in the rate of soil erosion (Garrison and Laliberte 2010). Phosphorous and nitrogen are important nutrients for the growth of aquatic plants and algae. The synthetic fertilizers and soils contain potassium (K). Thus, soil erosion and the input of commercial fertilizers in the watershed can be distinguished from the variation in the profile of potassium (K) (Garrison and Laliberte 2010). The urban runoff is mostly accompanied by zinc (Zn) as it is a component of galvanized roofs and tires

Table 6.1 Important chemical indicators of catchment or lake processes

Process	Chemical elements
Soil erosion	Aluminium, potassium, titanium
Synthetic fertilizer	Potassium
Urban	Zinc, copper
Ore smelting	Zinc, cadmium, copper
Nutrients	Phosphorus, nitrogen
Lake productivity	Organic matter

After Garrison and Laliberte (2010)

(Garrison and Laliberte 2007). The addition from the smelting of lead–zinc ores is seen in the zinc (Zn) profile of the cores (Dean 2002). Similarly, the use of soil modifications for the development of amenities is reflected in the calcium (Ca) profiles and the changes in the oxygen levels in the bottom waters are indicated by the manganese (Mn) profiles (Garrison 2008).

The accumulation rate gives information about the impact of watershed processes on the lake ecosystem (Garrison and Laliberte 2010). The calculation of the rate of accumulation of some geochemical elements of the NBL was done by adding the geochemical concentrations with the rate of sedimentation (Calculation by ^{210}Pb dating and CRS model). The elemental deposition in the NBL through time was determined from the accumulation rate. The accumulation rate of phosphorous was similar to the erosional indicators (Ti, Al) in the NBL. The phosphorus content has been progressively increased since ~1919 and reached its highest level in ~1998, with its further decline since ~2000 till present at the top of the core (Fig. 6.4). Therefore, it can be surmised that the phosphorous content of the NBL is mostly derived from the sediment and soils from the watershed and played an important role as a major source of the nutrient. The profile of phosphorous and nitrogen may not reflect the lakes eutrophication history due to diagenesis (Anderson and Rippey 1994 and Fitzpatrick et al. 2003) and the breakdown of phosphorous and nitrogen into the inorganic components with some of the material then recycled into the water column and removed out of the sediments (Garrison and Laliberte 2010). The calcium content of the NBL is similar to the titanium till the top of the core and indicates that the major part of it was derived from the sediments and soils of the watershed (Fig. 6.4). The geochemical elements such as silica, manganese and sodium also show a similar increasing trend like Ti and Al, which reflects that the major concentration of these elements is derived from the sediments and soils in the watershed of the NBL (Fig. 6.4).

The selected geochemical elements from the NBL are normalized to aluminium (Al) to distinguish the anthropogenic inputs other than mineral sediments. Thus, the factors which influence the lake in addition to sediment/soil input from the watershed can be known. The anthropogenic impact both from the concerned watershed and atmospheric transport through time on the sediments of the NBL have been observed from the elemental profiles normalized to Al (Fig. 6.5). The ratio of these elements would be the same throughout the core if soil particles/sediments were the only sources of potassium, as both aluminium (Al) and potassium (K) are found in sediments/soil particles in the watershed (Garrison and Laliberte 2007). There is a gradual depletion in the K: Al ratio in the NBL since ~1906–1966(Fig. 6.5). This decline in the ratio till ~1966 indicates that potassium was not mainly derived by the erosional processes in the watershed considering the increase in the concentration of erosional indicators (Ti and Al) (Fig. 6.4) and other anthropogenic sources in the form of potash fertilizers in the nearby agricultural lands of the NBL. The K: Al shows an increasing trend during ~1967–~1975 similar to the profile of Al. This indicates that K is an important source derived from soil particles or sediments from the catchment of the NBL. The phosphorous content shows a declining trend, while the K: Al ratio slightly increased during ~1976–1991 indicating the mixed source of K, i.e. from soil particles along with some anthropogenic origin. The K: Al ratio once again shows a rising trend during ~1992–1998 similar to the profile of K. This points to the increased maximum rate of soil erosion in the NBL watershed. From ~1999 till ~2009 the rise and decline in K: Al ratio indicates mix origin of K, i.e. soil particles and potash fertilizers used in the agricultural lands in the catchment of the NBL (Figs. 6.4 and 6.5).

Phosphorous and nitrogen are important nutrients for algal growth (Garrison and Laliberte 2010). The ratio of phosphorous to aluminium (P: Al) shows a continuous decline from ~1906 to ~1956, while the P profile shows a continuous rise in its content during this period. It shows that during this period P was mainly derived through the excessive use of synthetic fertilizers in the agricultural land of the catchment area of the NBL rather than soil particles mainly after the construction of the rockfill dam in ~1917. The P: Al ratio shows a peak during ~1957–1975 more or less similar to the P profile indicating its origin through soil particles or sediments of the catchment area of the NBL. There is little rise in the ratio of P: Al during ~1976–1991 in contrast to the major decline of the P profile suggesting the possible mixed origin of the P. The trend of the P: Al ratio has declined from ~1992 till ~1998 with the highest peak of the P profile. It indicates that most of the P was derived from the erosion of soils or sediments from the catchment of the NBL during this period. The P: Al ratio from ~1999 to ~2009 shows a mixed origin of the P (Figs. 6.4 and 6.5).

The Zn: Al ratio in the NBL core is highest post ~1917 coinciding with the year of the dam construction till ~1933 similar to the Zn profile indicating an increased rate of soil erosion in the catchment area (Figs. 6.4 and 6.5). The increased rate of soil erosion could be attributable to the clearing of forest area for the development of the agricultural fields in the catchment of the NBL. A more or less uniform profile of the Zn: Al ratio is observed from ~1934 to ~1991 pointing to a consistent moderate level of supply of Zn mainly through soil erosion in the catchment. The profile of the Zinc concentration remained highest from ~1992 to ~2009 with its highest content in ~1998. The Zn: Al ratio declined a little from ~1992 to ~1998 indicating some part of the Zn could have derived from a source other than the soil particles. This Zn could have possibly derived from unknown sources and transported through the air and precipitated in the NBL basin during this period. The increasing trend of the Zn: Al ratio post ~1999 similar to a higher Zn profile could be attributed to a rise in soil erosion in the catchment of the NBL (Figs. 6.4 and 6.5).

The manganese (Mn) preferentially migrates to iron once the bottom water loses oxygen (Engstrom et al. 1985). This manganese generally moves from the sediments into the deepest water causing the environment of manganese in the deeper waters sediments (Garrison 2008). The iron and manganese tend to remain in suspension for a long period with their preferential movement together (Jones and Bowser 1978). Thus, the ratio of iron to manganese (Fe: Mn) declines (Mn increases) with the loss of oxygen (Garrison 2008). The profile of iron (Fe) to manganese (Mn) in the core of the NBL shows an increasing trend from ~1906 to ~1933 (Fig. 6.5), which shows the existence of a fairly good oxygenated environment during this period. However, the ratio of Fe: Mn started showing a declining trend after ~1933 till ~2009. This indicates that the bottom water of NBL had started losing oxygen progressively after ~1933 and the lake bottom became more and more anoxic till ~2009. Thus, the hypolimnetic oxygen of the NBL has declined for the past 80 years indicating increased eutrophication in the lake till the present.

Paleo-Rainfall and Droughts

The Indian sub-continent experiences >80% of rainfall as a result of the seasonal monsoon, which is the lifeline of Indian agriculture (IPCC 2007). Therefore, it is vital to forecast the future climate, particularly rainfall trends considering global warming and its adverse effects (Church and white 2006). Hence, the investigation of the past monsoonal fluctuations is highly essential to forecasting rainfall and understanding past extreme weather events. The paleoclimatic studies in India particularly from the last ice age to the present add comprehensive knowledge on rainfall and droughts (Singhvi and Kale 2008). The different proxies have been used by the various workers to investigate past monsoonal fluctuations. The responses of the diatom community to climate change have been studied by several workers (Anderson et al. 2012; Chen et al. 2014). The other proxies used are chemical weathering intensity and geochemical ratios normalized with erosional indicators like Ti and Al (Colin et al. 1998; Achyuthan et al. 2007; Kotlia et al. 2010 and Veena et al. 2014).

The sediment accumulation rate of the NBL was increased after the construction of a small rockfill dam in ~1917 (Fig. 6.4). The sediment input in the NBL is dependent upon the amount of rainfall that occurred in its catchment area. A higher rainfall is coupled with an enhanced deposition of sediments into the lake. The reduced deposition of sediments in the NBL till ~1919 is coeval with calamitous and moderate drought events that occurred in major parts of India during 1918 and 1920, respectively (De et al. 2005; Fig. 6.4). The exceptional flood in India in ~1933 coincides with the initial rise in the concentration of the geochemical elements of the NBL (Fig. 6.4). The gradual rise in the accumulation of the majority of the geochemical elements was continued and reached peaked in ~1956. This continuous rise of the geochemical elements could be attributed to improved precipitation in the catchment of the NBL. There was a gradual decline in the geochemical content from ~1957 till ~1966. The maximum fall in the concentration of a geochemical element in ~1966 is again coinciding with the other moderate drought event in India in ~1965. Another rise in the geochemical content started in ~1967 and reached its peak in ~1975 (Fig. 6.4). Another decline in the concentrations of geochemical elements of the NBL was noticed after ~1976 and the same trend was continued till ~1991. Many parts of India experienced severe drought in 1987 (De et al. 2005). Thus, the central Indian region in general and the catchment area of the NBL, in particular, had suffered a major drought from ~1987 to ~1991. The highest level of deposition of the geochemical elements in the NBL was observed in ~1998 and possibly could be attributed to the occurrence of maximum rainfall or flood event in the eastern part of the Vidarbha region of Maharashtra, Central India (Fig. 6.4). The post ~1998 period showed a gradual reduction in the rainfall till ~2004 and the low rainfall condition prevailed up to ~2009 (Fig. 6.4). The peaks of the geochemical elemental content also suggest the increased soil erosion in the catchment of the NBL during ~1933, ~1956, 1975 and 1998, respectively (Fig. 6.4). The increased agricultural practices have led to deforestation in the NBL catchment resulting in a rise in soil erosion coupled with varying strength of rainfall in the monsoonal core zone of Central India.

The Chemical Weathering Intensity (CWI)

The oxide such as Al_2O_3, TiO_2 and SiO_2 is used as the most resistant and insoluble elements in the environment during weathering and erosion, whereas CaO, MgO and Na_2O are more soluble and mobile (Engstrom and Wright 1984; Veena et al. 2014; Garrison and Laliberte 2007). The chemical weathering intensity (CWI) is calculated to know the fluctuations in weathering intensity (Sun et al. 2009). The higher value CWI points to more weathering. This corresponds to wetter conditions and heavy monsoonal rainfall in the study area (Central India). The more intense development of anthropogenic carbonate in the soil of the sediment source area corresponds to higher CaO/MgO ratios in sediment, which points to warm climate with alternating wet and dry periods (Veena et al. 2014; Wang et al. 1990; Wang 1992).

The CWI and CaO/MgO ratio with the depth (cm) of the NBL core are compared for the past 100 years (Fig. 6.6). The CWI and CaO/MgO ratio started declining after ~1918 till ~1933 coinciding with the one calamitous drought event in ~1918 and a moderate drought event in ~1920 in many parts of India. The further decline in the ratio of the CWI and CaO/MgO after ~1920 till ~1933 points to a gradual reduction of rainfall (dry period) in Central India. The further decline in this ratio from ~1934 to ~1975 indicates the prevalence of low rainfall in the catchment of the NBL. The moderate drought event of ~1965 and the severe drought event of ~1972 in India (De et al. 2005) also corroborate our findings of low rainfall conditions from ~1934 to ~1975. The CWI, CaO/MgO ratio further shows a declining trend after ~1984 indicating the reduction in the rainfall till ~1991. The severe drought event in India during ~1987 also supports the present finding. The CWI and CaO/MgO ratios show increasing trends after ~1992 indicating moderate to high rainfall in Central India till ~2009 (De et al. 2005; Fig. 6.6).

Diatom Analysis

The significant shift of diatoms from benthic to planktonic assemblage has been attributed to climate warming (Chen et al. 2014). Lake sediments from stratification during summer and climate warming periods and thus more stratification would point to increased length and strength of warm/dry periods (Wang 2012). The small size and fast-growing planktonic diatoms in the sediment cores indicate longer and stronger thermal stratification (Smol et al. 2005; Rühland et al. 2010; Chen et al. 2014). The NBL falls in the tropical climatic region. Thus, the temperature variation in the region has not been significant since the last century. However, the diatom species may respond to fluctuations in the precipitation (rainfall) and dry periods (droughts) existing in the region. The changing patterns of planktonic to benthic diatom assemblage (P/B) are recorded in the NBL core from ~1906 to ~2012 (Fig. 6.7). The planktonic diatom *A. granulata* is dominate throughout the NBL core along with the appearance of another planktonic diatom taxa *D. stelligera* after ~2004. However, the comparative decline in abundance of *A. granulata* was associated with the increase in the concentration of benthic diatoms such as *R. musculus, R. gibberula* and *G. undulatum.* The concentration of *A. granulata* remained higher since ~1906 and more or less constant till ~1961. A major shift in speciation was seen at ~1920, ~1933, ~1966 and ~ 1983 till ~2005 with the preponderance of the benthic species *R. musculus* coeval with a decline in planktonic forms. The abrupt disappearance of *R. gibberula* was observed from ~1983 to ~2005. The nutrient input in response to climatic conditions and anthropogenic activities has played a vital role in the diatom shift of the NBL. The aridity/low rainfall mainly affects the water chemistry and inflicts salinity (Winder et al. 2009). *D. stelligera* indicates changing water levels and deforestation, while *A. granulata* exists in varying trophic status (Costa-Bödeker et al. 2012). *R. musculus* is a mesohalobus and alkalibiontic species found in the swamp, swampy lakes, springs and oozing areas (You et al. 2009). The major period of the profusion of *R. musculus* in the NBL core was during ~1919–1921, ~1933–1940, ~1956–1972 and ~ 1978–2004, respectively. Thus, the NBL water was mainly mesotrophic, Meso euhyaline and alkalibiontic. The longer dry/less rainfall period that prevailed in the NBL watershed could have possibly developed swampy conditions in the lake during these years. The existence of a low rainfall/dry period also corroborates several drought events recorded in India, for example, in ~1918, ~1920, ~1965, ~1972 and ~1987 (De et al. 2005; Fig. 6.7). *A. granulata* was predominant during ~1906–1914, ~1920–1922, ~1941–1943, ~1948–1956, ~1972–1976 and ~ 1973–1982. The increased concentration of *A. granulata* during these periods in the NBL mainly points to the accelerated soil erosion and flux, persistent wet period and increased nutrient levels. It also clearly signifies that the NBL was highly eutrophic with high phosphorus and nitrogen values. A decline in the *A. granulata* population is linked with an increase in the population of *R. gibberula* and *G. undulatum* also coincides with dry/low rainfall periods (De et al. 2005; Fig. 6.7). The major dominance of *R. musculus* was observed on the top of the core during ~1983–2006. The prevalence of the longer dry/low rainfall periods could have again led to swampy conditions in the NBL with the mesotrophic waters. *R. musculus* and *A. granulata* both decreased drastically at the top of the core during ~2007–2012, while other small benthic and planktonic forms became predominant during this period (Fig. 6.7). *D. stelligera* was the common abundant form observed during this period. Diatom assemblages dominated by *D. stelligera* indicate a thermally stable environment and alkaline conditions (Cardozo et al. 2014). The abundance of *D. stelligera* also indicates a decline in the nutrient concentration due to enhanced stratification in response to the warming trend in the region (Stone et al. 2010). Thus, the abundance of planktonic diatom species, *D. stelligera* during this period suggests low nutrient, mesotrophic, alkaline lake waters with comparatively dry/less rainfall period indicating changing water level and deforestation in the catchment of the NBL. Thus, differences in aquatic habitat, duration and strength of dry and wet periods could have resulted in the different diatom responses in the NBL.

Conclusions

Nutrients such as phosphorous are mostly derived from the sediment and soils from the watershed of the NBL. The concentration of geochemical elements like K, Si, Fe, Mn, Zn and Na are also mainly derived from the sediments and soils in the watershed of the NBL. From ~1906 to ~1956, P was mainly derived through the excessive use of synthetic fertilizers in the agricultural land of the catchment area of the NBL rather than soil particles mainly after the construction of the rockfill dam, i.e. ~1917. The hypolimnetic oxygen of the NBL has declined for the past 80 years indicating increased eutrophication in the lake till the present. The decline in the deposition of the geochemical elements till ~1919 was coeval with the calamitous and moderate drought events that occurred in major parts of India during 1918 and 1920. Similarly, the exceptional flood in India in ~1933 coincides with the initial rise in the concentration of the geochemical elements of the NBL. The gradual rise in the accumulation of the majority of the geochemical elements till ~1956 could be attributed to improved precipitation in the catchment of the NBL. The maximum fall in the concentration of a geochemical element in ~1966 again coincides with the other moderate drought event in India in ~1965. The central Indian region in general and the catchment area of the NBL, in particular, had suffered a major drought period from ~1987 to ~1991. The highest level of deposition of the geochemical elements in the NBL in ~ 1998 possibly could be attributed to the occurrence of maximum rainfall or flood event over Central India. The post ~1998 period showed a gradual reduction in rainfall till ~2004 and the low rainfall conditions existed up to ~2009. The CWI record of the NBL shows that the accumulation of the geochemical elements was mainly controlled by the changing climatic conditions, namely wet and dry periods in the catchment area along with the enhanced soil erosion caused by anthropogenic activities. The changes in the aquatic conditions are mainly caused by varying duration and strength of dry and wet periods over the central Indian region and enhanced anthropogenic activities such as deforestation, use of synthetic fertilizers, etc., in the catchment area and could have resulted in the diverse diatom responses in the NBL.

Acknowledgements SKH and SSH thank University Grants Commission (UGC), New Delhi for the funding to the department under UGC-SAP DRS- II (No.F.550/2/DRS-II/2016; SAP-I; 03/05/2016) and the Department of Science and Technology (DST), Government of India for providing financial support to the department under the DST-FIST I program.

References

Achyuthan H, Kar A, Eastoe C (2007) Late quaternary environmental changes and evolution of the Tal Chhapar salt lakes, Rajasthan. J Paleolimnol 38:493–507

Anderson NJ, Rippey B (1994) Monitoring lake recovery from point source eutrophication: the use diatom-inferred epilimentic total phosphorus and sediment chemistry freshwater biology. Hydrobiologia 32:625–639

Anderson NJ, Foy RH, Engstrom DR, Rippey B, Alamgir F (2012) Climate forcing of diatom productivity in low land, eutrophic lake: White Lough revisited. Freshw Biol 57:2030–2043

Batterbee RW, Charles DF, Dixit SS, Renberg I (1999) Diatom as indicators of surface water acidity. In: Stoermer EF, Smol JP (eds) . Sciences Cambridge University Press, pp 85–127

Batterbee RW, Jones VJ, Flower RJ, Cameron NG, Bennion H, Carvalho L, Juggins S (2001) Diatoms. In: Smol JP, Birks HJB, Last WM (eds) Tracking environmental change using Lake sediments: terrestrial, algal, and siliceous indicators, vol 3. Kluwer Academic Publishers, Dordrecht, pp 155–202

Bennion H, Wunsam S, Schmidt R (1995) The variation of diatom-phosphorus transfer functions: an example of Mondsee, Austria. Freshw Biol 34:271–283

Cardozo AZY, Gomes DF, Silva E, Da M, Duque SRE, Rangel JOC, Sifeddine A, Turcq B, Albuquerque ALS (2014) Holocene Paleolimnological reconstruction of high altitude Colombian tropical lake. Palaeogeogr Palaeoclimatol Palaeoecol:1–10. https://doi.org/10.1016/J.palaeo.2014.03.013

Chen C, Zhao L, Zhu C, Wang J, Jiang J, Yang S (2014) Response of diatom community in Lugu Lake (Yunnan–Guizhou plateau China) to climate change over the past century. J Paleolimnol 51(3):357–373. https://doi.org/10.1007/s10933-013-9760-4

Church J, White N (2006) A 20th-century acceleration in global sea-level rise. Geophys Res Lett. https://doi.org/10.1029/2005GL024826

Cohen AS (2003) Paleolimnology. Oxford University Press, 500p

Colin C, Kissel C, Blamart D, Turpin I (1998) Magnetic properties of sediments in the Bay of Bengal and the Andaman Sea: impact of rapid North Atlantic Ocean climatic events on the strength of the Indian Monsoon. Earth Planet Sci Lett 160:623–635

Costa-Bödeker S, Bennion H, Jesus TA, Albuquerque ALS, Figueira RCL, Bicudo DC (2012) Paleolimnologically inferred eutrophication of a shallow, tropical, urban reservoir in Southeast Brazil. J Paleolimnol 48:751–766

De US, Dube RK, Prakasa Rao GS (2005) Extreme weather events over India in the last 100 years. Indian Geophy Union 9:173–187

Dean W (2002) A 1500-year record of climatic and environmental change in Elk Lake, Clearwater County, Minnesota II: geochemistry, mineralogy, and stable isotopes. J Paleolimnol 27:301–319

DRM (2000) District Resource Map of Bhandara and Gondia Districts. Geological Survey of India

Engstrom DR, Wright HE (1984) Chemical stratigraphy of lake sediments as a record of environmental change. In: Lund H, Jm WC (eds) Lake sediments- environmental history. Leicester University Press, pp 11–67

Engstrom DR, Swain EB, Kingston JC (1985) A paleolimnological record of human disturbance from Harvey's Lake, Vermont: geochemistry, pigments, and diatoms. Freshw Biol 15:261–288

Fitzpatrick FA, Garrison PJ, Fitzgerald SA, Elder JF (2003) Nutrient, trace-element, and ecological history of Musky Bay, Lac Courte Oreilles, Wisconsin, as inferred from sediment cores. US Geologic Survey Water-Resource Investigation Report 02-4225: 141p

Garrison PJ (2008) Paleoecological study of Grindstone Lake, Sawyerl County. Wisconsin Department of Natural Resources, Bureau of Integrated Science Services 1–16

Garrison PJ, Laliberte GD (2007) Paleoecological study of Big Round Lake, Polk County. Wisconsin Department of Natural Resources, Bureau of Integrated Science Services 1–17

Garrison PJ, Laliberte G (2010) Paleoecological Study of lake Chetac, Sawyer Country. Wisconsin Department of Natural Resources, Bureau of Integrated Science Services 1–17

Gondia District Gazetteer (2013) Government of Maharashtra, India

Grimm E (1991) Tilia. Version 2.0. Illinois State Museum, Illinois

Humane SS, Humane SK, Juare S (2016) ^{210}Pb dating and trace metal accumulation study of sediment cores from Navegaon Bandh and Bodhalkasa Lakes, Gondia District, Maharashtra, India. Environ Earth Sci 75(7):1–26

IPCC (2007) Climate change 2007: the physical science basis. In: Solomon S, Qin D, Manning M, Chen Z, Marquis M, Averyt KB, Tignor M, Miller HL (eds) *Contribution of a working group to the fourth assessment report on the intergovernmental panel on climate change*. Cambridge University Press, 996p

John J (2012) Diatoms in the Swan River estuary, Western Australia: taxonomy and ecology. Koeltz Scientific Books, 456p

Jones BF, Bowser CJ (1978) The mineralogy and related chemistry of lake sediments. In: Lerman A (ed) Lakes: chemistry, geology, physics. Springer, New York, pp 179–235

Juggins S (2007) C2 version 1.5.0: a program for plotting and visualizing stratigraphic data. University of Newcastle

Karthick B, Hamilton PB, Kociolek JP (2013) An illustrated guide to common diatoms of Peninsular India. Gubbi labs, 206p

Kotlia BS, Sanwal J, Phartiyal B, Joshi LM, Trivedi A, Sharma C (2010) Late quaternary climatic changes in the eastern Kumaun Himalaya India as deduced from multi-proxy studies. Quat Int 213:44–55

Liu J, Zhangwen L, Huanjun Z, Han BP (2012) Hydrodynamic change recorded by diatoms in sediments of Liuxihe Reservoir, southern China. J Paleolimnol 47:17–27

Liu B, Williams DM, Tan L (2017) Three new species of *Ulneria* (Bacillariophyta) from the Wuling Mountains Area, China. Phytotaxa 306(4):241–258. https://doi.org/10.11646/phytotaxa.306.4.1

McFadden MA, Patterson WP, Mullins HT, Anderson WT (2005) Multi-proxy approach to long- and short-term Holocene climate-change: evidence from eastern Lake Ontario. J Paleolimnol 33:371–391

Metzeltin D, Lange-Bertalot H (2007) Tropical diatoms of South America, 2. In: Lange-Bertalot H (ed) Iconographia diatomologica, annotated diatom micrographs, vol 15. Koeltz Scientific Books, Sttutgart, 736p

Metzeltin D, Lange-Bertalot H, Garcia-Rodriguez F (2005) Diatoms of Uruguay. Iconographia Diatomologica 15:1–736

Meyers PA, Teranes JL (2001) Sediment organic matter. In: Last WM, Smol JP (eds) Tracking environmental changes using lake sediment – vol. 2: physical and geochemical methods, pp 239–270

Mullins HT, Patterson WP, Teece MA, Burnett AW (2011) Holocene climate and environmental change in Central New York (USA). J Paleolimnol 45:243–256

Rühland KM, Paterson AM, Hargan K, Jenkin A, Clark BJ, Smol JP (2010) Reorganization of algal communities in the Lake of the Woods (Ontario, Canada) in response to turn of – the-century damming and recent warming. Limnol Oceanogr 55:2433–2451

Schroeder L, Martin SC, Kerns GJ, McLean CE (2016) Diatom assemblages in a reservoir sediment core track land-use changes in the watershed. J Paleolimnol 55(1). https://doi.org/10.1007/s10933-015-9860-4

Singhvi AK, Kale VS (2008) Paleoclimate studies in India: last ice age to the present, IGBP-WCRP-SCOPE report series. Indian National Science Academy, New Delhi, p 4

Smol JP (2008) Pollution of lakes and Rivers: a Paleoenvironmental perspective, 2nd edn. Blackwell Publishing, Oxford, 383p

Smol JP, Wolfe AP, Birks HJB, Douglash MSV (2005) Climate-driven regime shifts in the biological communities of arctic lakes. Proc Natl Acad Sci U S A 102:4397–4402

Stone J, Westover K, Cohen A (2010) Late Pleistocene paleohydrography and diatom paleoecology of the central basin of Lake Malawi, Africa. Palaeogeogr Paleoclimatol Paleoecol 303:51–70

Sun Q, Wang S, Zhou J, Chen Z, Shen J, Xie X, Wu F, Chen P (2009) Sediment geochemistry of lake Daihai, north-Central China: implications for catchment weathering and climate change during Holocene. J Paleolimnol 43:75–87

Veena MP, Achyuthan H, Eastoe C, Farooqui A (2014) A multi-proxy reconstruction of monsoon variability in the late Holocene, South India. Quat Int 325:63–73

Wang SM (1992) Lake records of the Holocene climatic change in China. In: Shi YF (ed) The Holocene climatic optimum and associated environment in China. Science Press, Beijing (in Chinese), pp 146–152

Wang Q (2012) Environmental evolution of Lugu Lake, Yunnan and response to Southwest monsoon climate since LGM. Nanjing Institute of Geography and Limnology, CAS, University of Chinese Academy of Sciences, Nanjing China, PhD thesis, 128p

Wang SM, Tu YS, Wu RJ, Freng M (1990) The Daihai Lake: environment evolution and climate change. University of Science and Technology of China Press, Hefei, pp 1–191. (in Chinese)

Winder M, Reuter JE, Schladow SG (2009) Lake warming favours small-sized planktonic diatom species. Proc R Soc Britain Biol Sci 276:427–435

You Q, Liu Y, Wang Y, Wang Q (2009) Taxonomy and distribution of diatoms in the genera *Epithemia* and *Rhopalodia* from the Xinjiang Uygur Autonomous Region, China. Nova Hedwigia 89(3–14):397–430

Chapter 7
Anthropogenic Interference and Climatic Changes Influencing Precipitation, Sediment Load and Sand Deposition by the Phalgu River in and Around Jahanabad District, Bihar: A Case Study

Akhouri Bishwapriya, Abhishek Kumar Chaurasia, and V. Aneesh Kumar

Abstract The peninsular bound rivers in the state of Bihar join the median principal drainage, River Ganga along its south bank flowing from south to north. These emanate mostly from the rocky uplands of the Chotanagpur Gneissic Complex, a few from the Vindhyans, Bihar Mica Belt, Gondwanas and other geological formations. Amongst them, River Falgu has its emergence from the Chotanagpur Gneissic Complex and is a promising repository for sand used for construction purposes. Understandably, there is a composite licence awarded by the state government to harness the sand potential of the river along its fertile stretches which is sometimes unchecked from the point of view of mining complying with the requisites of existing guidelines. The region which experiences hot and humid climate conditions has suffered irreversible vagaries in the past few decades with a conspicuous decrease in the overall precipitation, thus influencing the groundwater conditions. The recent past is replete with climatic surprises from unexpected rainfall in a couple of seasons to intimidating drought-like conditions in others. Undoubtedly, these events cannot be segregated from short-term climatic changes which are now affecting different parts of the country with a certain amount of regularity. This has influenced the sedimentation pattern and carrying capacity of the river. The aspect of sedimentation in a stretch of one of the peninsular bound rivers viz. Phalgu, originating from Chotanagpur Gneissic Complex (CGC) and debouching into the River Ganga near Patna district, forms the soul of the present study in light of fluctuation in sedimentation pattern, anthropogenic interference (particularly aberrations in sand mining), the effect of cross drainage structures (viz. barrage, bridge, weir) and diversion of the streamflow for various irrigational purposes which together or in isolation have greatly influenced the ecological health of the river.

A. Bishwapriya (✉) · A. K. Chaurasia
Geological Survey of India, Patna, India

V. Aneesh Kumar
Central Ground Water Board, Patna, India
e-mail: aneeshkv-cgwb@gov.in

N. Khare (ed.), *Science, Policies and Conflicts of Climate Change*, Springer Climate,
https://doi.org/10.1007/978-3-031-16254-1_7

The study has been carried out on 11 no Sand Mining ‘Ghats’ officially identified as lease areas for minor minerals in Jehanabad district, Bihar along the course of River Phalgu. Studies reveal that anthropogenic activities are significantly affecting the river bed morphology by influencing sediment supply and transportation in the river basin. Transportation of sediment is the predominant work of the river through various hydrological processes such as a solution, suspension, saltation and traction under highly fluctuating hydrodynamic conditions, but decreased precipitation has affected the carrying capacity of the river. This chapter deals with the tight rope walking in arriving at the best possible balance resulting from environmental safety guideline procedures involved in mining, safeguarding the ecology of the river amidst challenges and ways to minimize the unprecedented effects of short-term climatic change. It also highlights aspects like drivers responsible for the reduced flow, erratic modes of sand deposition vis-a-vis undisciplined extraction and recommendations towards sustainable exploitation in tune with the expectation and guidelines served by the Hon’ble National Green Tribunal in this case along Phalgu River, in Jehanabad district, Bihar.

Keywords Fluvial ecosystem · River adjustment · Channel evolution · Land dereliction · Global positioning system

Introduction

The recent past is replete with incidences of disaster arising out of climatic surprises that create havoc and wide-scale catastrophe in their aftermath. The flood disaster of Himachal Pradesh and Uttarakhand (2013), the incessant rainfall leading to widespread inundation in Chennai and Kerala (2015 and 2018, respectively) and the recent abnormally high rainfall within the span of 3–4 days in parts of Bihar (2019), changed the perspective with which such vagaries in climatic condition need to be seen. Undoubtedly, these events cannot be segregated from short-term climatic changes which are now affecting different parts of the country with a certain amount of regularity.

The aspect of sedimentation in a stretch of one of the peninsular bound rivers viz. Phalgu, originating from the zone of rocky upland comprising Chotanagpur Gneissic Complex (CGC) and ultimately debouching into the river Ganga near Patna district, forms the soul of the present study in light of fluctuation in sedimentation pattern, anthropogenic interference (particularly aberrations in sand mining), the effect of cross drainage structures (viz, barrage, bridge, weir) and diversion of the streamflow for various irrigational purposes which together or in isolation have greatly influenced the ecological health of the river. The study has been carried out on 11 no Sand Mining ‘Ghats’ officially identified as lease areas for minor minerals in Jehanabad district, Bihar along the course of River Phalgu.

Climatic change in this part of the country in terms of rainfall has conspicuously experienced a sustained shift in precipitation from high to low and excessive dependence on groundwater resulting in over drafting has influenced the fluctuation in sedimentation pattern and carrying capacity of the river.

Anthropogenic activities are significantly affecting the river bed morphology by influencing sediment supply and transportation in a river basin. The fluvial ecosystem like others is vulnerable to changes emerging from the increased human occupation of an area and climatic changes (Cincotta et al. 2000). The variations in sedimentation regime also affect the ecological integrity of the river channel by altering the channel's stability, bed and bank erosion, and aquatic habitats of the riverine ecology (Comiti et al. 2011). Sediment extraction from river beds is one of the detrimental anthropogenic activities responsible for altering the river bed morphology and ecological degradation (Petit et al. 1996; Arnaud et al. 2015; WyZga 2007). The impact of sediment flux and extraction has been widely studied in different parts of the world (Gaillot and Piegay 1999; Batalla et al. 2006; Surian 1999; Scorpio et al. 2015). In some cases, sediment extraction is an important driving force for river adjustment and channel evolution (Warner 2000). Gaillot and Piegay (1999) have studied sedimentation patterns and the impact of sediment lifting. The impact of sediment abstraction and sediment fluctuation is also dependent on the magnitude of the anthropogenic intervention and the timescale of the sediment extraction activity (Surian and Rinaldi 2003; Belletti et al. 2015). In the monsoon phase, most of the Indian rivers obtain a high amount of water mass (Singh 1996; Gupta 1997; Goswami 1998; Kale et al. 1997; Liebault and Piegay 2002; Starkel et al. 2008; Wiejaczka 2016; Ayaz et al. 2018). Transportation of sediment is the predominant work of the river during monsoon through various hydrological processes such as a solution, suspension, saltation and traction under highly fluctuating hydrodynamic conditions and sediment loads.

This paper deals with the tight rope walking in arriving at the best possible balance resulting from *environmental safety guideline procedures involved in mining, safeguarding the ecology of the river amidst challenges and ways to minimize the unprecedented effects of short-term climatic change*. It also highlights aspects like drivers responsible for the *reduced flow, erratic modes of sand deposition vis-a-vis undisciplined extraction and recommendations towards sustainable exploitation in tune with the expectation and guidelines served by the Hon'ble National Green Tribunal* in this case along Phalgu River, in Jehanabad district, Bihar.

About the Drainage

The Phalgu River, which flows past Gaya, India, in the state of Bihar, is a sacred river for Hindus and Buddhists. The name Phalgu is believed to be a combination of *phala* (merit) and *gau* (wish-fulfilling cow). The etymology implies that the river manifests the highest power of piousness and merit. In ancient scriptures, it is called the Nairañjanā in Sanskrit. Hindu pilgrims from different parts of India gather on the banks of the river during monsoon (September – October) for a religious ritual – 'Pitrapaksh Tarpan' when pilgrims take a dip in the seasonal holy river 'Phalgu' (Vikram Kumar 2019).

Phalgu River is formed by the confluence of Lilajan (*also called Niranjan or Nilanjan*) and the Mohana, two large hills originated streams each of which is over 250 m wide. The confluence is marked ~3 kilometres d/s of Bodh Gaya. The higher stream flows further to the north, past the town of Gaya, where it attains a width of over 820 m (Vikram Kumar 2019). Phalgu River is known for having minimal surface flow and most of the water content is known to occur superficially.

Diversion of water through manmade canals/distributaries by means of water interception civil structures like weirs, barrages and dams for irrigation has rendered the active channel of the stream almost devoid of water. Excessive dependency on surface flow and somewhat poor management of water resources are also responsible factors for depleting water content.

The short-term climatic changes (severity in heat spells and reduced or unpredictable precipitation) have also influenced the surface runoff in the river channel. The decreased surface runoff in turn is increasing the pressure on the available freshwater resources in south Bihar which is also a phenomenon in several rain-starved states of India. Water levels are falling to the tune of 4–6 mm/year in the Indus Basin and the Ganges-Brahmaputra Basin (Vikram Kumar 2019). The disappearance or drying up of River Saraswati due to the effect of climatic changes and tectonism is an enigma environmentalists are dealing with. In the state of Uttar Pradesh, the Gomti River has dried up in its upper catchment as the river has lost its tributaries in almost all the districts it flows through. Likewise, the Behta River (UP) has shrunk to ponds. Streams like Bhainsi and Tareuna in Shahjahanpur district UP have turned dry to the extent that for ~16 km in the upstream reaches, no signs of the river are visible. Chhota and Andhra Chhoha in the Lakhimpur area have almost dried up owing to the acute fall of groundwater levels in nearby areas. With a baseline water stress score of 4.15, India's Palar River is placed at 15th rank on WRI's list of the world's most water-stressed rivers (Vikram Kumar 2019). Nayyal is another river in Tamil Nadu that is endangered along with the waterways of Chennai (*Cooum and Adyar River).*

India accounts for about 17% of the world's population but has only 4% of the world's freshwater with the per capita annual availability of 1544 cubic meters placing it as an already water-stressed country, inching rapidly towards severe water scarcity. The downwards trends in the availability of fresh water in rivers have become a serious concern globally posing questions about sustainability. Experts have predicted that water availability will be one of the major challenges faced by human society and the lack of water would be a key factor limiting development. (Vikram Kumar 2019).

Location and Study Area

The Phalgu River is a dominantly south to northbound drainage flowing through the middle of Gaya district; Bihar (Fig. 7.1); however, it is pushed to the eastern fringe when it enters Jehanabad district. The higher-order derives its name from the two

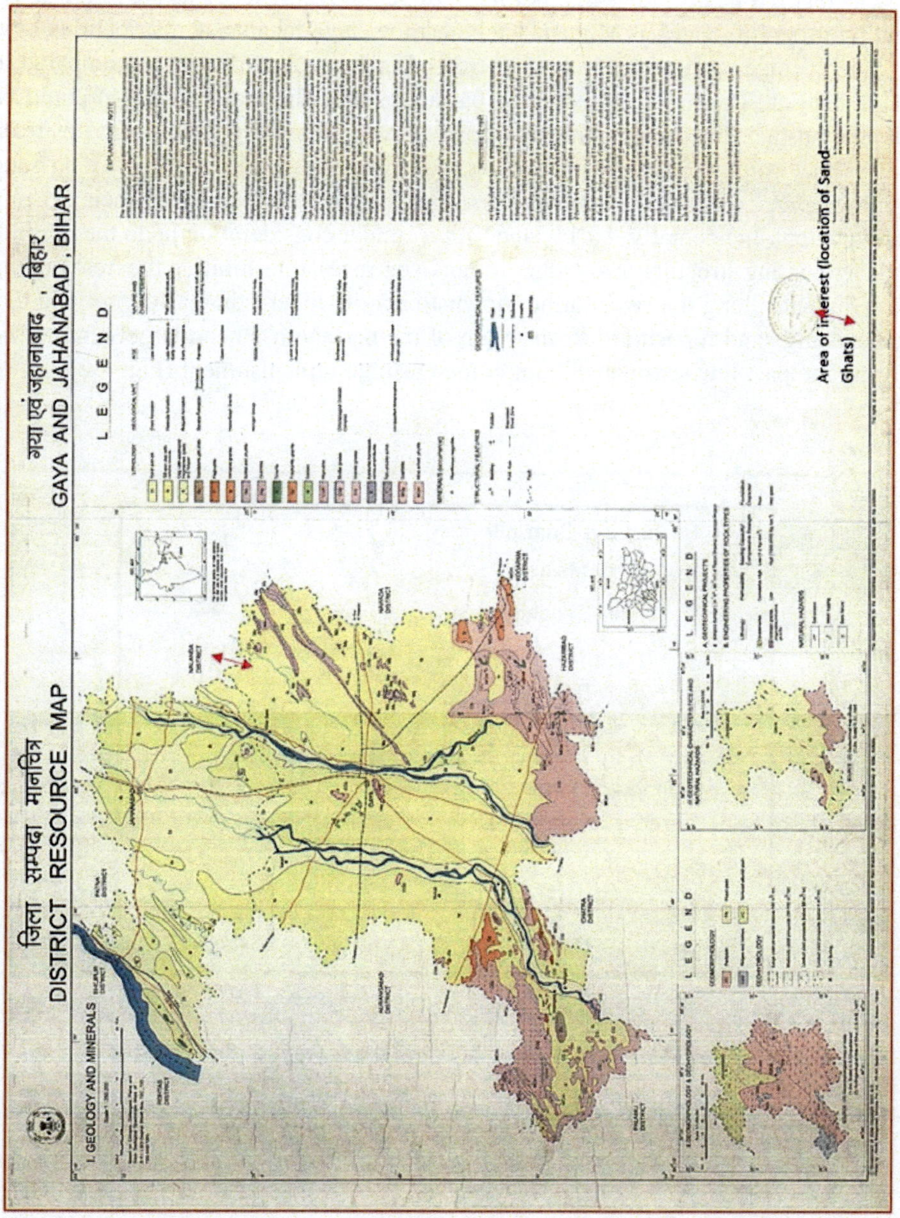

Fig. 7.1 District resource map showing location of the area of study

tributaries Lilajan and Mohana River which join to make the higher order drainage (Phalgu River). The Lilajan (flowing SW-NNE) and Mohana River (flowing SE to NNW) emanate from the rocky uplands of Chotanagpur Gneissic Complex in the south and make a confluence, immediately south-east of Gaya township. Further downstream, it nearly flows in a northerly direction rechristened as the Falgu River and it further bifurcates as Mohana Nadi near the Gaya Jehanabad district boundary.

The average span of the river is decreased in Jehanabad district as compared to Gaya district. The newly constructed barrage near Udherasthan (~2016) could be another obstacle intercepting the runoff and allowing more deposition to its upstream unless the water is released in the downstream areas. In this stretch, 11 no sand mining '*ghats*' identified as mining lease areas, have been studied for their existing sand content. Incidentally, 10/11 sand '*ghats*' under consideration lie to the north of the engineering structure near Udherasthan (a barrage), thus limiting the fresh supply of sediments along the river further downstream of Udherasthan also implying that the existing sand repository downstream of the barrage has been largely harnessed during the past few seasons without commensurate replenishment (Fig. 7.2).

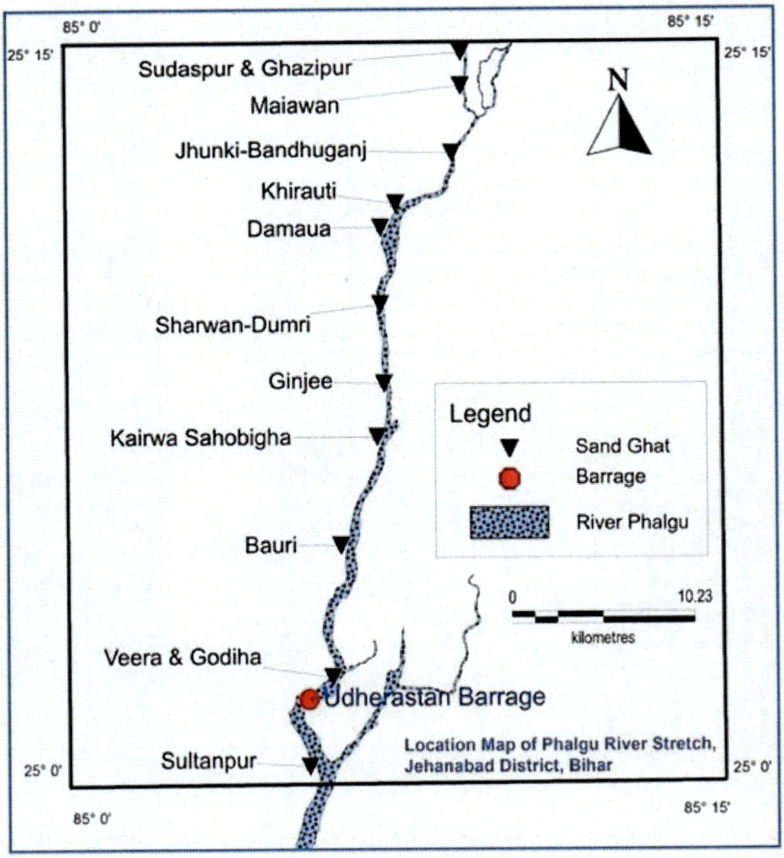

Fig. 7.2 Study area map

Geology

The districts of Gaya and Jehanabad have about 80% of its surface area under cover of alluvium. The southern part of the Gaya district constitutes mainly Precambrian rocks while the northern part together with the present territorial boundary of the Jehanabad district is represented by quaternary sediments. The rocky upland comprising hilly tract as well as inselbergs in the southern periphery is composed of Chotanagpur Gneissic Complex (mainly granites and its variants) along with older metamorphic comprising mica schist, quartz mica schist and phyllite. Migmatites and augen gneiss are also present.

Amphibolites, hornblende schist, epidiorites, etc., are exposed along hills north-east of Gaya. The younger quartzites and Mica schists of Munger Group of Middle Proterozoic age are exposed east of Gaya along NE-SW trending ridges. Intrusive granites and pegmatites are present along the eastern border of the district. Gondwana Group of Rocks in limited exposure is present in the southern part of the district.

The quaternary sediments overlying the Precambrian rocks are expressed as alluvial uplands characterized by fluvial imprints like levees, channel bars, etc., the northern part of the district is covered by older flood plain sediments which have a marginally higher relief as compared to younger flood plain sediments.

The whole of Jehanabad district and the northern part of Gaya district contain a high yielding aquifer zone while the southern part of Gaya district is covered with hard rock or a thin cover of alluvium with the low yielding aquifer. The high yielding aquifer dwindles in thickness in the terrain south of the main railway line passing through Gaya town. In the blocks south of Gaya town viz. Dobhi, Bodh Gaya, Wazirganj, etc., high to moderate yielding zones are found in pockets where aquifers occur within the shallow zone of 30 m in the form of saturated sand extending widely on either side of the course of Phalgu, Morhar and other perennial rivers.

Material and Methodology

River channel morphology results from a complex interplay of tectonics, geomorphic and anthropogenic processes that occur in a basin at various spatial and temporal scales (Achyuthan 2003; Lbisate et al. 2011; Wolf and Faust 2016; Prizomwala et al. 2015). Scientific observations have been taken at independent sites (11 no ghats) on the parameters viz. river morphology & flow regime, identification pillars (if any), over-excavation/exploitation (if any) in the area, impact assessment, proximity to major engineering structures, land dereliction, bank erosion and/or other salient features, the status of the available sand horizon.

To assess the combined impact of the geomorphic processes and anthropogenic activities on river morphology, Global Positioning System (GPS), Google Earth (GE) imagery, and ecological species assessment has been tentatively made using a

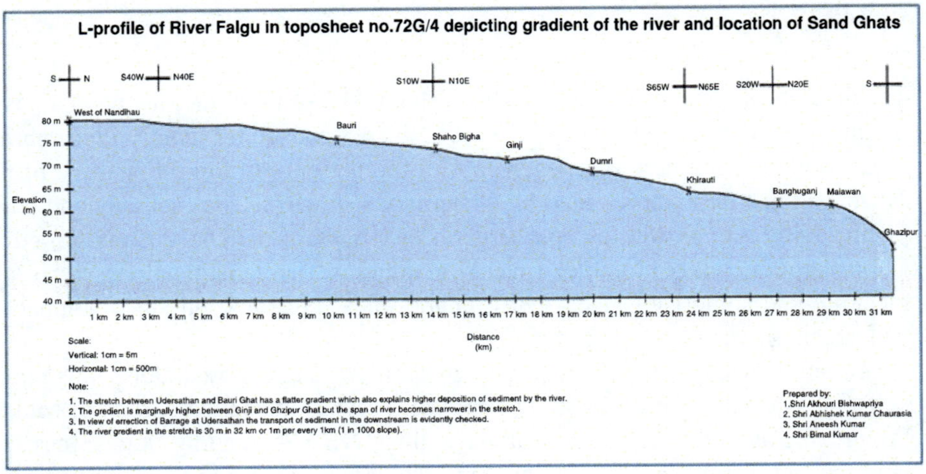

Fig. 7.3 Longitudinal profile of the area

measuring tape and graduated bamboo stakes. The data in regard to longitudinal river profile (Fig. 7.3), stream water velocity, river bank breaches, etc., was obtained using high-resolution satellite data supported by detailed field measurements.

With an exaggerated scale along the Y-axis symbolizing the height of the terrain, the drop in gradient could be depicted. It also iterates the fact that the gradient for this peninsular bound river is almost negligible which further establishes the fact that any deposition of sand in a colloidal state will depend on the available water in the river and its carrying capacity.

As indicated in the flowchart in Fig. 7.4, sediment extraction from the river regime has offsite and onsite impacts. Though offsite impacts only relate to transport-related problems, the latter may result in depletion and sometimes day-lighting of groundwater table, increased salinity and turbidity. On a parallel flow, excavation of sediments in an unsystematic manner may result in bank erosion, bed degradation, caving phenomenon, definite bed lowering, channel instability, widening to even change in watercourse within the river system. In the present case, bed degradation, bank erosion, caving, the threat to embankments, civil structures, widening and migration of pits and change in watercourse within the banks has been observed and documented.

Analysis & Discussion

River morphology studies suggest that spatial variations in climatic conditions, geomorphology, downstream changes in bedrock lithology, tectonic processes and anthropogenic activities along a river stretch can lead to steeper, or flatter longitudinal river profiles (Selander 2004; Hossain et al. 2013; Besné and Ibisate 2015; Yanagida 2016; Žibret and Žibret 2017).

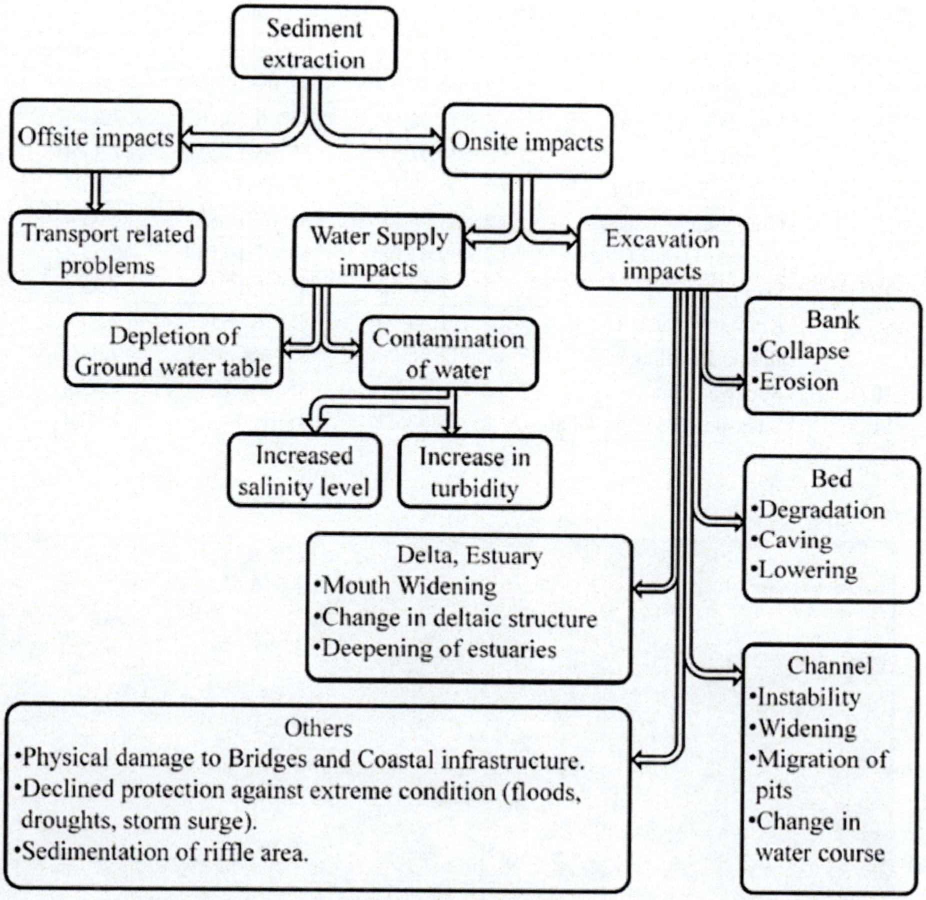

Fig. 7.4 Sediment extraction impacts

A site investigation was done on the Phalgu River and the site coordinates are given in Table 7.1 below.

Observations at Individual Sand 'Ghats'

Location 1: Sultanpur: The river flows in the N20°E direction. The location is slightly upstream of the bifurcation of Phalgu River into Mohana River (Fig. 7.5a–f). The span of the river (bank to bank) is ~550–600 m. The river channel had water present in it but not in flowing condition. Conspicuous excavations on the ground as a result of quarrying were noticed mostly in irregular fashion forming pits as deep as 7–8 ft. below bed level (Fig. 7.5f).

Compared to the general river bed profile, a relief difference of about 3–4 m within the channel has been observed implying that sand has been drafted with norm

Table 7.1 Location details of sand mining ghats

S. no.	Location name	Latitude	Longitude	Toposheet no.
1.	Sultanpur Ghat	25°00′01.02”	85°05′30.47″	72G/4
2.	Veera Godiha Ghat	25°02′08.59″	85°06′00.70″	72G/4
3.	Bauri Ghat	25°04′47.54″	85°06′13.55″	72G/4
4.	Kairwa Saho Ghat	25°06′59.15″	85° 07′04.58″	72G/4
5.	Ginji Ghat	25°08′04.28″	85°07′14.83″	72G/4
6.	Sharvan Dumri Ghat	25°09′41.08″	85°07′10.51″	72G/4
7.	Damuo Ghat	25°11′14.66″	85°07′12.17″	72G/4
8.	Khirauti Ghat	25°11′44.00″	85° 07′ 33″	72G/4
9.	Jhunki Bandhuganj	25°12′46.00″	85° 08′ 50.6″	72G/4
10.	Maiawan Ghat	25°14′ 6.56″	85° 09′ 3.5″	72G/4
11.	Sudaspur-Ghazipur Ghat	25°15′23.47″	85°08′54.46	72G/4

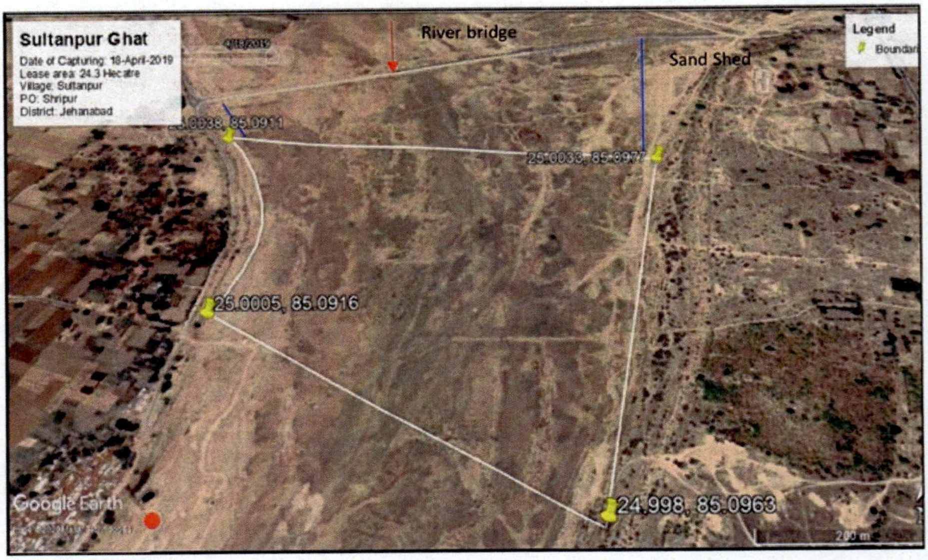

Fig. 7.5 (**a**) The Google Image is of 18 April 2019. Within the lease area, prima facie, sand horizon can be observed but on the date of observation the entire stretch has water cover. The flow path is mainly towards the middle and right bank. The banks do not provide confirmatory evidence on depth of sand; however, a few available sections on the left bank indicate pre-sense of more than 3 m sand column. (**b**) The excavated section is within an active channel of the river and the height of the person is 6.1 ft, implying that the height of the section is nearly 10 ft. or > 3 m. Moreover, the coordinates of this location falling much outside the lease area indicating the possibility of excavation outside the lease area. (**c**) Irregular pits observed along the Phalgu River channel (Near Sultanpur). (**d**) Sand mixed with clay within active channel (Phalgu River; Near Sultanpur). (**e**) Thick column of piled up sand within river section (Phalgu River; Near Sultanpur). (**f**) Irregular shaped pits indicating quarry activity in the river section (Near Sultanpur)

Fig. 7.5 (continued)

violation of going beneath 3 m. Within the river bed also, irregular puddles of water were noticed (Fig. 7.5f). Thus, overexploitation could be in practice. Several isolated water pools have been created within the river bed giving rise to undulating topography leading to some amount of land dereliction within the active channel (Fig. 7.5e). This is the only site that lies upstream of the barrage at Udherasthan, implying that it has a good chance of being replenished during the monsoon period. Understandably, voluminous sand quantity but more on the bank fringes is still available at the site.

Location 2: Veera and Godhiya: The river flows in the N30°E direction. The location is downstream of the barrage location (1.2 km) on Phalgu River near Udherasthan (Fig. 7.6a). The span of the river (bank to bank) is ~300 m. The river channel at the time of observation had extensive water but was not in flowing condition. No Identification pillars for exploration were observed on the ground at the time of observation. The barrage at Udherasthan is ~1.2 km in upstream (Fig. 7.6b). The

Fig. 7.6 (**a**) The Google image is of 18 April 2019. Within the lease area, prima facie, sand horizon cannot be observed but on the date of observation the stretch has water cover. The flow path is mainly towards the middle and right bank. The banks do not provide confirmatory evidence on depth of sand; however, a few available sections on the left bank indicate pre-sense of sand mixed with clay. (**b**) Upstream view of the barrage at Udherasthan from Veera-Godhiya Ghat. (**c**) Observation taken from right bank of the river. A distinct embankment on the right bank acting as a levee with a height of about 1–2 m. A path developed for vehicular movement observed (Near Veera Godiya; River Phalgu). (**d**) The ghat at the time of observation appeared sand starved although heaps of sand on the right bank of the Phalgu River at levee portion were observed (Near Veer Godiya)

eastern face of the area under exploitation is 280–373 m from the bridge on Garbhi Nadi. Sand is sparsely present on the ground but doesn't appear to be of high quality and is mixed with appreciable clay (Fig. 7.6c, d). Dark bands of the intervening clay sequence can be observed. Depth of sand can only be ascertained by a scout or exploratory drilling/auguring.

Location 3: Bauri: The river flows in a northerly direction at the site. The span of the river (bank to bank) is ~300 m (Fig. 7.7a). The river bed is dry and wet with extensive grass growth (~5–6 ft) (Fig. 7.7b). Small puddles of the water zone are present and they could be covering the areas of the river channel from where sand is derived. At the site, the area shows signatures of past quarrying activity, and it is observed that sand has been derived from the beyond the defined boundary (Fig. 7.7e). The river bed is eroded and there is unmistakable evidence of unsystematic quarrying. Irregular fashioned pits are observed though the depth of pits is less than 1 m (Fig. 7.7f), but land dereliction is evident. Gully erosion and subsequent sand retrieval are also present. A fresh column of riverine sand (> = 7–8 ft) is observed in the river section but it is not universally present. An appreciable amount of sand has been extracted from the location.

Location 3: Bauri: The river flows in a northerly direction at the site. The span of the river (bank to bank) is ~300 m (Fig. 7.7a). The river bed is dry and wet with extensive grass growth (~5–6 ft) (Fig. 7.7b). Small puddles of the water are present and they could be covering the areas of the river channel from where sand is derived. At the site, the area shows signatures of past quarrying activity, and it is observed that sand has been derived from the beyond defined boundary (Fig. 7.7e). The river bed is eroded and there is unmistakable evidence of unsystematic quarrying. Irregular fashioned pits are observed though the depth of pits is less than 1 m (Fig. 7.7f).

Location 4: Khairwa Sahu Bigha: The river flows in the N10° E direction. The site area lies along the left bank of the river. The span of the river (bank to bank) is ~250 m (Fig. 7.8a). The river channel at the time of observation had a pool of stagnant water (not in flowing condition). No active mining or quarrying activity was observed (Fig. 7.8b, c). Extensive vegetation (mainly tall grasses) was observed within the river section (Fig. 7.8c). The left bank area is generally dry and undulating leading to the abutments. No prima facie signs of land dereliction or bank erosion. On the left bank covering the site area, the sand on the surface is minimal. The surface appears to be clayey/muddy.

Location 5: Ginjee: The river flows in a northerly direction. The mining area occupies the centre of the river and the right bank. The span of the river (bank to bank) is ~180 m (Fig. 7.9a). The river bed has extensive grass growth (~4–5 ft) (Fig. 7.9e). Small puddles in the water zone are present and they could be covering the areas of the river channel from where sand is derived. No active mining or quarrying activity is evident. There is sand presence on the banks of the

Fig. 7.7 (**a**) Sand distribution on river bed near Bauri Ghat, Jehanabad. (The quarry area demarcated on the map with coordinates). (**b**) Thick grass growth along river bed. Signatures of past quarrying in the area evident (Near Bauri; Phalgu river). (**c** and **d**) Evidence of (**c**) land dereliction owing to sand extraction; (**d**) the relief difference depicted by the section also indicates sand extraction at different levels. (**e**) Signatures of vehicular movement in study area (Bauri Ghat). (**f**) Irregular excavated pits resulting from unscientific extraction of sand (Bauri Ghat)

Fig. 7.7 (continued)

river but as such, it is not superficially present in the river section. The river bed is sand starved, eroded and there is unmistakable evidence of past quarrying. Puddles of water suggest quarrying from these zones (Fig. 7.9f). Dereliction is evident within the river bed.

On the surface, the sediment is mostly clayey or muddy (minimal sand). One striking feature is that in the past year's imageries, the presence of sand could be made out but the replenishment has not been commensurate with the exploitation. For instance, in an image dated 7 March 2014, the sand repository in the river bed was extensive (Fig. 7.9a). Subsequently, in the image dated 16 November 2015, the flow was found confined to the right bank and in channels through the centre of the river, but overall a marked decrease in the sand volume has been observed (Fig. 7.9b). Again, in the image dated 22 January 2017, there is a further decrease in the volume of sand and the flow of the water has now shifted more towards the left bank (Fig. 7.9c). Thus the excavation has visibly affected the channel flow. Finally, in the image dated 18 April 2019, the river channel appears flattened and the river starved of its sand content as compared to earlier temporal images (Fig. 7.9d).

Location 6: Sharwan Dumri Ghat: The river flows in a northerly direction. The area of study occupies the major portion of the right bank, taking a convex loop towards the left bank at its northern extremity and tapers considerably at its southern end, towards the right bank giving it a sickle shape. The span of the river (bank to bank) is ~300 m (Fig. 7.10a). The river bed has extensive grass growth (~4–6 ft) (Fig. 7.10c). Small but extensive puddles of stagnant water zone are present and they could be covering the areas of the river channel from where sand is derived (Fig. 7.10c, f, h). The river channel is more clayey but the sediment beneath the water column is sand. Sand is also exposed on the river bed at a few locations (Fig. 7.10b). No active mining or quarrying activity was observed. The area had signatures of past quarrying activity. There is sand present on the bank of the river and is also observed in the levee cut section with an intervening sequence of clay and silt. The bank has an elevation difference of 3–4 m from the river bed (Fig. 7.10e).

Fig. 7.8 (**a**) The Google image is of 18 April 2019. Within the lease area, prima facie, sand horizon can be observed but on the date of observation the entire stretch has stagnant water. The banks do not provide confirmatory evidence on depth of sand; neither are there cut sections to ascertain that. (**b**, **c**) No active mining or quarrying activity is observed. (**b**) Extensive vegetation (mainly tall grasses) observed within the river section. (**c**) The water in the channel may be filling up earlier formed pits

On the right bank, there is a good quantity of sand available but with an intervening sequence of clay. There is a second tier of abutment also within the river channel. The right bank has been exploited leaving an exposed section of 4–5 feet. The exposed section has ~70 cm of clay, followed by 1.2 m of sand which has probably been exploited as inferred from site observation. The dereliction is evident within the banks (Fig. 7.10g). The river bed is eroded and there is evidence of past quarrying. Puddles suggest quarrying from these zones. Water pools are observed with stagnant water within the river channels. Sand is available on the abutments mixed with clay. Sand is also available within the river channel as observed at a few locations but spatial consistency and depth continuity cannot be exactly ascertained.

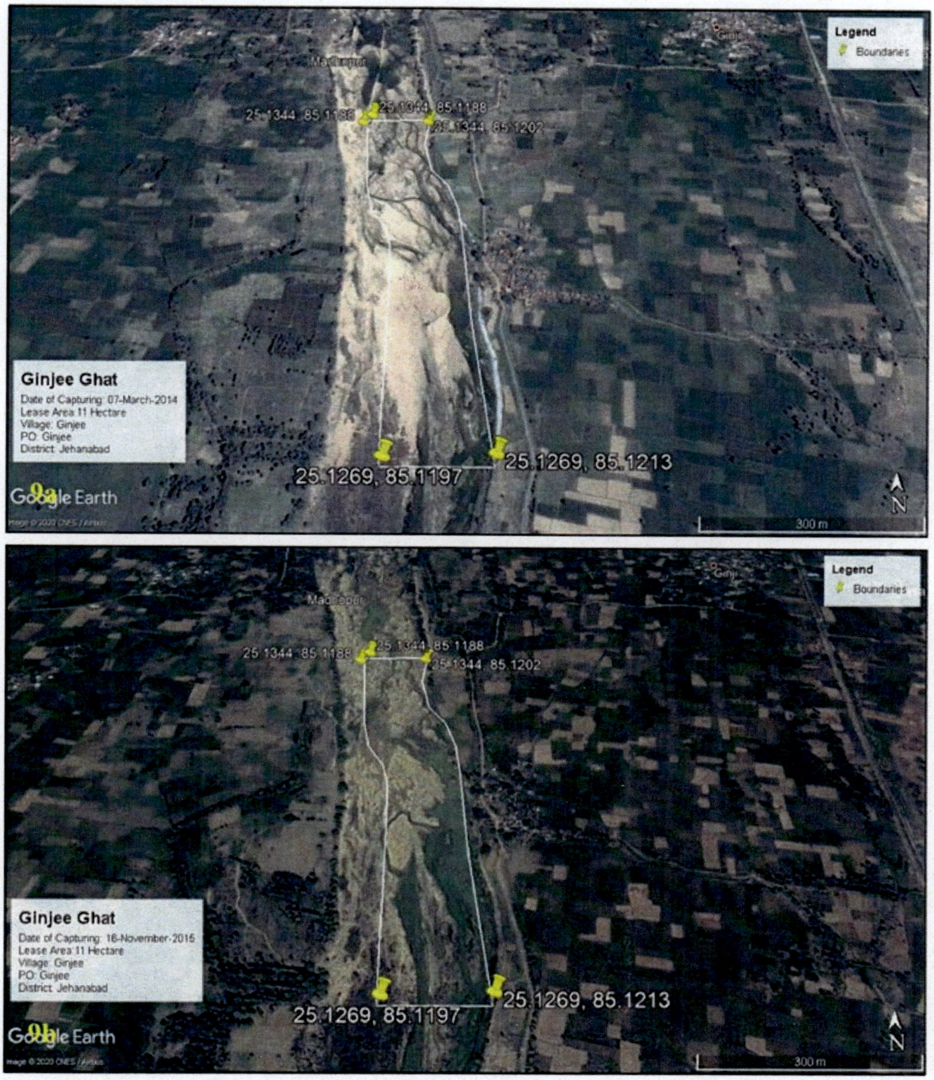

Fig. 7.9 (**a**) Around Ginjee Ghat, in an image dated 7 March 2014, the sand repository in the river bed was extensive. (**b**) Subsequently, in the image dated 16 November 2015, the flow was found confined to the right bank and in channels through the centre of the river, but overall a marked decrease in the sand volume has been observed. (**c**) In the image dated 22 January 2017, there is further decrease in the volume of sand and the flow of the water has shifted more towards the left bank. Thus, the excavation has visibly affected the channel flow. (**d**) Finally, in the image dated 18 April 2019, the river channel appears flattened and the river starved of its sand content as compared to earlier temporal images. (**e**) River bed under dense vegetation comprising grasses and shrubs. Generally sand. (**f**) Puddles of water within river bed indicating excavation. (**g**) Films of water-covered areas within the river bed. (**h**) Intermittent stretches of water covered zones on the river bed

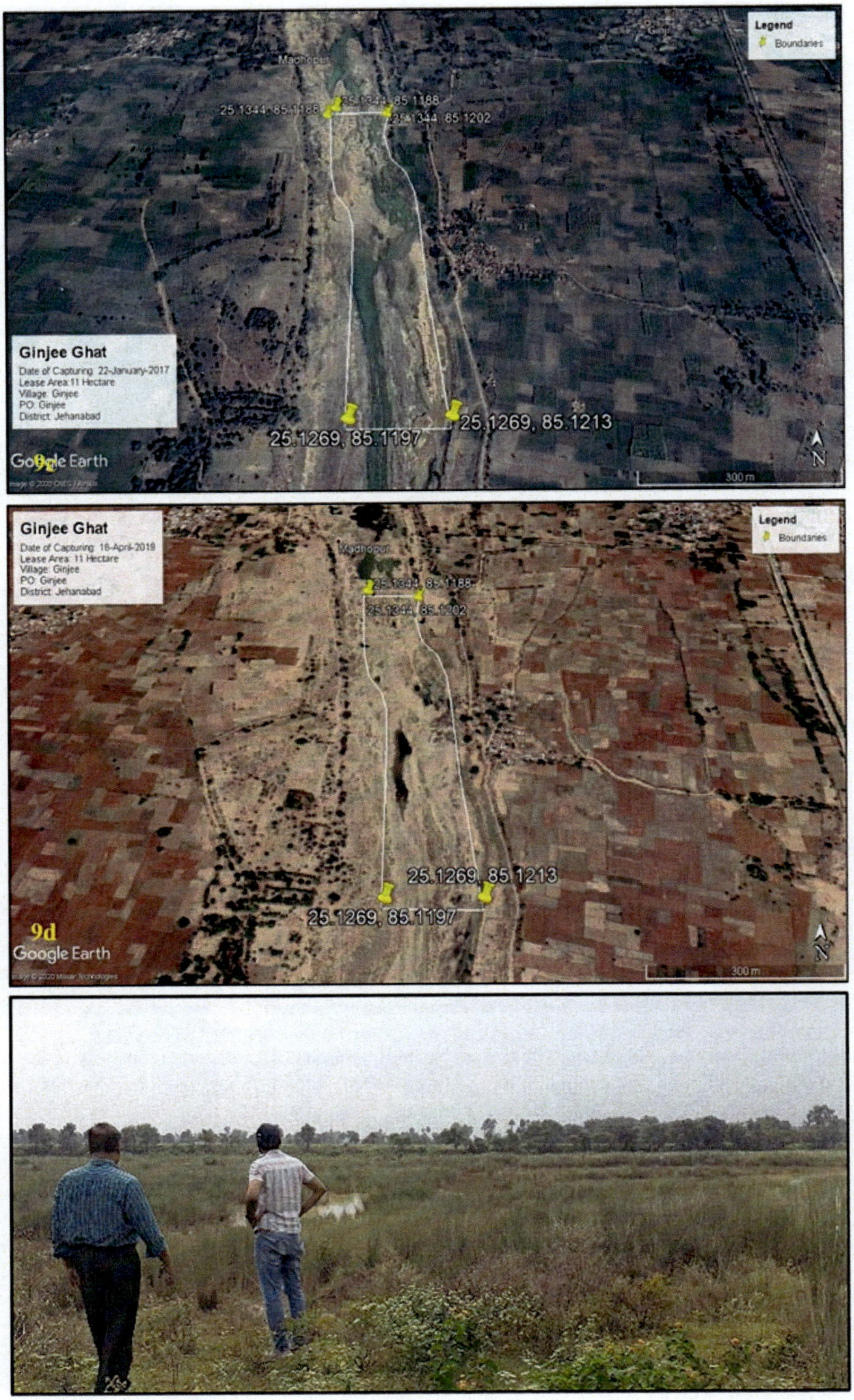

Fig. 7.9 (continued)

Fig. 7.9 (continued)

Location 7: Dumao Ghat: The river flows in the N10°E direction. The area occupies a major portion of the left bank, extending to the centre of the river. The area has a roughly rectangular shape. The span of the river (bank to bank) is ~450 m

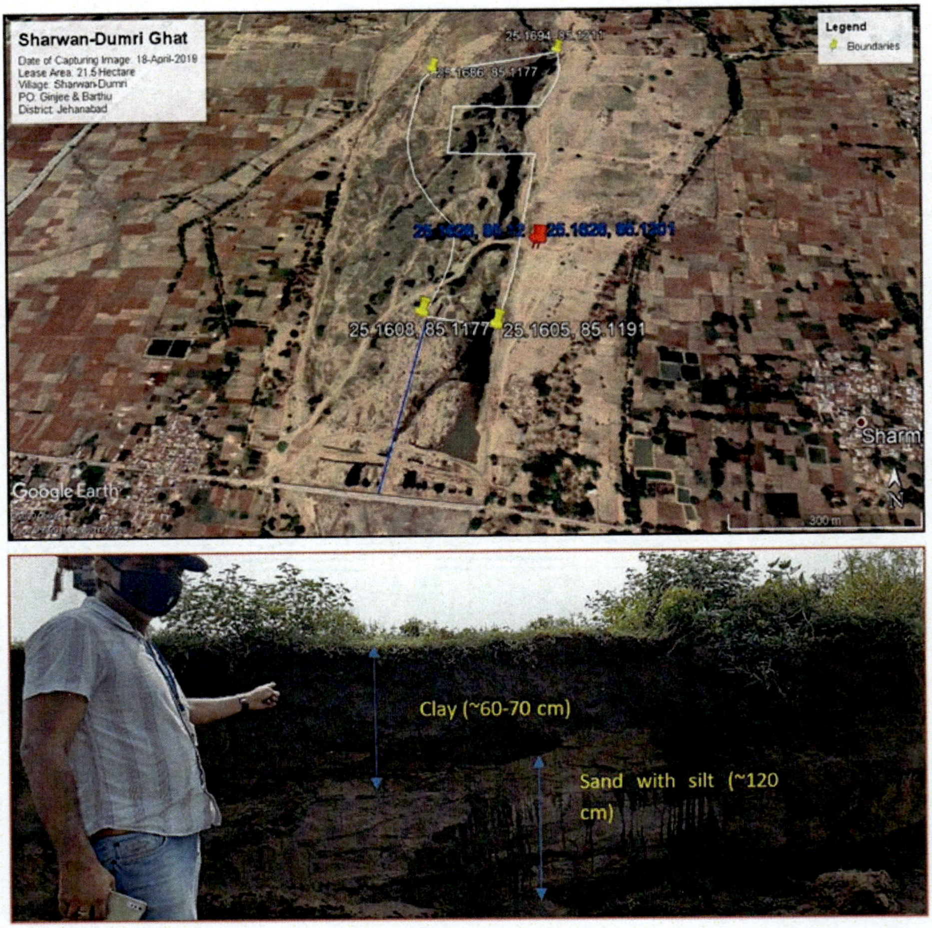

Fig. 7.10 (**a**) The Google image is of 18 April 2019. The image shows the lease area boundary. However, independent observation locations (coordinates indicated in blue) suggest that there is some quarrying activity outside the boundary limits. It needs to be ascertained whether the river bank is included in the mining lease area, as the location in red is outside the river regime. The banks do not provide confirmatory evidence on depth of sand, but there are cut sections to ascertain that. (**b**) The top soil comprising clay has also been exploited along with the sand which comprises the underlying sequence of strata. A thin layer of clay follows the sand layer immediately beneath. (**c**) Wild grass growth within the river channel. Pools of stagnant water observed. (**d**) Land dereliction. Sand column mixed. (**e**) Land dereliction due to exploitation resulting in loss of one tier of surface on the banks. (**f**) Water pools of stagnant water created due to land erosion. (**g**) Riverbank aggradations and extensive. (**h**) Water pools observed in river section. On the periphery lies a veneer of the sand horizon. Sand may have been derived from these zones

(Fig. 7.11a). Observation has been made from the left bank. The river bed (more towards the middle portion) has extensive grass growth (~4–6 ft). The left bank area is adversely affected by signatures of earlier mining leaving pools of stagnant water, river bed erosion and dereliction (Fig. 7.11d). The flow regime appears to be affected in segments. The area had fresh signatures of quarrying activity. GPS

Fig. 7.10 (continued)

Fig. 7.10 (continued)

Fig. 7.11 (**a**) The Google image is of 18 April 2019. The image shows the lease area boundary. However, independent observation locations (coordinates indicated in blue) suggest that there is some quarrying activity outside the boundary limits. This precisely is the observation of the committee as these quarry locations lie on the banks leading to land dereliction. The banks do not provide confirmatory evidence on depth of sand, but there are cut sections to ascertain that. (**b**) Pools of water created as a result of pit formation and unsystematic mining. (**c**) Vertical cut section of the bank for sand exploitation. (**d**) Evident sand quarrying from river bed affecting river bed topography. (**e**) After sand removal, the clay layer has been exposed along the banks

Fig. 7.11 (continued)

coordinates suggest that mining has taken place outside the lease boundary as well as suggesting anthropogenic interference outside norms. There is extensive sand present on the bank of the river and is also observed in the levee cut section with an intervening minor sequence of clay and silt (Fig. 7.11c). The left bank has been exploited unsystematically outside the river channel regime with apparently a lot of material being excavated from the area, leaving the area disturbed and adversely aggraded creating an undulating to irregular surface topography.

The river bed is aggraded, eroded and there is evidence of past quarrying. Irregular small and large pits have been created making water covered zones suggesting quarry operations in these areas. On the left bank, a continuous sequence of sand is observed. River bed profile, topography and flow regime are disturbed due to the unsystematic derivation of sand. The topsoil has also been exploited as could be made out from the section. Sand is available on the abutments mixed with marginal clay. Sand is also available within the river channel as observed at a few locations with substantial depth continuity (Fig. 7.11c).

Location 8: Khirauti Ghat: The river flows in the N60°E direction. The area is located at a point where the river takes a turn towards the NE direction. The span of the river (bank to bank) is ~320 m (Fig. 7.12a). The river bed is absolutely dry and tall grasses are observed (Fig. 7.12b). Evidence of sand quarrying is observed within the boundary limit towards the upstream portion. An appreciable amount of sand is observed on the right bank of the river (Fig. 7.12c). The area had signatures of past quarrying activity. Sand is present on the bank of the river which is also clearly observed in the river section. The levee portion of the apartment has been quarried for sand recovery. The bank is disturbed owing to land erosion. Uneven pits are also present in the upstream part, from which sand might have been derived (Fig. 7.12e). An extensive sand deposit is observed on the right bank of the river. Though clay admixture with sand is not ruled out.

Location 9: Jhunki-Bandhuganj Ghat: The river flows in the N20°E direction. The lease area is located on the right bank of the river. The span of the river (bank to bank) is ~110 m (Fig. 7.13a). The river bed is dry and in patches wet with extensive grass growth (~5–6 ft). An appreciable volume of water is observed in the river channel from where sand is derived (Fig. 7.13b). The river had substantial water accumulated due to the retention of water by the Mandai weir downstream (about 1.2 km from the Banduganj Bridge). No active mining or quarrying activity is observed. However, a huge pile of stacked sand is observed on the right bank of the river almost at the mid-point of the area (Fig. 7.13c). The stacked repository also included reddish clay which may also have been exploited as a by-product. Sand mixed with clay (inferior quality sand) has also been observed in the area.

Location 10: Maiwan Ghat: The river flows in a northerly direction. The span of the river (bank to bank) is ~60 m (Fig. 7.14a). The river has water (not in flowing condition). The right levee portion is at a height of 4–5 m from the river bed (Fig. 7.14b). Sand may have been derived from water-covered areas of the river channel. There are signs of sand excavation observed within the area, as observed

Fig. 7.12 (**a**) Google Image of the lease area obtained on 18th April 2019 indicates a flat river bed with dying remnants of sand. Understandably, after mining/quarry operation, not much sand has been replenished on the river bed. The red pinned locations are the ones having remnants of quarrying and also shown to the committee. The same indicates quarrying outside lease boundary. (**b**) Near Khirauti Ghat, river bed is dry with tall grasses. Evidence of sand quarrying is observed within the boundary limit towards the upstream portion. (**c**) Appreciable amount of sand is observed on the right bank of the river. Sand is present on the banks and in the river section. (**d**) The levee portion of the abutment has been quarried for sand recovery. The bank is disturbed owing to land erosion as visible in the photograph. (**e**) Uneven pits are also present in the upstream part, from which sand might have been derived

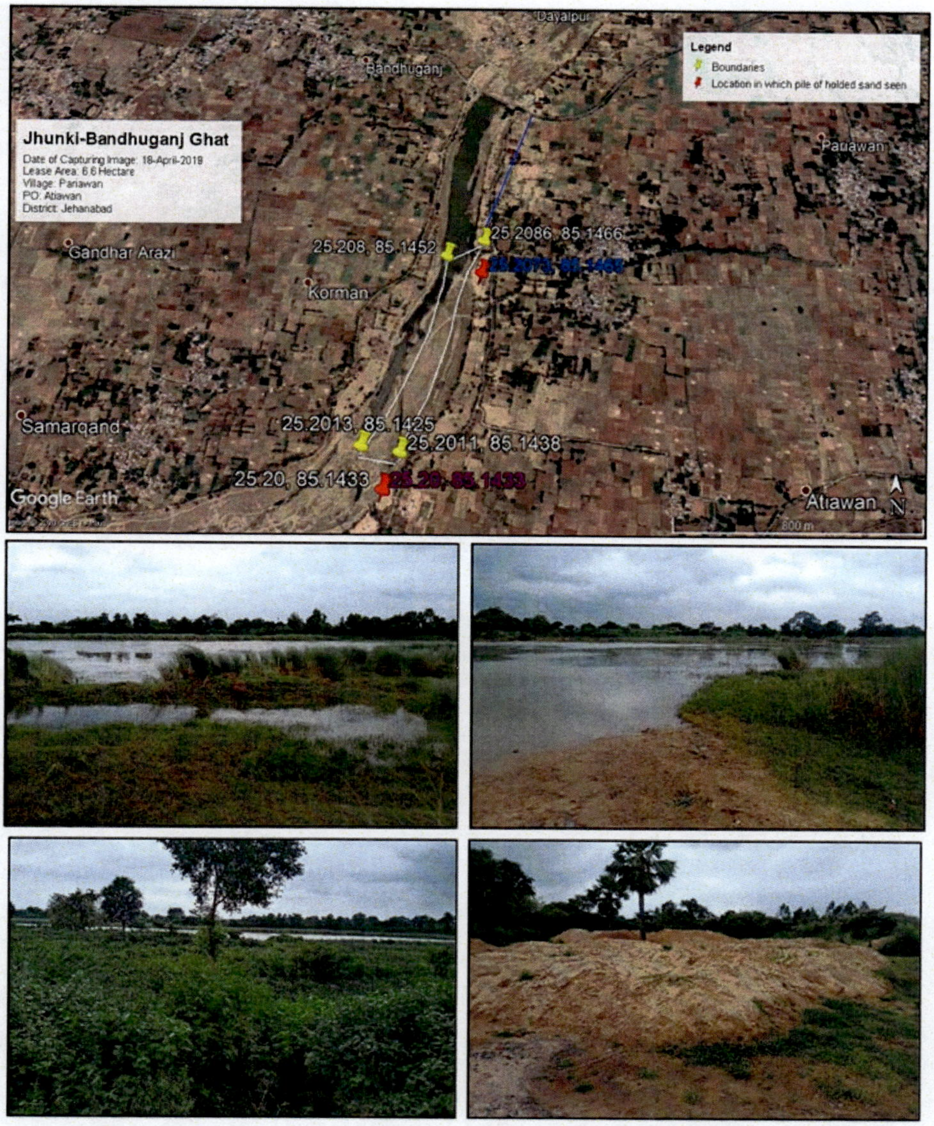

Fig. 7.13 (**a**) Google Image of the lease area obtained on 18th April, 2019 indicates a flat river bed with dying remnants of sand. Understandably, after mining/quarry operation, not much sand has been replenished on the river bed. The water flow is confined to the left bank although on the date of observation, the entire river section was under water on account of retention of water by the Mandai weir further downstream. (b) The Falgu river at Jhunki sand ghat is full of water and there is extensive foliage on the banks. (c) Extensive foliage on the right bank and heaps of sand stacked along the river in the mid-portion of the lease area. Heaps of sand were observed

in the exposed cut sections. Sand is also present in the river bed (Fig. 7.14e). Extensive slope failure and deep gully erosion have been observed on the right bank of the river, which might have developed during the exploitation (Fig. 7.14f). The left bank of the river has a relatively lower relief difference, as

Fig. 7.14 (**a**) Two comparative Google images are depicted for comparison. In the figure above, the red pinned locations are the ones shown to the committee. They fall well outside the lease boundary area suggesting exploitation outside the lease area. The fig below shows the distance of the southern boundary from the Mandhai weir (the stone throw distance is ~550 m). (**b**) Exploitation along banks is leading to slope failure of the levee portion as depicted in the image around Maiwan Ghat on the Phalgu River. (**c**, **d**) Water within river channel observed from the river road bridge. The arrows indicate the flow direction. (**e**) Sand exposed on river bed mixed with clay. The river bridge has been seen further downstream. (**f**) Gully erosion along the right bank of the Phalgu River probably aggravated by human interference

Fig. 7.14 (continued)

compared to the right bank. No signs of sand quarrying are observed on the left bank. The cut section on the right bank has greater clay content as compared to sand. However, an intervening sequence of sand is present within clay layers.

Location 11: Sudaspur-Ghazipur Ghat: The river flows in a northerly direction at this location. The span of the river (bank to bank) is ~120 m (Fig. 7.15a). The river is completely dry and thick and a potential sand column in the river bed is observed. A fresh layer of good quality coarse to medium-grained riverine sand covering the entire river bed is present (Fig. 7.15b). The thickness of river bed sand is 4 to 6 ft. The appearance of the sand layer provides unmistakable evidence that no quarrying has been done at this location. The fertility of the ecology, sand column and its spatial continuity is preserved.

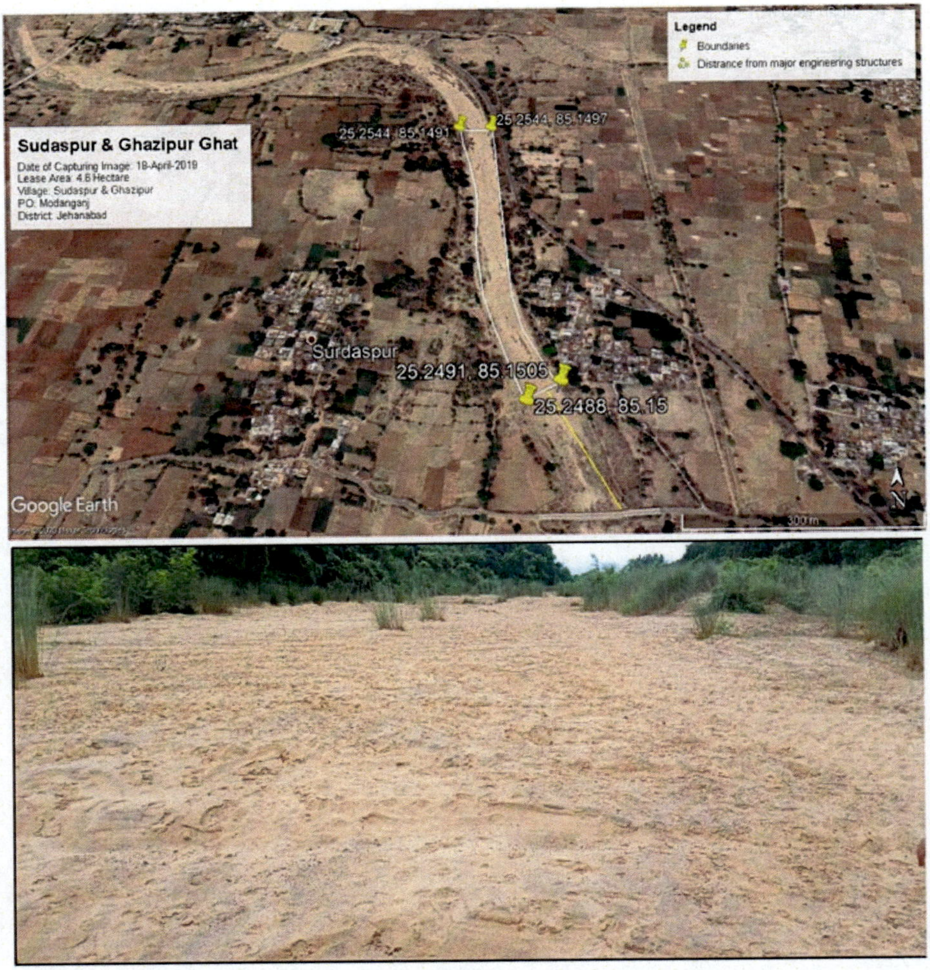

Fig. 7.15 (**a**) The Google Image dated 18 April 2019 is worth a million words. The sand layer on the river bed is intact and is an eye opener to the contrast earlier dated (archive) images indicate in regard to the health of the river in absence of anthropogenic interference. The fertility of the river in terms of sand content at other sites is either lost or is severely affected as compared to this site. (**b**) Ocean of orange to pink coloured sand on the river bed with the extensive spatial extent and the river bed depicting no signs of any quarrying in the past

Timeline Comparison of the Area

Temporal data for the ghats has been compiled to see the changes between 2014 and 2019 which give unmistakable evidence of the changing sedimentation pattern in the context of minor minerals (sand). The wavering water flow path as seen in various images also conforms to the erratic removal and deposition of sand for the period under consideration (Table 7.2).

Table 7.2 Timeline comparison of area boundary

Sl. No.	Location	Image as on 07 March 2014	Image as of 22 January 2017	Image as on 18 April 2019
1	Sultanpur			
2	Veera & Godiha			
3	Bauri			

(continued)

Table 7.2 (continued)

Sl. No.	Location	Image as on 07 March 2014	Image as of 22 January 2017	Image as on 18 April 2019
4	Kairwa Sahobigha			
5	Ginjee			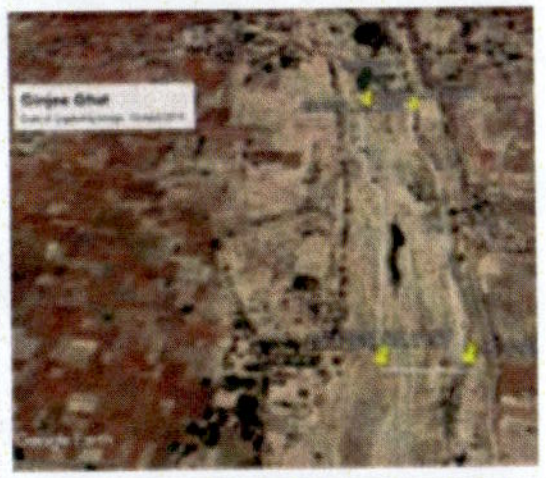
6	Sharwan Dumri			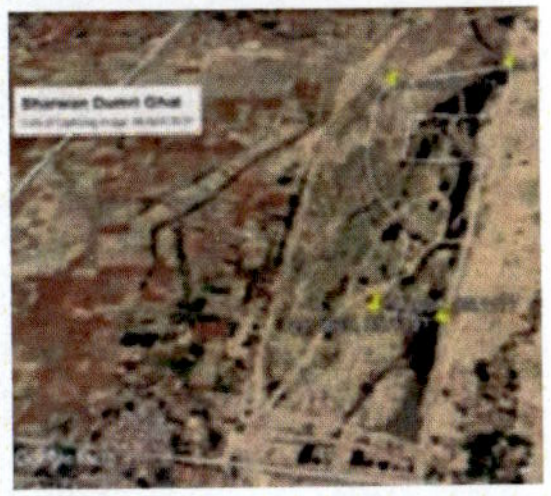

7	Damaua			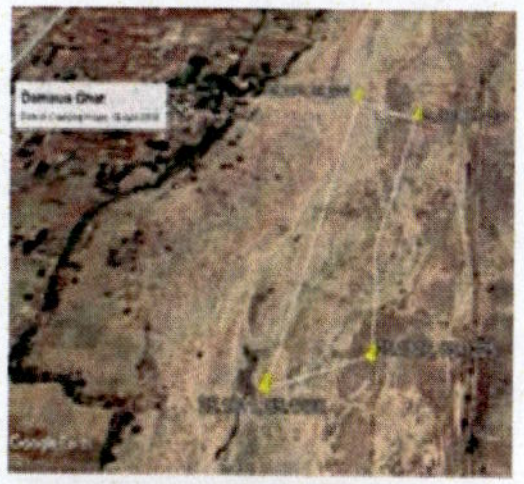
8	Khirauti			
9	Jhunki Bandhuganj			

(continued)

Table 7.2 (continued)

Sl. No.	Location	Image as on 07 March 2014	Image as of 22 January 2017	Image as on 18 April 2019
10	Maiwan		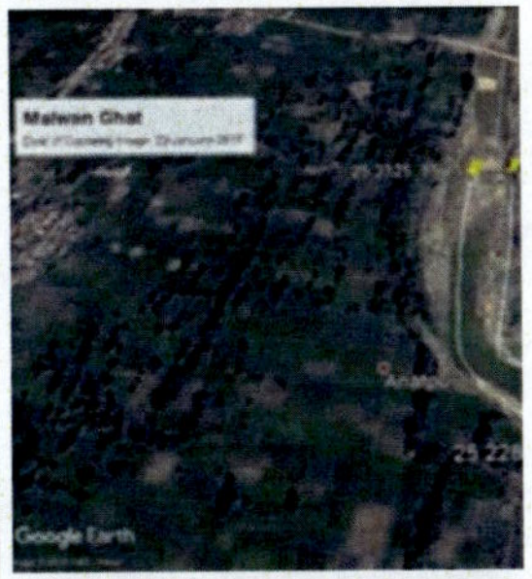	
11	Sudaspur Ghazipur			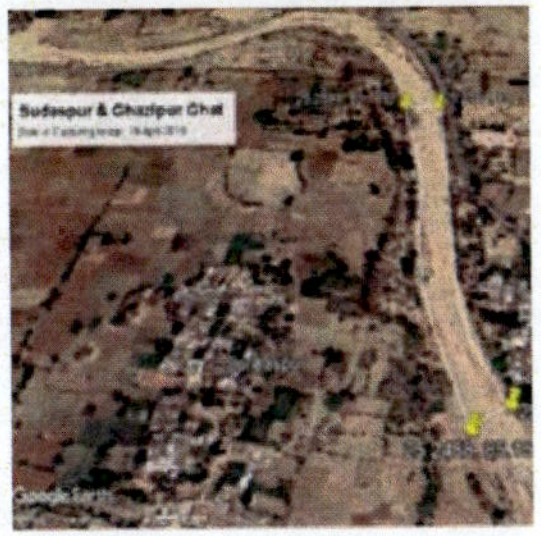

Rainfall analysis: There are several studies available on the observed trends and variability of rainfall with extreme rainfall events, which are based on data for the past 100 years or more (Warwade et al. 2018; Guhathakurta et al. 2015; Guhathakurta et al. 2011; Guhathakurta and Rajeevan 2008 etc.). In the observed rainfall variability and changes over the state of Bihar, the analysis of rainfall patterns, trends and variability has been done based on the data for the past 30 years (1989–2018). The recent data of the past three decades has helped to present an idea of the recent changes to the state authorities for climate change adaptation and management (IMD 2020).

Figures 7.16 and 7.17 show the time series of annual rainfall in mm for June, July, August, September and the onset of the southwest monsoon season, respectively. The trend lines are also displayed for each series. The southwest monsoon season rainfall shows a significant decreasing trend, but neither monthly rainfall nor annual rainfall shows any significant increasing/decreasing trend. Although, a decreasing trend is observed in all the southwest monsoon months along with a decrease in the seasonal and annual mean.

During the past 30 years, the highest rainfall in June, July, August and September has been recorded in the years 2008 (291.8 mm), 2007 (548.9 mm), 1998 (377.1 mm) and 2016 (357.9 mm), respectively and the lowest rainfall of June, July, August and September have been recorded in the year 2009 (69.4 mm), 2013 (158.7 mm), 2016 (150.9 mm) and 1992 (86 mm), respectively. The highest southwest monsoon rainfall of 1349.3 mm and annual rainfall of 1516.8 mm were recorded in the year 2007 and the lowest southwest monsoon rainfall of 539.5 mm and annual rainfall of 632.5 mm in 2010 (IMD 2020).

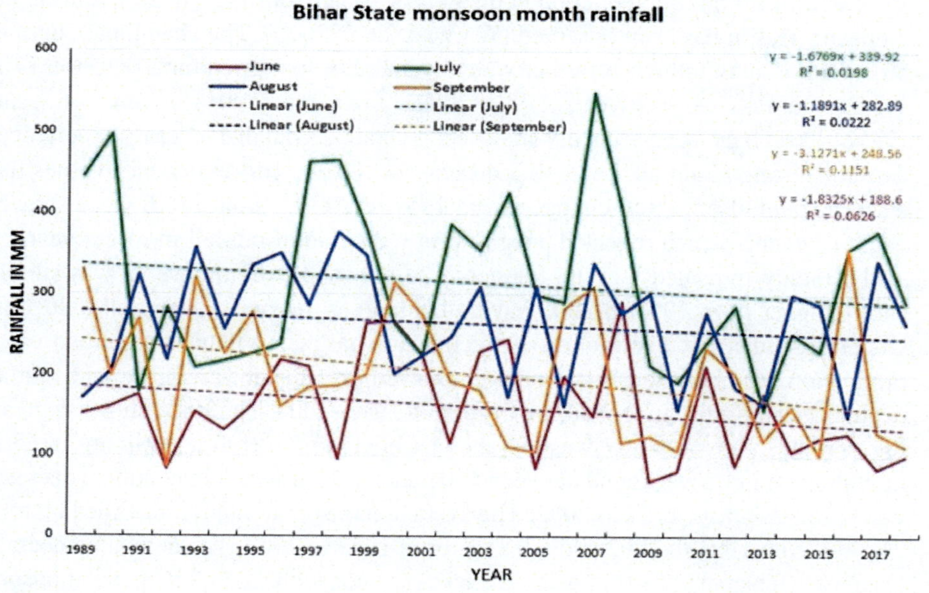

Fig. 7.16 Time series of rainfall in mm for June, July, August, September and trends

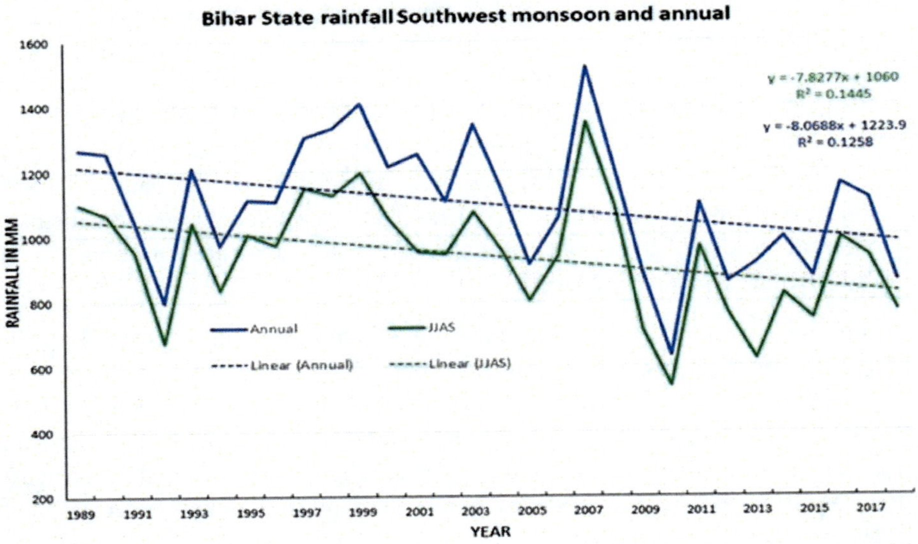

Fig. 7.17 Time series of rainfall in mm for the southwest monsoon season and annual trends

If we observe the linear trend in Figs. 7.16 and 7.17, a conspicuous decline in the annual precipitation in the past three decades is observed, which has several spikes of annual lows vis-a-vis highs as compared to the average annual precipitation. This establishes the fact that a shift in the climatic regime has been experienced by most districts in Bihar which has undoubtedly affected the surface runoff in major streams and also the groundwater recharge conditions within confined (shallow–deep) aquifers.

A significant rainfall reduction in districts of Bihar from 1957 to 2002 during the monsoon season has been observed (Warwade et al. 2018). The changing pattern of precipitation may influence further water availability for agricultural purposes. The average rainfall of 102 years (1901–2002), 56 years (1901–56) and 46 years (1957–2002) have been critically analysed. Overall, 26 out of 102 years experienced extreme negative rainfall, i.e. >10% decrease while 24 years experienced a positive event of rainfall of >10% change. From 1957 to 2002, 19 out of 46 years showed negative events which reflected a significant reduction of rainfall in recent years.

The study indicated that the frequency of negative extreme events was significantly higher from 1957 to 2002 having continuous 10 years of rainfall reduction. Uncertain and erratic distribution of precipitation has been a major limitation to crop production. Annual monsoonal rainfall showed significant variability and rainfall decline was most prominent during the period 1992–2002. The study shows decreasing rainfall trends at all 37 weather stations in Bihar. Overall a declining trend in rainfall magnitude was found between 1901 and 2002. The resultant annual decrease was −18.92% from 1957 to 2002. The average monsoonal rainfall declined up to −21.84% from 1901 to 2002 and −0.18% from 1957 to 2002. The annual percentage change in Jehanabad district was −28.69%, in monsoon (−34.93%), pre-monsoon (30.71%), and post-monsoon (7.86%) and winters (7.47%) from 1957 to 2002.

The declining trend in the past decade has changed the overall scenario of rainfall and the rainfall variability may impact future climatic conditions bringing about a conspicuous shift in the wake of climate change (Warwade et al. 2018).

Groundwater Status

In 2019, the groundwater level for Jehanabad District was 3.38 meters below ground level. Though the groundwater level of the district fluctuated substantially in recent years, it tended to decrease from 2015 to 2019 ending at 3.38 meters below ground level in 2019. In Jehanabad district in 2015–2016, 29% of wells showed a rise in water table, whereas 71% showed a drop during pre-monsoon, whereas in post-monsoon season 100% of wells showed a drop in water level (GWYB, Bihar; 2015–16). It has been observed that though the piezometric surface usually restores its average depth level, it is now the pressure on deeper confined aquifers that is causing a decline in groundwater repository with increasing population, tube well revolution, limited surface repositories and uncertain avenues of recharge.

Conclusions

The chapter brings to light the shifting climatic regime during the past three decades resulting in increasing temperatures and decreasing precipitation. The shift in the mood of the regional climate has given rise to unpredictability in weather conditions viz. overall precipitation, erratic and incessant rains having a bearing on the surface runoff, sedimentation pattern within the active channel of the river, groundwater condition and likewise.

The role of anthropogenic activities in affecting the ecological status of the rivers cannot be understated. The norms of mining and the guidelines therein thus need to be strictly followed for optimizing the natural resources in the present case, for example, sand (minor mineral) which has a huge economical utility.

The comparative temporal study also highlights the fate of a lease area continually worked out for its sand content and those allowed to replenish with time. The Google Image dated 18 April 2019 in regard to Sudaspur- Ghazipur sand *ghat* is worth a million words. The sand layer on the river bed is intact and is an eye-opener to the contrast earlier dated (archive) images show in regard to the health of the river in absence of anthropogenic interference. The fertility of the river in terms of sand content at other sites along the Phalgu River is either lost or is severely affected as compared to this site.

Whether the quantity of minor minerals extracted from the leased out area is replenished in totality is not known. In view of the erection of the barrage at Udherasthan, the release of water downstream is regulated. This needs to be clarified with a seasonal study carried out by the Irrigation department. Few of the sand ghats looked starved of the requisite sand content indicating reduced replenishment overexploitation.

Stray pits of irregular size and fashion have been observed which generally do not comply with planned excavation. The formation of benches of a height of 1 m and width of 0.9 m to prevent slope failure was not readily observed.

Comparative timeline imageries for the sand lease area have been studied in detail, and it is observed that there is a diminishing sediment load in the river channel from 2014 to 2019. It implies that the rejuvenation/replenishment has not been at par with the exploitation which is a critical environmental aspect to be addressed in regard to the effect of mining on the river.

As per State Level Environment Impact Assessment Authority (SEIAA) guidelines, Mining shall be done in layers of 1 m depth to avoid the ponding effect and after the first layer is excavated, the process is to be repeated for the next layers. However, this approach to excavation was not observed, rather the formation of irregular pits, a few shallow to even deep have affected the land topography and profile leading to the ponding effect.

The present excavation area is from the active river channel. But at a few sand ghat locations, the buffer of 3 m has not been maintained from the river bank.

Recommendations as a road map for preventive measures include:

- The absence of a distinct/clear boundary will leave loose ends to prevent rightful exploitation. Flagging the boundary of the area of interest is paramount to mining under open cast conditions as per norms.
- Assessment of the mineable sand in the defined area may be made mandatory. As the depth of interest doesn't go beyond 2–3 m, the same can also be ascertained by improvised hand auguring by physically lowering hollow pipes of 1 m, 2 m and 3 m to obtain core samples following which grain size analysis can yield the percentage of sand particles.
- Ground Penetration Radar is a non-invasive, non-destructive geophysical technique very useful for deciphering shallow subsurface conditions and may be used as a constructive tool for sub-surfaced condition investigation.
- Random drone aided surveys could be useful to check over-exploitation if any and maintain regular monitoring.
- The river carries sand to provide the resource and thus it is necessary to ascertain the rate of replenishment of the mineral. A regular replenishment study needs to be carried out to keep a balance between deposition and extraction.
- Mining within or near the river bed has a direct impact on the stream's physical characteristics, such as channel geometry, bed elevation, substratum composition and stability, in-stream roughness of the bed, flow velocity, discharge capacity, and sediment transport capacity, turbidity, temperature, etc. Alteration or modification of the above attributes may cause an impact on the ecological equilibrium of the riverine regime, and disturbance in channel configuration and flow paths. This may also harm instream biota and riparian habitats. Thus, staggered replenishment surveys must be carried out to ascertain the extent of sediment revival.

Here also services of geoscientific and engineering-based scientific organizations can be sought.

- There is a need to develop a Standard Operating Procedure (SOP) that would be terrain specific for harnessing minor minerals through systematic river auditing. This will have room for a mechanism to evaluate loss to the ecology and to recover the cost of restoration of such damage through well-devised mechanisms.
- As per State Level Environment Impact Assessment Authority (SEIAA) guidelines, no quarrying is to be done within 5 m from both banks of the river, the quarrying is to be prohibited within 100 m u/s and d/s from any dam, weir or diversion structure erected for irrigation purposes. No quarrying would be permitted within a 46 m distance from the flood control embankment. The quarrying of sand is to be prohibited within 300 m on both sides of the railway bridge or bridge falling under the National Highway/State Highway and also 100 m on both sides of any other bridge. These guidelines are relevant to the area of study; however, the exhaustive stipulations are documented in the SEIAA document, Bihar. *Enforcement & Monitoring Guidelines for Sand Mining* from the Ministry of Environment, Forest and Climate change, 2020 brings out the norms for strict adherence.

Acknowledgements The authors wish to express their gratitude to *Shri R.S. Garkhal*, Director General, GSI for extending support to carry out the work. Special thanks to *Dr Ranjit Rath*, DG, GSI, (formerly) for his continued support and encouragement. *Dr Rajesh Asthana*, Dy. D.G., State Unit Bihar (formerly), Eastern Region, GSI, has been the main motivating factor for realizing this publication and without his support, it would not have been feasible. Gratitude goes to *Shri Sanjay Kumar Dutta*, Deputy Director General (Additional Charge), State Unit Bihar, GSI, Patna who has been a source of real encouragement and support towards all academic endeavours.

Special thanks to the District Administration, Jehanabad district, Bihar, DGM, Government of Bihar, Director, CGWB and Shri Bimal Kumar, Superintending Engineer, Irrigation Department, Govt. of Bihar for their contribution at various stages of work and final accomplishment.

References

Achyuthan H (2003) Bedrock Channel incision in the arid tracts of Jaisalmer, Rajasthan. Proc Indian Natl Sci Acad Part A 69(2):241–250

Arnaud F, Piegay H, Schmitt L, Rollet A, Ferrier V, Beal D (2015) Historical geomorphic analysis (1932–2011) of a by-passed river reaches in process-based restoration perspectives: the old Rhine downstream of the kembs diversion dam (France, Germany). Geomorphology 236 (Supplement C):163–177. http://www.ciencedirect.com/science/article/pii/S0169555X15000926

Ayaz S, Biswas M, Dhali MK (2018) Morphotectonic analysis of alluvial fan dynamics: a comparative study in Spatio-temporal scale of Himalayan foothill, India. Arab J Geosci. Springer. https://doi.org/10.1007/s12517-017-3308-2

Batalla R, Vericat D, Martínez T (2006) River-channel changes downstream from dams in the lower Ebro River. Z Geomorphol 143(Suppl. B):1–14

Belletti B, Dufour S, Piégay H (2015) What is the relative effect of space and time to explain the Braided River width and island patterns at a regional scale? River Res Appl 31:1–15. https://doi.org/10.1002/rra.2714

Besné P, Ibisate A (2015) River channel adjustment of several rivers reaches on Ebro basin. Quat Int 364:44–53

Cincotta RP, Wisnewski J, Engelman R (2000) Human population in the biodiversity hotspots. Nature 404:990–992

Comiti F, Da Canal M, Surian N, Mao L, Picco L, Lenzi M (2011) Channel adjustments and vegetation cover dynamics in a large gravel-bed river over the last 200years. Geomorphology 125:147–159

Gaillot S, Piegay H (1999) Impact of gravel-mining on stream channel and coastal sediment supply: example of the Calvi Bay in Corsica (France). J Coast Res 15(3):774–788

Goswami DC (1998) The fluvial regime and flood hydrology of the Brahmaputra River, Assam. In: Kale VS (ed) Flood studies in India, vol 41. Geological Soc India Memoir, Bangalore, pp 53–76

Guhathakurta P, Rajeevan M (2008) Trends in rainfall pattern over India. Int J Climatol 28:1453–1469

Guhathakurta P, Sreejith OP, Menon PA (2011) Impact of climate changes on extreme rainfall events and flood risk in India. J Earth Syst Sci 120(3):359–373

Guhathakurta P, Rajeevan M, Sikka DR, Tyagi A (2015) Observed changes in southwest monsoon rainfall over India during 1901–2011. Int J Climatol 35:1881–1898

Gupta S (1997) Himalayan drainage patterns and the origin of fluvial megafans in the Ganges foreland basin. Geology 25:11–14

Hossain MA, Gan TY, Baki ABM (2013) Assessing morphological changes of the Ganges River using satellite images. Quat Int 304:142–155

India Meteorological Department (2020) Observed Rainfall Variability and Changes over Bihar State, Met Monograph No.: ESSO/IMD/HS/Rainfall Variability/04(2020)/28

Kale VS, Hire P, Baker VR (1997) Flood hydrology and geomorphology of monsoon dominated rivers: the Indian Peninsula. Water Int 22:259–265

Lbisate A, Ollero A, Diaz E (2011) Influence of catchment processes on fluvial morphology and river habitats. Limnetica 30(2):169–182

Liebault F, Piegay H (2002) Causes of 20th-century channel narrowing in mountain and Piedmont Rivers of southeastern France. Earth Surf Process Landform 27:425e444

Petit F, Poinsart D, Bravard J-P (1996) Channel incision, gravel mining and bedload transport in the Rhone River upstream of Lyon, France ("Canal de Miribel"). Catena 26(3):209–226. http://www.sciencedirect.com/science/article/pii/034181629500047X

Prizomwala SP, Gandhi D, Ukey VM, Bhatt N, Rastogi BK (2015) Coastal boulders as evidences of high-energy marine events from Diu Island, west coast of India: storm or palaeotsunami? Nat Hazards 75(2):1187–1203

Scorpio V, Aucelli PP, Giano SI, Pisano L, Robustelli G, Rosskopf CM, Schiattarella M (2015) River channel adjustments in southern Italy over the past 150 years and implications for channel recovery. Geomorphology 251:77–90

Selander J (2004) Chapter 4: Influences on river morphology in a sediment-dominated system, University of Oregon B.S. thesis, Eugene, Oregon, Chapter Four, pp 1–13

Singh M (1996) The Ganga River: fluvial geomorphology, sedimentation, processes and geochemical studies. Ruprecht-Karls- Universitat, Heidelberg

Starkel L, Sarkar S, Soja R, Prokop P (2008) Present-day evolution of the Sikkimese Bhutanese Himalayan Piedmont. In: Geographical studies, vol 219. Institute of Geography and Spatial Organization PAS, Warsaw

Surian N (1999) Channel changes due to river regulation: the case of the Piave River, Italy. Earth Surf Process Landf 24:1135–1151

Surian N, Rinaldi M (2003) Morphological response to river engineering and management in alluvial channels in Italy. Geomorphology 50(4):307–326

Vikram Kumar (2019) The mighty Phalgu River dries up. Acta Sci Agric 3(9):01–02

Warner RF (2000) Gross channel changes along the Durance River, southern France, over the last 100 years using cartographic data. Regul Rivers Res Manag 16(2):141–157

Warwade P, Tiwari S, Ranjan S, Chandniha SK, Adamowski J (2018) Spatio-temporal variation of rainfall over Bihar State, India. J Water Land Dev 36:183–197. https://doi.org/10.2478/jwld2018-0018

Wiejaczka L (2016) Riverbeds level changes in the margin and foreland of the Darjeeling Himalaya during the years with a normal monsoon rainfall. In: Singh RB, Prokop P (eds) Environmental geography of South Asia, advances in geographical and environmental sciences. https://doi.org/10.1007/978-4-431-55741-8_5

Wolf D, Faust D (2016) River braiding caused by rapid floodplain deformation–insights from Holocene dynamics of the Jarama River in Central Spain. Quat Int 407:126–139

WyZga B (2007) 20 a review on channel incision in the polish Carpathian rivers during the 20th century. In: Habersack H, Piegay H, Rinaldi M (eds) Gravel-bed Rivers VI: from process understanding to river restoration, Developments in earth surface processes, vol 11. Elsevier, pp 525–553

Yanagida M (2016) Landform development of mountains and rivers in Japan. Quat Int 397:223–233

Žibret G, Žibret L (2017) River gradient anomalies reveal recent tectonic movements when assuming an exponential gradient decrease along a river course. Geomorphology 281:43–52

Part IV
Land (Geology, Archeology/Paleo/Tectonics)

Chapter 8
Impact of Contrasting Climate on the Intensity of Chemical Weathering and Maturity of Sediments in Different Latitudes: A Geochemical Study from Peninsular and Extra Peninsular India

Ishfaq Ahmad Mir

Abstract Chemical weathering and sediment maturity under contrasting climatic conditions from tropical peninsular and temperate extra peninsular parts of India are investigated. Rocks of peninsular India are moderate to intensively weathered, while rocks of extra peninsular India are moderately weathered. High chemical weathering conditions in south India appear to result from climate forcing related to high rainfall and temperature. The distribution of C-values and Sr/Cu ratios collectively indicate a generally humid climate in peninsular India and a semi-humid to dry climate in extra peninsular India. There is an increase in humidity from higher to lower latitudes. The compositional maturity of the sediments is evaluated by using the ICV and SiO_2/ $Al2O_3$ ratio. The results collectively indicate that the sediments in the peninsular area are mineralogically more mature than the sediments of the extra peninsular area. There is a decrease in sediment maturity from higher to lower latitudes meaning a high degree of clays and mineralogical maturity.

Keywords Geochemistry · Sediments · Climate · Weathering · Maturity

Introduction

Geochemical investigations carried out from stream and slope wash sediments are used in understanding the geology of the catchment area, mineralization, environment, agriculture, land use studies, and in understanding the sedimentary processes like weathering, erosion, sediment sources, depositional mechanisms and tectonics (Mir and Mir 2019). Under the National Geochemical Mapping (NGCM) program, the Geological Survey of India (GSI) is carrying out geochemical mapping of the entire country intending to generate a multi-elemental database finding use in multi

I. A. Mir (✉)
Geological Survey of India, State Unit: Karnataka and Goa, Bengaluru, Karnataka, India

N. Khare (ed.), *Science, Policies and Conflicts of Climate Change*, Springer Climate,
https://doi.org/10.1007/978-3-031-16254-1_8

geoscientific applications. In pursuance of this program in the year 2014–15 and 2017–18, geochemical mapping was carried out on a 1:50,000 scale in SOI toposheet nos. 43 J/11 and 43 J/12 (part) covering Bandipora–Ganderbal areas, Kashmir valley, western Himalaya and in toposheet no. 48I/11 covering parts of Belgaum, Dharwad and Uttar Kannada districts of Karnataka state of peninsular India, respectively.

The climatic conditions under which weathering and erosion take place influence the sediment composition (Pattan et al. 2012; Mir et al. 2013). Chemical weathering brings changes in compositions, mineral phases and textural features compared to the fresher parental rock and is mainly related to geology, vegetation, temperature, precipitation and geomorphological conditions (Scarciglia et al. 2008). Over a longer geological time scale, chemical weathering is influenced by tectonism and climate and is an important feature of the global element cycles (Giovanoli et al. 1988). Labile minerals, like feldspars, are easily chemically weathered under hot and humid conditions in lower latitudes while physical weathering is more intense under semi-arid and glacial conditions in higher latitudes. The geochemistry of the weathered rocks gives important information about the degree of weathering recorded in different geochemical proxies such as the Chemical Index of Alteration (CIA; Nesbitt and Young 1982), Chemical Index of Weathering (CIW; Harnois 1988), Plagioclase Index of Alteration (PIA; Fedo et al. 1995), the Weathering Index of Parker (WIP; Parker 1970), the ratio of Rb/Sr (McLennan et al. 1993); climatic conditions recorded by geochemical proxies of C-value (Zhao et al. 2007), and Sr/Cu (Meng et al. 2012); and the sediment maturity recorded in proxies of the Index of Composition Variability (ICV; Cox et al. 1995), and the ratio of Si/Al (Potter 1978). In the present study, an attempt is made to understand the impact of contrasting climatic conditions on the degree of chemical weathering and sediment maturity in peninsular and extra peninsular India by investigating the geochemical characteristics of the stream and slope wash sediments. The Dharwad-Belgaum area from the peninsular part of Karnataka state and the Bandipora–Ganderbal area from the extra peninsular part of the union territory of Jammu and Kashmir within the republic of India are selected for this study.

Previous Work

Only a limited number of geochemical studies of sediments have been carried out in the valley of Kashmir (VOK). Kapoor (1963) carried out geochemical studies of Lidder stream sediments. Rashid et al. (2013) studied the geochemistry of the Wular Lake sediments. Bhat and Bhat (2014) carried out geochemical studies of the recent sediments of the Pulwama district to understand the mineralogy, environment and sedimentary processes. Mir and Jeelani (2015) studied the geochemical characteristics of the bottom sediments of the River Jhelum to understand the pollution and other geochemical characteristics of river sediments. The sediments of the Manasbal and Wular lakes have been studied recently by Rashid et al. (2015) for environmental aspects. Babeesh et al. (2017) carried out a geochemical study of the Manasbal lake sediments to understand the sedimentary processes such as weathering, provenance, sorting, maturity and the tectonic setting of the catchment area. Shah et al. (2017) studied the physicochemical characteristics of the Wular lake sediments to understand the geological processes in the lacustrine setup. Mir and Mir (2015, 2019) carried out inorganic geochemical spatial distribution studies from the Bandipora–Ganderbal area to understand the mineralization, geo-environment, geo-hazards, provenance, weathering, sorting and maturity of recent sediments.

In the Belgaum–Dharwad area, there are only a limited number of geochemical studies of sediments carried out. Bruce Foote (1876, 1888) had described the major lithology of the Dharwar Group of rocks as schistose. Slater (1906) described these rocks as pseudo-greywacke associated with beds of magnetite quartzite and mica schists. Gururaja Rao and Sinha (1976) mapped a large area in the Dharwar district and described the metasedimentary sequence of greywacke, shale, ferruginous shale, phyllite, quartz-mica schist as geosynclinal flysch sediments. Iyengar (1976) mapped the stratigraphy, structure and correlation of the Dharwar Supergroup. Hegde and Chavadi (2008) studied the geochemistry of the late Archaean meta-greywacke from the Western Dharwar Craton for understanding implications for the provenance and nature of the Late Archaean crust. Bhat et al. (2019) studied the geochemical composition of stream and slope wash sediments of the Belgaum, Dharwad and Uttar Kannada areas to understand the mineralization, geo-environment, geo-hazards, provenance, weathering, sorting and maturity of recent sediments.

Study Area

The extra peninsular study area covers parts of the Bandipora and Ganderbal districts of Kashmir valley, India. The study area lies between 34′10° to 34′30°N latitudes and 74′30° to 74′45°E longitudes. The area of investigation covers an area of about 800 km^2 within toposheet number 43 J/11 and part of 43 J/12 (Fig. 8.1). The topography of the study area consists of flat to hilly areas. The N-NE hilly area reaches the height of 5142 meters above sea level (masl) whereas the W-SW flat area drops to an elevation of 1550 masl. The Jhelum River and its tributaries like Sindh, Erin, Madhumati and Ningli form the major drainage network of the study area. The study area also consists of many water bodies and wetlands. Wular and Manasbal are the major lakes, whereas the Haigam, Takia, Tulmul and Naugam are the other wetlands of the area.

The peninsular study area covers the parts of the Belgaum, Dharwad and Uttar Kannada districts in the Western Ghats of Karnataka, India. The study area is bounded by 15°15′ to 15°30′N latitudes and 74°30′ to 74°45′E longitudes. The area of investigation covers an area of about 800 km^2 within the toposheet number 48I/11 (Fig. 8.1). The mapped area is an NW-SE trending hilly terrain with steeply rising hills and narrow valleys towards the western side and plain land towards the eastern side, forming a part of Western Ghats. The highest elevation is 891 masl and

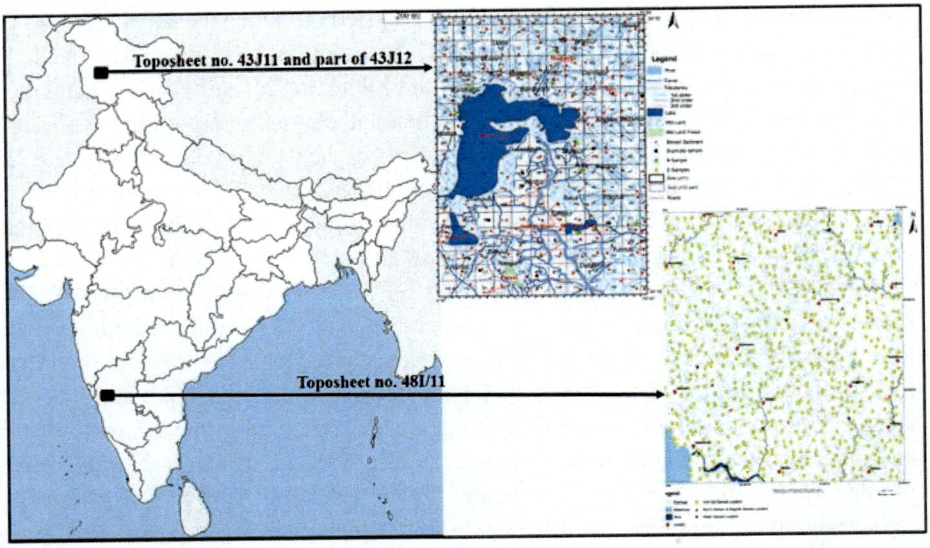

Fig. 8.1 Location map of the study areas. The location of sampling points from the Bandipora–Ganderbal area (extra peninsular India) is shown in the top right and the Belgaum–Dharwad–Uttar Kannada area (peninsular India) is shown in the bottom right side of the figure

the lowest elevation is 526 masl. The area is mainly drained by the Kali River and its tributaries, Barchi, Pandhari, Karkia, Tatti and Gundoli Hallas.

Geological Setting

The extra peninsular study area is covered on the northern side by the Pohru group of rock formations, including limestone, siltstone, shale and arenite of the Cambrian–Silurian and agglomeratic slates of the Upper Carboniferous age (Singh 1982). The Panjal volcanic and Triassic limestone formations are exposed on the NE and NW sides, whereas the Quaternary Karewa deposits and recent alluvium are present in the SE, S and SW. Panjal volcanics are of the Permian age and consist mainly of distinctly layered andesitic to basaltic lava flows. Triassic limestone consists of a thick sequence of compact blue limestone, slates and dolomites (Wadia 1931). The low-lying area is filled with sediments of the Karewa group composed of fine silty clays with sand and boulder gravel conglomerates and recent alluvium (De Terra and Paterson 1939) made up of fine particles of silt and clay and coarser particles of sand and gravel. The geological map of the study area modified after Thakur and Rawat (1992) is shown in Table 8.1.

The peninsular study area lies in Western Dharwar Craton (WDC) of the Indian shield (Swami et al. 1976) and consists of gneissic rocks of the Peninsular Gneissic Complex (PGC); quartzite, limestone, phyllite, argillite, banded iron formation (BIF), Dharwar supergroup schist belts of the late Archean age, dolerite intrusives

Table 8.1 Stratigraphic set up of study area, SOI toposheet no. 43 J11 and part of 43 J12, after Thakur and Rawat (1992)

Age	Group	Formation	Lithology
Holocene	–	Recent Deposit	Alluvium, Scree
Lower Pliocene-Pleistocene	Karewa	Dilpur	Loess, Silt
Upper Permian-Triassic	Vihi	Khrew	Limestone, Dolomite, Quartizite
Permian		Panjal Volcanics	Andesite, Basalt
Upper Carboniferous		Agglomeratic Slate	Pebbly Slate, Quartz Arenite, Congalomerate
		Trehgam	Limestone, Siltstone, Shale
Camberian-Silurian	Pohru	Natnus	Shale, Siltstone, Arenite

Table 8.2 Stratigraphic set up of study area, SOI toposheet no. 48I/11, after Murthy and Mehrotra (1979)

Age	Supergroup	Group	Formation	Lithology
Palaeo-Proterozoic		Younger Intrusives	Basic Intrusives	Dolerite
Archean	Dharwar	Chitradurga	Ranibennur	Quartz-Sericite Schist, Quartzite, Argillite.
			Joldal	Banded Iron Formation, Argillite, Limestone, Manganiferous Phyllite.
			Jhandimatti	Ortho-Quartzite.
	Peninsular Gneissic Complex (Basement)			Granite Gneisses-I type.

of the Palaeo-Proterozoic age. The PGC forms much of the WDC and is made up of tonalitic-trondhjemitic gneiss with many inclusions of older sedimentary and igneous rocks. The geological map of the study area after Murthy and Mehrotra (1979) is shown in Table 8.2.

Climate

The extra peninsular Bandipora–Ganderbal area has a temperate climate and is classified as 'Cfb' as per the Köppen climate classification (Koppen 2011). The climate here falls under the sub-Mediterranean type with four distinct seasons (Mir and Mir 2019). The climate is hot in summer, cold in winter and mild in spring and autumn seasons. The hottest month is July with an average temperature of 21.3 °C and the coldest month is January with a minimum temperature of 0.3 °C (https://en.climate-data.org). Precipitation occurs throughout the year and no month is particularly dry. Precipitation in the winter months usually falls in the form of snow. In winters, precipitation occurs from the western disturbances originating from the Mediterranean Sea and during the summer season precipitation occurs from south-

Materials and Methods

The sample collection from both peninsular and extra peninsular areas was carried out by following the guidelines as per the standard operating procedure (SOP 2014) of the Geological Survey of India (GSI) for the National Geochemical Mapping (NGCM) mission. For sampling purposes, the 2x2 gridded toposheets were used as the base maps. Stream sediment (SS) samples were collected from the first and second second-order streams and in the area devoid of streams, slope wash (SW) sampling was carried out. The fine-to-medium grained sediments (~5 kg) were collected at each sampling site, and actual coordinates and other field-related information such as lithology, land-use land cover, sample material type and other details were recorded. A total of 200 sediment samples along with 10 duplicate samples for cross-check analysis were collected from both of the study areas. The samples were air-dried, deplumed, powdered and sieved to 120-micron mesh size. The samples were then coned, quartered and packed into two sets each of 500 g quantity. One set was submitted to GSI, Chemical Laboratory, Lucknow (northern region) and Hyderabad (southern region) for chemical analysis, whereas the other set has been stored in the sample repositories maintained at GSI, Jammu and GSI Bengaluru offices.

The major oxides and a few trace elements of the composite samples were measured by M/S Panalytical; MAGIX, 2.4 KW Sequential XRF Spectrometer instruments at GSI chemical laboratories located at Lucknow and Hyderabad. The details are discussed in Mir and Mir (2019) and Bhat et al. (2019). The detection limit is 0.1% for major elements and 1.00 ppm for trace elements. Total iron is determined as Fe_2O_3. The precision of the instrument is ±10%. A few trace and rare earth elements (REE) were analysed using a Perkin Elmer Sciex ELAN 6100 inductively coupled plasma-mass spectrometer (ICP-MS) facility of GSI chemical laboratories at Lucknow and Hyderabad. The fusion method was performed for analysis. The details are discussed in Bhat et al. (2019). The method has quantification limits from 0.1 to 1 ppm and the precision of the instrument is ±10%. Standard reference material (SRM) (GSD10) with known element concentrations was analysed after each batch of 20 samples for accuracy and duplicate samples after each batch of 10 samples were analysed for repeatability. The accuracy of the measurement is ±3% and the precision is ±5%. Geochemical data processing, statistical analysis and plotting were carried out by using the M/S Excel and Origin Pro 2016 software.

The Plagioclase Index of Alteration (PIA; Fedo et al. 1995) and the ratio of Rb/Sr (McLennan et al. 1993) have been calculated to understand the intensity of chemical weathering in the study area; climatic conditions in the area have been calculated from geochemical proxies of C-value (Zhao et al. 2007) and Sr/Cu (Meng et al. 2012); and the sediment maturity is calculated from the proxies of the Index of Composition Variability (ICV; Cox et al. 1995), and the ratio of Si/Al (Potter 1978).

$$PIA = [(Al_2O_3 - K_2O)/(Al_2O_3 + CaO + Na_2O - K_2O)] \times 100 \quad (8.1)$$

$$C - value = \Sigma\,(Fe + Mn + Cr + Ni + V + Co)/\Sigma\,(Ca + Mg + Sr + Ba + K + Na) \quad (8.2)$$

$$ICV = Fe_2O_3 + K_2O + Na_2O + CaO^* + MgO + TiO_2/Al_2O_3 \quad (8.3)$$

CaO* represents the amount of CaO added in the silicate fraction of the rock (Zhao et al. 2007).

Results

The geochemical results of major and trace elements of the surface sediments of the Bandipora–Ganderbal area are discussed in Mir and Mir (2019) and the results from the Belgaum–Dharwad–Uttar Kannada area are discussed in Bhat et al. (2019). The descriptive statistics of weathering indices, climate proxies and sediment maturity indicators used in this study are given in Table 8.3. In the northern Bandipora–Ganderbal area CIA varied from 26.41 to 77.79 with a mean of 65.29, PIA varies from 24.76 to 85.59 with a mean of 70.63, Rb/Sr varies from 0.10 to 1.48 with a

Table 8.3 Descriptive statistics of weathering indices, climate proxies and sediment maturity indicators used in this study, CIA: *Mir and Mir (2019), #Mir et al. (2021)

	Standard deviation	Variance	Skewness	Kurtosis	Minimum	Maximum	Mean
Bandipora area data							
CIA*	11.25	126	−1.69	2.47	26.41	77.79	65.29
C-Value	0.11	0.01	1.20	3.72	0.25	1.01	0.50
Sr/Cu	4.74	22.53	4.66	28.56	1.75	44.7	5.08
SiO_2/Al_2O_3	0.44	0.19	−0.96	0.91	2.94	5.14	4.37
ICV	0.51	0.26	3.05	10.19	0.91	3.88	1.29
PIA	14.50	210	−1.47	1.60	24.76	85.59	70.63
Rb/Sr	0.29	0.08	−0.49	0.02	0.10	1.48	0.84
Dharwad area data							
CIA#	5.07	25.71	0.24	−0.13	67.69	94.01	80.76
C-Value	0.43	0.19	3.29	14.71	0.18	3.41	0.70
Sr/Cu	3.11	9.70	1.37	1.72	0.25	15.14	4.50
SiO_2/Al_2O_3	0.98	0.96	−0.26	−0.13	1.77	6.82	4.66
ICV	0.16	0.02	1.04	3.91	0.45	1.64	0.78
PIA	5.16	26.68	−0.09	−0.65	74.58	97.26	86.95
Rb/Sr	0.49	0.24	1.28	1.28	0.13	2.47	0.70

mean of 0.84, C-value varies from 0.25 to 1.01 with a mean of 0.50, Sr/Cu varies from 1.75 to 44.7 with a mean of 5.08, ICV varies from 0.91 to 3.88 with a mean of 1.29 and SiO_2/Al_2O_3 varies from 2.94 to 5.14 with a mean of 4.37. In the southern Belgaum–Dharwad–Uttar Kannada area CIA varied from 67.69 to 94.01 with a mean of 80.76, PIA varies from 74.58 to 97.26 with a mean of 86.95, Rb/Sr varies from 0.13 to 2.47 with a mean of 0.70, C-value varies from 0.18 to 3.41 with a mean of 0.70, Sr/Cu varies from 0.25 to 15.14 with a mean of 4.50, ICV varies from 0.45 to 1.64 with a mean of 0.78 and SiO_2/Al_2O_3 varies from 1.77 to 6.82 with a mean of 4.66.

Discussion

Sediment geochemistry based on major and trace elements is the most practicable for defining the weathering degree of source sediment to interpret the weathering intensity, sediment maturity and climatic conditions. In this work, surface sediment geochemical data of granitic terrain of peninsular India (Belgaum–Dharwad–Uttar Kannada area of Karnataka) were compared to those of basaltic and carbonate terrain of extra peninsular India (Bandipora–Ganderbal area of Kashmir) to evaluate the generalized picture of weathering history in the drainage basins from lower and higher latitudes instead of direct and quantitative proxies. Degree of weathering, sediment maturity and climate of the area were determined based on selected major and trace element proxies. The geochemical weathering parameters (PIA and Rb/Sr), sediment maturity parameters (ICV and SiO_2/Al_2O_3) and climate reconstruction parameters (C-value and Sr/Cu) for the considered area are given in Table 8.3.

Weathering Variations

The litho-units in the study areas experienced diverse weathering conditions recorded by CIA (derived from Mir and Mir 2019; Mir et al. 2021), PIA and Rb/Sr (Fig. 8.3). The Belgaum–Dharwad–Uttar Kannada area shows medium to high CIA values (67.69 to 94.01; Table 8.3), and the area has experienced a moderate to very intensive chemical weathering alteration. The Bandipora–Ganderbal area shows low to medium CIA values (26.41 to 77.79; Table 8.3), the area has experienced a low to moderate chemical weathering alteration, some parts have not experienced any chemical weathering as evidenced by CIA values of <50. Similarly, PIA and Rb/Sr values from this study (74.58 to 97.26 and 0.13 to 2.47; Table 8.3) for the Belgaum–Dharwad–Uttar Kannada area confirm moderate to very intensive chemical weathering alteration. On the other hand, the Bandipora–Ganderbal area shows low to moderate PIA and Rb/Sr values (24.76 to 85.59 and 0.10 to 1.48; Table 8.3) confirming low to moderate chemical weathering alteration. PIA values in unaltered rocks are between 0 and 50 and for altered rocks are between

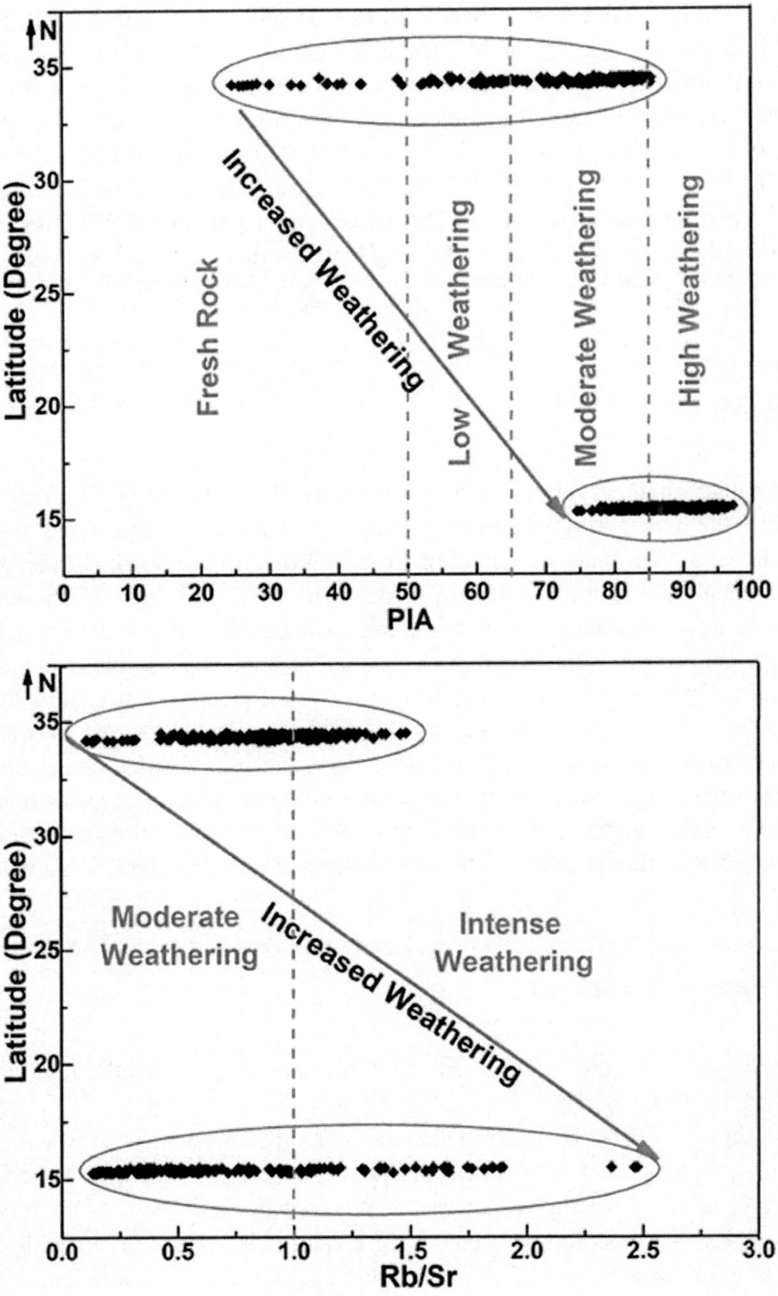

Fig. 8.3 Comparison of PIA (left) and Rb/Sr ratio (right) from different latitudes of India as a proxy for variation in chemical weathering intensity

50 and 100. The PIA index is used to evaluate the degree of weathering of plagioclase minerals in host rocks. The PIA values of the Belgaum–Dharwad–Uttar Kannada area are between 74.58 and 97.26 and for the Bandipora–Ganderbal

area sediments are between 24.76 and 85.59, indicating extra peninsular sediments contain better-preserved plagioclase than the peninsular sediments. PIA values are higher than CIA values in sediments from both areas due to differential rates of K-feldspar and plagioclase weathering (Goldich 1938). Strong chemical weathering of rock increases the Rb/Sr ratio; therefore, ratios >1 are indicators of intense chemical weathering (McLennan et al. 1993). The Rb/Sr values of the Belgaum–Dharwad–Uttar Kannada area are between 0.13 and 2.47 and for the Bandipora–Ganderbal area sediments are between 0.10 and 1.48, supporting the low to moderate chemical weathering for extra peninsula sediments and moderate to high chemical weathering for peninsula sediments of granitic rocks of south India.

Climate Conditions

Latitudinal differences in climate are an important factor for the variations in the intensity of chemical weathering of the source area for the Indian peninsular and extra peninsular sediments. The weathering index values of the peninsular sediments are higher, and the climate is likely more humid in the Belgaum–Dharwad–Uttar Kannada area in comparison with the Bandipora–Ganderbal area. The composition and distribution of a few major and trace elements in the sediments can show the climatic variations (Hu et al. 2017). Mn, Fe, Cr, Co, Ni and V are relatively enriched in sediments under humid conditions and Ca, Mg, K, Na, Sr and Ba are relatively enriched under arid conditions (Cao et al. 2012). Zhao et al. (2007) successfully performed the C-value (Eq. 8.2) as a proxy for climate change. The C-values for the Bandipora–General area sediments range from 0.25 to 1.01 and for the Belgaum–Dharwad–Uttar Kannada area sediments range from 0.18 to 3.41 (Table 8.3). The C-value results reflect a generally semi-arid to the semi-moist climate in the Bandipora–Ganderbal area and semi-arid to the moist climate in the Belgaum–Dharwad–Uttar Kannada area (Fig. 8.4). There is an increase in humidity from higher to lower latitudes. The climate in tropical areas is more humid than in temperate areas. Similarly, the Sr/Cu ratio has been used for climatic studies of different regions (Meng et al. 2012). A low Sr/Cu ratio indicates a warm and humid climate, whereas a high Sr/Cu ratio indicates a cold and semi-humid climate (Jia et al. 2013). The ratio of Sr/Cu between 1.3 and 5.0 indicates a warm and humid climate, whereas the ratio of Sr/Cu >5.0 indicates a hot to cold and arid to semi-arid climate (Lerman 1978). The Sr/Cu ratios of the Bandipora–General area sediments range from 1.75 to 44.7 and for the Belgaum–Dharwad–Uttar Kannada area sediments range from 0.25 to 15.14 (Table 8.3). This distribution of Sr/Cu ratios indicates dry to semi-arid and cold to wet climate conditions in extra peninsular India and semi-arid to humid and warm climate conditions in peninsular India (Fig. 8.4). From the Sr/Cu plot, it is also evident that there is an increase in humidity from higher temperate to lower tropical latitudes. The C-values and Sr/Cu ratios collectively indicate generally humid climate conditions in southern India and semi-humid climate in northern India. The climate proxies based on C-values and Sr/Cu

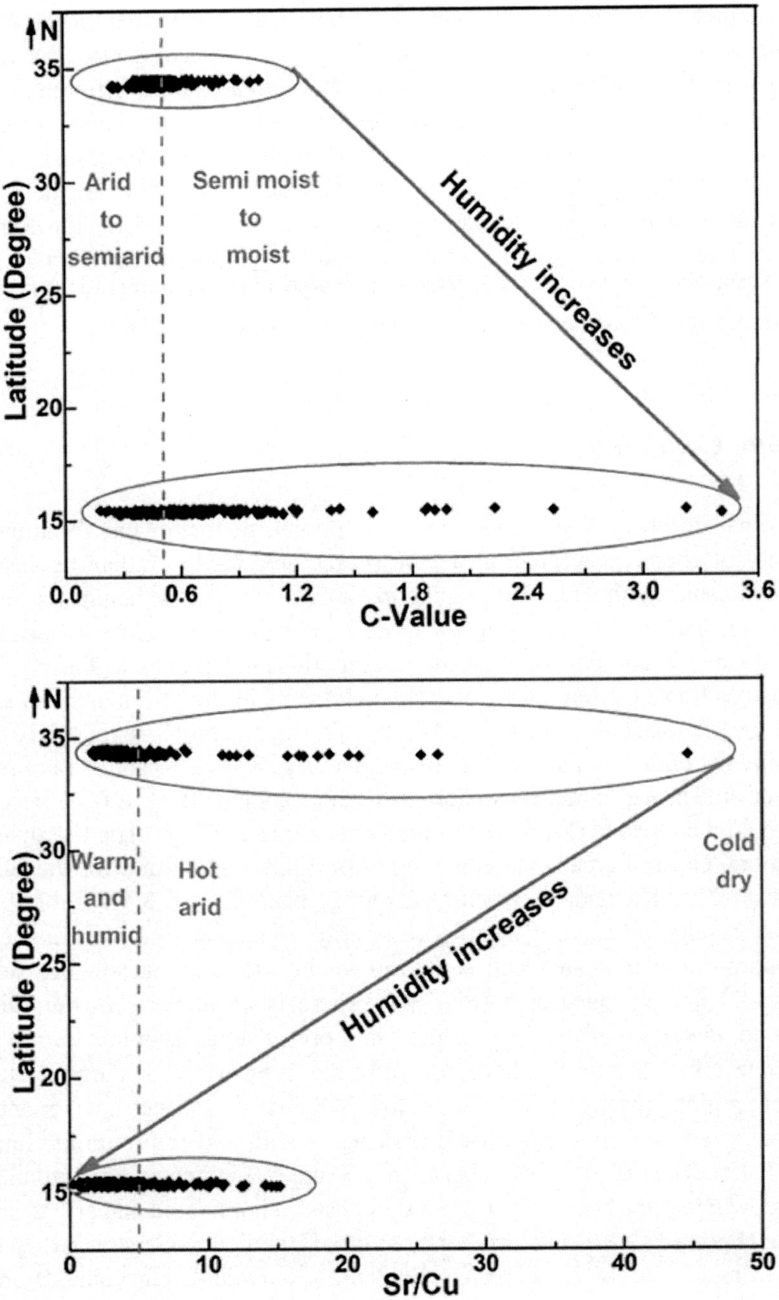

Fig. 8.4 Comparison of C-value (left) and Sr/Cu ratio (right) from different latitudes of India as a proxy for variation in climate

match very well with the meteorological data of the area (Fig. 8.2, https://en.climate-data.org). The impact of varying climates is very well reflected in the weathering indices also (Fig. 8.3), confirming a very important role of regional climatic conditions on the intensity of chemical weathering in the different parts of the world.

Sediment Maturity

The compositional maturity of the sediments is evaluated by using the Index of Compositional Variability (ICV; Cox et al. 1995) and the SiO_2/Al_2O_3 (Potter 1978) ratio. An approach to assessing detrital mineralogy is to use the ICV (Eq. 8.3). Matured sediments with more clayness show lower ICV values (<1.0), such sediments are mostly derived from a cratonic environment (Cox et al. 1995). The results of the ICV in the Bandipora–Ganderbal area range between 0.91 and 3.88 and in the Belgaum–Dharwad–Uttar Kannada area range between 0.45 and 1.64 (Table 8.3) indicating that the sediments in the peninsular area are mineralogically more mature than the sediments of extra peninsular area (Fig. 8.5). In this study, there is a decrease in ICV values from higher to lower latitudes meaning a high degree of clays and mineralogical maturity. Tropical area sediments are rich in clay minerals, due to the alteration of feldspars into clay minerals under hot and wet climatic conditions, and have lower ICV values (Cox et al. 1995) than sediments rich in non-clay silicate minerals of temperate area sediments. Immature sediments have high ICV values (>1.0) indicating a single weathering cycle, whereas the mature sediments have low ICV values (<1.0) suggesting multiple cycles of chemical weathering, erosion and redeposition (Cox et al. 1995). Highly weathered sediments show low ICV values and high PIA values (Figs. 8.5 and 8.3). The maturity of the sediments is also reflected by the ratio of SiO_2/Al_2O_3 (Potter 1978). The higher the SiO_2 content in sediments, the lower the degree of clays. A high ratio of SiO_2/Al_2O_3 indicates mineralogically mature sediments and a low ratio indicates chemically immature sediments. The Belgaum–Dharwad–Uttar Kannada area has SiO_2/Al_2O_3 values between 1.77 and 6.82 (Table 8.3), indicating that the sediments are formed under humid conditions tending towards increasing chemically maturity. The Bandipora–Ganderbal area shows SiO_2/Al_2O_3 values between 2.94 and 5.14 (Table 8.3), suggesting that the sediments are formed under semi-arid to semi-humid conditions tending towards decreasing chemically maturity. High ratios indicate peninsular sediments are mineralogically mature, while low ratios indicate extra peninsular sediments are chemically immature (Fig. 8.5). The ratio between SiO_2 and Al_2O_3 is also used in the Ruxton Ratio (RR; Ruxton 1968). Low RR values suggest intense chemical weathering, whereas high RR values suggest moderate to weak chemical weathering. In our study, there is a decrease in the ratio of SiO_2/Al_2O_3 from higher temperate to lower tropical latitudes indicating an increase in chemical weathering, humidity, sediment maturity and clay mineral development (Fig. 8.5).

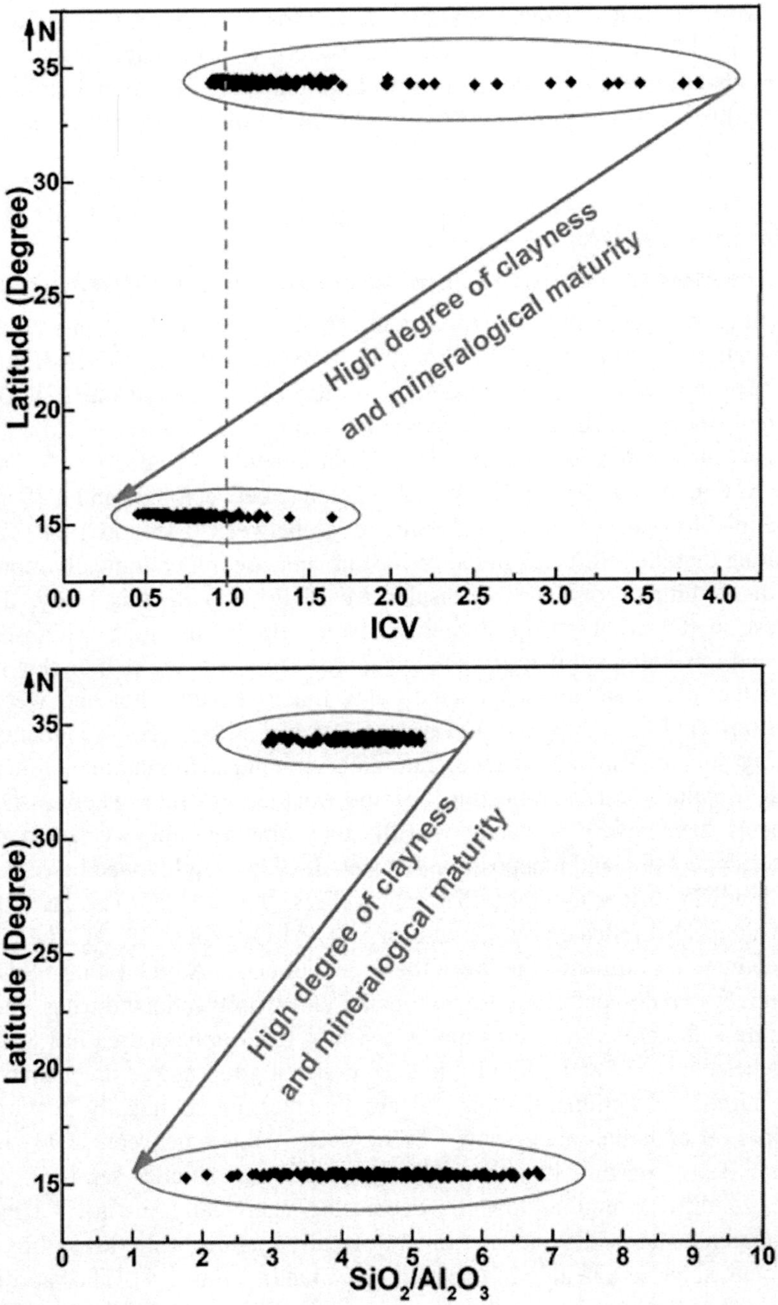

Fig. 8.5 Comparison of ICV (left) and SiO_2/Al_2O_3 ratio (right) from different latitudes of India as a proxy for variation in mineralogical maturity and clay mineral formation

Discussion on Relationships Between Chemical Weathering, Sediment Maturity and Climatic Conditions

In this study weathering indices, temperature proxies and sediment maturity indices reflect a generalized picture of weathering history in the drainage basins from lower and higher latitudes instead of direct and quantitative proxies. The tropical climatic conditions in the Belgaum–Dharwad–Uttar Kannada area (south India) and temperate climatic conditions in the Bandipora–Ganderbal area (north India) (Figs. 8.2 and 8.4) influence the variations in the chemistry of sediments. The basaltic and carbonate terrain of Bandipora–Ganderbal area under climate characterized by moderate precipitation and low temperature (10.8 °C, yearly average) experienced low to moderate chemical weathering but relatively intensive physical weathering (<50 PIA and CIA values), whereas the granitic terrain of the Belgaum–Dharwad–Uttar Kannada area developed under climate characterized by high precipitation and temperature (25.15 °C, yearly average) experienced moderate to intense chemical weathering (>50 PIA and CIA values). The chemical weathering and sediment maturity processes are faster at higher temperature and moisture conditions (Binkley and Fisher 2019). High precipitation and temperature in tropical areas promote the dissolution of minerals by supplying more moisture to mineral surfaces and accelerating the net dissolution reactions forward (Ferrier et al. 2012). Thus, heavy rainfalls coupled with higher temperatures increase the rate of chemical weathering. This phenomenon is supported by many works from climatically diverse areas of the world (Bluth and Kump 1994; Rasmussen et al. 2011). Other factors such as lithology, physical erosion rates, vegetation, land use–land cover, and anthropogenic activities also influence the rate of chemical weathering and erosion (Ferrier et al. 2012). There is a very good correlation between chemical weathering indicators (CIA, PIA, Rb/Sr, Fig. 8.3); mean annual precipitation (Fig. 8.2) and geochemical climate proxies (C-value, Sr/Cu, Fig. 8.4); and sediment maturity proxies (ICV, SiO_2/Al_2O_3, Fig. 8.5) all collectively indicating an increase in the degree of chemical weathering and maturity of sediments with an increase in precipitation and temperature.

Factors Controlling Weathering

The composition of the rocks and minerals and the climate of an area are the major factors controlling the weathering rates. In a particular region under a uniform climate, different rock types weather at different rates (Raymo and Ruddiman 1992). Under a uniform climate, granitic rocks weather more slowly than basalt and limestone because basalt and limestone are softer and it is easy for weathering agents to impact and manipulate their structures (Daniel et al. 2016). In the present study, granitic rocks of southern India are more highly weathered than the basaltic and carbonate rocks of the Kashmir valley. Here the climate plays a more important role in weathering rate than lithology. The climate of a region is determined by the temperature and precipitation. At high temperatures and precipitation, chemical reactions become more intense and increase weathering (White and Blum 1995) than in cold and dry climates (Curtis 1990). A warmer climate also supports vegetation growth which also helps in weathering. In southern India, a hot and humid climate and dense evergreen vegetation weather the granites more than the basalt and limestone of Kashmir valley. The CIA, PIA and Rb/Sr content of stream sediments in southern and northern India have good correlation with temperature, precipitation and latitude reflecting the dominant control of the south-west monsoon climate on chemical weathering in southern India and western disturbances climate in the Kashmir valley. Li and Yang (2010) had cautioned against using weathering indices as a direct and quantitative proxy for evaluating the intensity of chemical weathering in the continents. In this study, weathering indices, temperature proxies and sediment maturity indices reflect a generalized picture of weathering history in the drainage basins from lower and higher latitudes instead of direct and quantitative proxies. We cannot deny the role of anthropogenic activities in enhancing the human-induced weathering in the study area, but further investigations are needed to understand the role of human activities in increasing the rate of weathering.

Conclusions

This work presents the processes of chemical weathering and sediment maturity under contrasting climatic conditions from different latitudes in the tropical (Belgaum–Dharwad–Uttar Kannada area of Karnataka) and temperate (Bandipora–Ganderbal area of Kashmir valley) parts of India. The main findings of this work are summarized below:

(a) In this study weathering indices, temperature proxies and sediment maturity indices reflect a generalized picture of weathering history in the drainage basins from lower and higher latitudes instead of direct and quantitative proxies.
(b) Chemical weathering conditions have been characterized in terms of variations of CIA, PIA and Rb/Sr of granitic terrain of peninsular and basaltic-carbonate terrain of extra peninsular India. All weathering indices closely resemble each other and show variations mainly related to the different climatic conditions. Rocks of the Belgaum–Dharwad–Uttar Kannada area of Karnataka state are moderate to intensively weathered, while rocks of the Bandipora–Ganderbal area of Kashmir valley is moderately weathered. High chemical weathering conditions in south India appear to result from climate forcing related to high rainfall and temperature.
(c) The distribution of C-values and Sr/Cu ratios collectively indicate a generally humid climate in south India and a semi-humid to dry climate in north India. There is an increase in humidity from higher to lower latitudes. The climate proxies based on C-values and Sr/Cu match very well with the meteorological

data of the area. The impact of varying climates is very well reflected in the weathering indices confirming a very important role of regional climatic conditions on the intensity of chemical weathering in the different parts of the world.

(d) The compositional maturity of the sediments is evaluated by using the ICV and SiO_2/Al_2O_3 ratio. The results collectively indicate that the sediments in the peninsular area are mineralogically more mature than the sediments in the extra peninsular area. There is a decrease in sediment maturity from higher to lower latitudes meaning a high degree of clays and mineralogical maturity.

Acknowledgements The author wishes to thank the Director-General, GSI, for devising the National Geochemical Mapping program and for providing the necessary facilities to carry out this work. IAM thanks the Additional Director General, GSI, Hyderabad (SR) and Lucknow (NR), and Deputy Director-General, State Unit: Karnataka and Goa and State Unit: Jammu and Kashmir for providing logistic and financial support to carry out this work. I.A.M. is also thankful to the staff of the geochemical laboratory of GSI, Hyderabad and Lucknow for sample analysis.

References

Babeesh C, Lone A, Achyuthan H (2017) Geochemistry of Manasbal Lake sediments, Kashmir: weathering, provenance and tectonic setting. J Geol Soc India 89:563–572

Bhat AA, Bhat NA (2014) Report on geochemical mapping of Kashmir nappe and Karewa group of rocks covering parts of Anantnag, Baramulla and Srinagar districts, Jammu & Kashmir in toposheet no. 43K13 and 14, Geological Survey of India, SU: J&K, NR, Field Season 2013–2014

Bhat AA, Mir IA, SreePrabash M, Sridhar V, Maruthi KV (2019) Report on geochemical mapping in toposheet no. 48I/11 in Belgaum, Dharwad and Uttar Kannada districts, Karnataka, Geological Survey of India, SU: K&G, SR, Field Season 2017–2018

Binkley D, Fisher RF (2019) Ecology and management of forest soils, 5th edn. Wiley-Blackwell, p 456

Bluth GS, Kump LR (1994) Lithologic and climatic controls of river chemistry. Geochim Cosmochim Acta 58:2341–2359

Bruce Foote R (1876) The geological features of the South Maharatta country and adjacent districts. Mem Geol Surv India 12:1

Bruce Foote R (1888) Dharwar system, the chief auriferous series in south India. Rec Geol Surv India XXI:40

Cao J, Wu M, Chen Y, Hu K, Bian LZ, Wang LG, Zhang Y (2012) Trace and rare earth element geochemistry of Jurassic mudstones in the northern Qaidam Basin, Northwest China. Chemie Erde-Geochem 72:245–252

Cox R, Lowe DR, Cullers RL (1995) The influence of sediment recycling and basement composition on evolution of mudrock chemistry in the southwestern United States. Geochim Cosmochim Acta 59:2919–2940

Curtis CD (1990) Aspects of climatic influence on the clay mineralogy and geochemistry of soils, palaeosols and clastic sedimentary rocks. J Geol Soc 147:351–357

Daniel EI, Jeremy KC, Seulgi M, Dana LT, Jens H, Chamberlain CP, Kate M (2016) Differential weathering of basaltic and granitic catchments from concentration–discharge relationships. Geochim Cosmochim Acta 190:265–293

De Terra H, Paterson TT (1939) Studies on the ice age in India and associated human cultures, vol 493. Carnegie Inst, Washington, DC, pp 1–354

Fedo CM, Nesbitt HW, Young GM (1995) Unraveling the effects of potassium metasomatism in sedimentary rocks and paleosols, with implications for paleoweathering conditions and provenance. Geology 23:921–924

Ferrier KL, Kirchner JW, Finke RC (2012) Weak influences of climate and mineral supply rates on chemical erosion rates: measurements along two altitudinal transects in the Idaho batholith. J Geophys Res 117:20–26

Giovanoli R, Schnoor JL, Sigg L, Stumm W, Zobrist J (1988) Chemical weathering of crystalline rocks in the catchment area of acidic Ticino Lakes, Switzerland. Clay Clay Miner 36:521–529

Goldich SS (1938) A study in rock-weathering. J Geol 46:17–58

Gururaja Rao TP, Sinha PA (1976) Geology of parts of Kalghatgi, Shiggaon and Hubli taluks, Dharwar District, Karnataka state (unpublished Progress report of GSI for FS-1975-76)

Harnois L (1988) The CIW index: a new chemical index of weathering. Sediment Geol 55:319–322

Hegde VS, Chavadi VC (2008) Geochemistry of late Archaean metagreywackes from the Western Dharwar craton, South India: implications for provenance and nature of the late Archaean crust. Gondwana Res 15:178–187

Hu J, Li Q, Song C, Wang S, Shen B (2017) Geochemical characteristics of the Permian sedimentary rocks from Qiangtang Basin: constraints for paleoenvironment and paleoclimate. Terr Atmos Ocean Sci 28:271–282

Iyengar S (1976) The stratigraphy, structure and correlation of the Dharwar Supergroup, GSI Misc. Publication No. 23, Part-II, pp 415–456

Jia J, Liu Z, Bechtel A, Strobl SAI, Sun P (2013) Tectonic and climate control of oil shale deposition in the upper cretaceous Qingshankou formation (Songliao Basin, NE China). Int J Earth Sci 102:1717–1734

Kapoor HM (1963) A note on the geochemical study of the stream sediments from Liddar river, Anantnag district Jammu & Kashmir state; Geological Survey of India, Operation: J&K, NR, Field Season 1962–1963

Koppen W (2011) The thermal zones of the earth according to the duration of hot, moderate, and cold periods and the impact of heat on the organic world. Meteorol Z 20:351–360

Lerman A (1978) Lakes: chemistry, geology, physics. Springer, New York

Li C, Yang S (2010) Is chemical index of alteration (CIA) a reliable proxy for chemical weathering in global drainage basins? Am J Sci 310:111–127

McLennan SM, Hemming S, McDaniel DK, Hanson GN (1993) Geochemical approaches to sedimentation, provenance, andtectonics. Processes controlling the composition of clastic sediments. Geol Soc Am Special Paper 284:21–40

Meng Q, Liu Z, Bruch AA, Liu R, Hu F (2012) Palaeoclimatic evolution during Eocene and its influence on oil shale mineralisation, Fushun basin, China. J Asian Earth Sci 45:95–105

Mir RA, Jeelani G (2015) Textural characteristics of sediments and weathering in the Jhelum River basin located in Kashmir Valley, western Himalaya. J Geol Soc India 86:445–458

Mir IA, Mir RA (2015) Report on geochemical mapping of Srinagar, Pampur and Ganderbal areas of Srinagar and Ganderbal districts, Jammu and Kashmir in toposheet nos. 43J/11 and part of 43J/12, Geological Survey of India, SU: J&K, NR, Field Season 2014–2015

Mir IA, Mir RA (2019) Geochemistry of surface sediments in parts of Bandipora-Ganderbal areas, Kashmir valley, western Himalaya: implications for provenance and weathering. J Earth Syst Sci 128(223):1–16

Mir IA, Pattan JN, Matta VM, Banakar VK (2013) Variation of paleo-productivity and terrigenous input in the eastern Arabian Sea during the past 100 ka. J Geol Soc India 81:647–654

Mir IA, Bhat AA, SreePrabash M, Sridhar V, Maruthi KV (2021) Surface sediment geochemistry for understanding the recent sedimentary environment in northwestern Karnataka, South India. Geosci J (under review)

Murthy TSS, Mehrotra SC (1979) Geology of manganiferous Formations in parts of North Kanara and Belgaum districts, Karnataka. Geological Survey of India, Unpublished Report (FS 1976–77)

Nesbitt HW, Young GM (1982) Early proterozoic climates and plate motions were inferred from the major element chemistry of lutites. Nature 299:715–717

Parker A (1970) An index of weathering for silicate rocks. Geol Mag 107:501–504

Pattan JN, Parthiban G, Gupta SM, Mir IA (2012) Fe speciation and Fe/Al ratio in the sediments of southeastern Arabian Sea as an indicator of climate change. Quat Int 250:19–26

Potter PE (1978) Petrology and chemistry of modern big river sands. J Geol 86(4):423–449

Rashid AS, Ather M, Farid AK (2013) Sediment water interaction at higher altitudes: examples from the geochemistry of Wular Lake sediments, Kashmir Valley, northern India. Procedia Earth Planet Sci 7:786–789

Rashid SA, Ganai JA, Masoodi A, Khan FA (2015) Major and trace element geochemistry of lake sediments, India: implications for weathering and climate control. Arabian J Geosci 8:5677–5684

Rasmussen C, Brantley S, Richter D, Blum A, Dixon J, White AF (2011) Strong climate and tectonic control on plagioclase weathering in granitic terrain. Earth Planet Sci Lett 301:521–530

Raymo ME, Ruddiman WF (1992) Tectonic forcing of late Cenozoic climate. Nature 359:117–122

Ruxton BP (1968) Measures of the degree of chemical weathering of rocks. J Geol 76:515–527

Scarciglia F, De Rosa R, Vecchio G, Apollaro C, Robustelli G, Terrasi F (2008) Volcanic soil formation in Calabria (southern Italy): the Cecita Lake geosol in the late quaternary geomorphological evolution of the Sila uplands. J Volcanol Geotherm Res 177:101–117

Shah RA, Achyuthan H, Lone AM, Ramanibai R (2017) Diatoms, spatial distribution and physicochemical characteristics of the Wular lake sediments, Kashmir valley, Jammu and Kashmir. J Geol Soc India 90:159–168

Singh I (1982) Report on petrological, volcanological and stratigraphical studies of Panjal Volcanics in Liddar and Sind valleys and Bandipora area, Anantnag, Srinagar and Baramulla districts, Jammu & Kashmir State; Geological Survey of India, Operation: J&K, NR, Field Season 1981–1982

Slater HK (1906) Memories of Mysore geological department, vol I

Standard Operation Procedure for National geochemical mapping & quality management. Revised and updated 2014, Mission-I: Baseline Geosciences Data Generation of Geological Survey of India

Swami NJ, Ramakrishnan M, Viswanatha MN (1976) Dharwar stratigraphic model and Karnataka craton evolution. Geol Surv India Rec 107:149–175

Thakur VC, Rawat BS (1992) Geological map of the Western Himalaya; Survey of India

Wadia DN (1931) The syntaxis of NW Himalaya, its rocks, tectonics and orogeny. Rec Geol Surv India 65:189–220

White AF, Blum AE (1995) Effects of climate on chemical weathering rates in watersheds. Geochimica et Cosmochimica Acta 59:1729–1747

Zhao Z, Zhao J, Wang H, Liao J, Liu C (2007) Distribution characteristics and applications of trace elements in Junggar Basin. Nat Gas Explor Dev 30:30–33

Chapter 9
Manganese Mineralization in Manganiferous Quartzite in the Boringpadar-Amath Area, Eastern Ghats Mobile Belt, Odisha, India: Implication for Climatic Changes

Pawan Kumar Yadav, Manorama Das, and Sradhanjali Subhadarshini

Abstract The chapter brings to light an array of manganese mineralization with promising values, hosted in manganese ore and manganiferous quartzite and predicts the responsible climatic environment for the formation of manganese based on field observations, ore petrographic studies and geochemical inputs. Two bands of manganese ore and four bands of manganiferous quartzite are mapped to the southeast of Boringpadar and northeast of the Amath area belonging to the Eastern Ghats Mobile Belt, India. The manganese ores are represented by bluish-black to brownish-black, massive to foliated, metallic luster, nodular and botryoidal forms which are mainly bedded and fragmental types. The ore bodies are strata bound occurring in between calc-silicate granulite and garnetiferous quartzite which defines a characteristic stratigraphic horizon. Manganiferous quartzite is hard, white to black, highly jointed, fractured and brecciated which occurs at the contact of khondalite. Braunite, bixbyite, jacobsite, manganite, pyrolusite and todorokite are primary minerals that are associated with host rocks like quartzite and these minerals occur as inclusions and anhedral in shape. Secondary minerals are formed by the process of colloidal influx or metasomatism psilomelane-cryptomelane, pyrolusite and goethite are secondary minerals which occur in higher quantities. A relatively high temperature metamorphic jacobsite-bixbyite-braunite assemblage, low temperature hydrothermal pyrolusite-psilomelane-cryptomelane assemblage and supergene pyrolusite-manganite assemblage are recorded which were formed by recycling of manganese in different stages of mineralization. Replacement and relict textures are common between braunite-manganite-pyrolusite, jacobsite-braunite-pyrolusite and pyrolusite-psilomelane-cryptomelane-goethite. A geophysical investigation has also been carried out in the area to know the subsurface continuity of manganese mineralization. In Boringpadar,

P. K. Yadav (✉) · M. Das
Geological Survey of India, SU, Bihar, Patna, India
e-mail: pawankumaryadavgsi@gmail.com

S. Subhadarshini
Geological Survey of India, SU, Odisha, Bhubaneswar, India

N. Khare (ed.), *Science, Policies and Conflicts of Climate Change*, Springer Climate,
https://doi.org/10.1007/978-3-031-16254-1_9

the apparent resistivity values vary from 10 to 240 ohm-m which displays low resistivity, and the apparent chargeability contour map shows high anomaly, indicative of a mineralized zone. In Amath, the apparent resistivity value varies from 20 to 140 ohm-m and the apparent chargeability contour map shows a high anomaly. The pseudo-depth sections were prepared over geophysical traverses to examine the subsurface distribution of apparent chargeability and apparent resistivity indicating mineralization prospects. This mineralized body starts almost from the surface and the contours of this anomaly zone open downward which is suggestive of the depth persistence of the causative source below 35 m. The analytical results of bedrock samples show high values of Mn, Fe, SiO_2, Al_2O_3, P, S and CaO. The values of Mn and Fe(T) ranged from 0.05 to 28.18 wt. % and 1.87–56.31 wt. %, respectively. In pitting/trenching samples, the Mn value ranges from 0.05 to 27.77 wt. % and the Fe (T) value ranges from 2.11 to 46.85 wt. %. The highest values of Mn come from manganese ore of the Amath area and maximum values of Fe(T) were recorded from manganese ore of the Boringpadar area. The above-mentioned data indicate that the manganese ores and manganiferous quartzite might be formed in the freshwater environment.

Keywords Manganese ores · Manganiferous quartzite · Mineralogy · Eastern Ghats Mobile Belt (EGMB) · India

Introduction

Manganese (Mn) is a silver metallic element with an atomic number of 25 which is not found as an element in nature. Manganite (MnO), purpurite ($MnPO_4$), pyrolusite (MnO_2), rhodonite [(Mn, Fe, Mg, Ca, (SiO_3)], rhodochrosite ($MnCO_3$) and sugilite [$KNa_2(Fe, Mn, Al)_2\ Li_3Si_{12}O_{30}$] are manganese minerals. Manganese mineral is also found in association with mineraloids like psilomelane and wad and is an important element for steel and non-alloying industries (Gandhi 2010; Singh and Biswas 2017). In Archean greenstone belts, the manganese mineralization occurs predominantly with the association of Banded Iron Formation (BIF) which is mainly reported from the Barberton greenstone belt of South Africa, the Sebakwian-Bulawayan-Shamvaian belt of Zimbabwe, the Isua Formation of Greenland, the Superior and Slave provinces of Abitibi belt, Canada, the Rio das Velhas deposit of Brazil, the Yilgarn and Pilbara blocks of Western Australia, the Bababudan and Chitradurga belt of South India and the Iron Ore Group (IOG) of the Singhbhum Craton (Goodwin 1973; Ramakrishnan et al. 1976; DeWit et al. 1980; Anhaeusser and Wilson 1981; Chadwick et al. 1981a, b; Condie 1981; Hallberg and Glikson 1981; Roy 1981; Dimroth et al. 1982; Windly 1982; Gross 1986; Schidlowski 1988, 1993; Machado and Carnerio 1992; Myers and Kröner 1994; Saha 1994; Teixeira et al. 1996; Martin et al. 1997; Ghosh et al. 2015). Continental crustal growth, decreasing mantle heat flux leading to diminished outgassing, along with suitable tectonic evolution, sedimentary environment, and biosphere–atmosphere–hydrosphere interaction has usually been cited as a source of initiation of sedimentary

Mn deposition during the late Archaean (Acharya et al. 1994; Roy 2000). The manganese accumulations predominantly occur in the Archean and Proterozoic periods and few occurrences were also noticed in the Phanerozoic Era. The Archaean–Proterozoic transition in the Earth system involved the rapid burial of organic matter, increased weathering on tectonically amalgamated supercontinents, reduction of reduced gases from a gradually oxygenated mantle, and the net increase in photosynthetic O_2 (Roy 2006; Mishra 2015). The formation of manganese rocks and ores on many geological boundaries was closely connected with global climatic, tectonic restructurings, and biological events (Kuleshov 2017). However, manganese rocks and ore were also derived from volcanogenic and/or hydrothermal and freshwater sources which are also reported from different parts of the world (Acharya et al. 1994; Roy 2000, 2006; Mishra 2015).

In Odisha, the manganese ore is found to occur in three stratigraphic horizons and widely spatially distributed, i.e. Iron Ore Group in Bonai-Keonjhar belt, Gangpur Group in Sundergarh, and Eastern Ghats Mobile Belt in Rayagada-Kalahandi-Bolangir belt. The geology, classification, mode of occurrence and genetic aspects of manganese ore deposits of the Jamda-Koira belt have been dealt with by many workers viz. Former 1909; Spencer 1948; Sen 1951; Ray 1954, 1955; Engineer 1956; Prasad Rao and Murty 1956; Mohapatra and Bagchi 1961; Mookherjee 1966; Basu 1969; Murty and Ghosh 1971; Ajmal 1990; Mishra 1994; Dasgupta et al. 1999; Mohapatra et al. 1989, 2009, 2010 and Mishra et al. 2016. The Jamda-Koira belt of the Noamundi basin contains more than 130 million tonnes of total manganese ore reserve (Ghosh et al. 2015). Manganese ores are predominantly reported in the Vizianagaram-Visakhapatnam sector in Andhra Pradesh belonging to the Eastern Ghats Mobile Belt (EGMB) (Straczek and Krishanswamy 1956; Krishna Rao 1963; Rao 1969; Siddiquie and Raza 2008; Siddiquie and Shaif 2015; Siddiquie et al. 2015). In contrast, only a few such occurrences have been reported from the EGMB in Odisha by the Geological Survey of India (Jayaram 1956; Jena and Devdas 1994; Dash et al. 2005; Yadav et al. 2017).

The study aimed to assess the manganese mineralization and predict the responsible climatic environment for the formation of manganese based on field observations, ore petrographic studies and geochemical inputs in Boriningpadar and Amath areas in the Eastern Ghats Granulite Belt belonging to the Eastern Ghats Mobile Belt, Odisha, India.

Geological Setting

The Eastern Ghats Mobile Belt encompassing an area of over 50,000 sq. km broadly trends NE-SW and extends from Brahmani River in Odisha to Ongole in Andra Pradesh for a distance of over 900 km with a maximum width of 300 km in Odisha in the north. It tapers down to a few tens of km in width in Andra Pradesh. The EGMB is a distinct geological entity and is known for its regionally developed granulite facies lithopackage, intense polyphase ductile deformation, metamorphism, migmatization and Meso to Neoproterozoic magmatism (Ramakrishna et al. 1998). It consists of granulite facies metapelitic and meta-psammatic gneisses, mafic and felsic orthogneisses/granulites and S-type granites. Massif-type anorthosite-leuconorite complexes and alkaline complexes are significant products of Meso to Neoproterozoic magmatism. In addition, the Eastern Ghats Granulite Belt has drawn the attention of various workers to study its stratigraphy, structure and tectono-thermal aspects (Murty and Ghosh 1971; Narayanswami 1975; Chetty and Murthy 1994; Bhattacharya et al. 1998; Ramakrishna et al. 1998) except a few exploration programmes for its economic mineral association such as graphite, manganese and bauxite, etc., mainly by the GSI. Ramakrishna et al. (1998) have divided EGMB into four longitudinal lithozones based on the dominant litho-suite present, viz. Eastern Khondalite Zone (EKZ), Central Charnockite-Migmatite Zone (CMZ), Western Charnockite Zone (WCZ) and Western Khondalite Zone (WKZ). In the northwestern corner of the belt in the Bolangir-Khariar region, the WKZ has possibly moved westward to produce a nappe like structure overriding and resting over the passive margin of Bastar Craton due to the westerly verging Vamsadhara, Nagavalli and Mahanadi shear zone (Ramakrishna et al. 1998). The prominent NE-SW trending Tel lineament running sub-parallel with the Tel River course possibly separates the main belt from the nappe block. Former (1909) and Pascoe (1950) were the pioneers who proposed the stratigraphic classification for the Eastern Ghats.

The manganese ore bodies are confined to the Khondalite Group of rocks of the Eastern Ghats Granulite Belt in Odisha. The prominent known areas are the 32 km long Kutinga-Nishikhal zone in Rayagada district and the 30 km long Kanaital-Uchhabapalli zone in Bolangir district. All the important manganese deposits of the region are found confined to these two small narrow zones as proved by exploration activities of GSI for manganese (Jena and Devdas 1994; Dash et al. 2005; Dash and Behera 2009). The manganiferous zone/horizon is characterized by the association of quartzite ± garnet (psammitic facies) and calc-silicate granulite (calcareous facies) with manganese ore bodies lying in between the two along with garnetiferous quartzo-feldspathic sillimanite schist/gneiss ± graphite (psammo-pelitic facies). In general, manganese ore bodies are intimately associated with calc-silicate granulite with profuse quartz veins in the vicinity of garnetiferous quartzo–feldspathic sillimanite schist/gneiss ± graphite. The litho members constitute a narrow but persistent manganiferous zone/horizon within the Khondalite zone. The zones occur as roof pendants, cuspate and lensoidal bodies, keels and often as scattered outcrops on coarse-grained, mesocratic, inequigranular and coarsely porphyritic granitoid but without the signature of mechanical stirring. In domal outcrops, granitoids and gneisses contain all the litho-variants of the manganiferous zone/horizon as enclaves usually in the proximity of the rich deposits of manganese. The present area (Boringpadar-Amath block) forms a part of the Western Khondalite Zone in the proximity of the Tel lineament zone. The dominant rock type exposed in this block is calc-silicate granulite, quartz + K-feldspar + sillimanite + garnet + biotite ± graphite (khondalite), Mn ore/manganiferous quartzite/ferruginous quartzite and garnetiferous quartzite (Khondalite Group), garnetiferous granite-gneiss and megacrystic garnetiferous granite-gneiss.

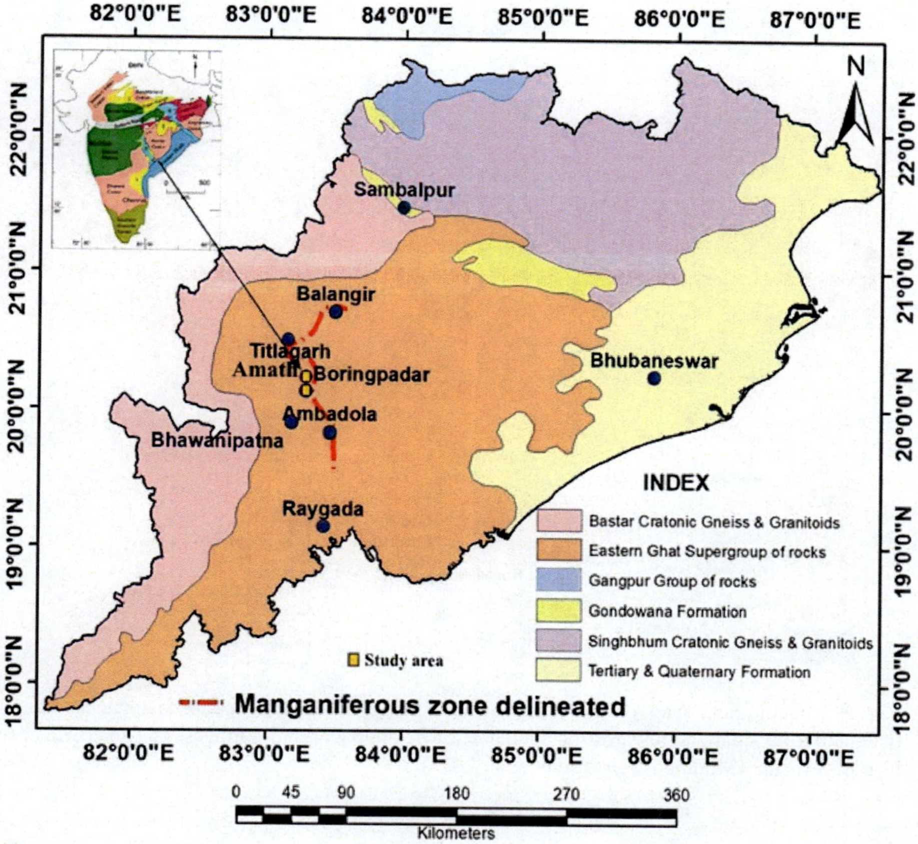

Fig. 9.1 Generalized geological map of the Eastern Ghats Mobile Belt, Odisha showing the location of the study area. (Modified after Ramakrishna et al. 1998)

In this chapter, two bands of manganese ore (200 × 50–70 m; 250 × 40 m) and four bands of manganiferous quartzite (150 × 10 m; 230 × 10 m; 150 × 5 m; 100 × 5 m) have been delineated in detail for the first time to the southeast of Boringpadar and northeast of Amath area in the Eastern Ghats Mobile Belt, Kalahandi district, Odisha (Fig. 9.1).

Geology of the Study Area

The study area forms a part of the Western Khondalite Zone (WKZ) in the proximity of the Tel lineament zone. The dominant rock type exposed in this block is calc-silicate granulite, quartz + K-feldspar + sillimanite + garnet + biotite ± graphite (khondalite), Mn ore/manganiferous quartzite/ferruginous quartzite and garnetiferous quartzite (Khondalite Group), garnetiferous granite-gneiss and megacrystic garnetiferous granite-gneiss (Fig. 9.2; Yadav et al. 2017).

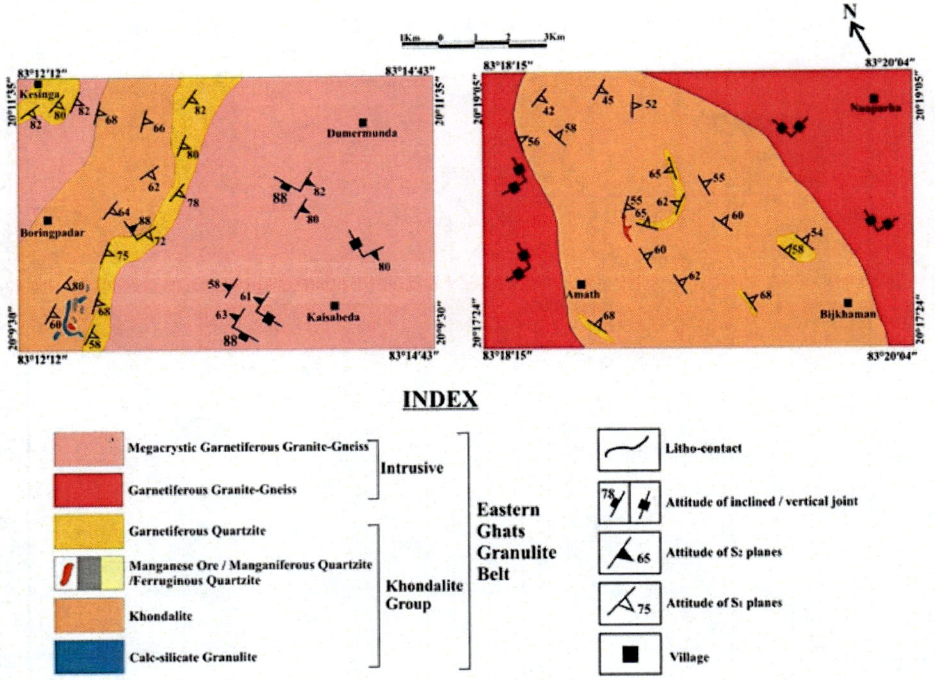

Fig. 9.2 Geological map of the study area showing the occurrence of manganese ore and manganiferous quartzite towards the southeast of Boringpadar and northeast of Amath Kalahandi district, Odisha. (Modified after Yadav et al. 2017)

Two bands of manganese ore have been observed in the mapped area near Boringpadar and Amath in parts of toposheets no. 64P/4 and 64P/7 (Fig. 9.2). The first band of manganese ore is mapped in the western flank of Sundari Dongri hill, southeast of Boringpadar which occurs in the lower flank of the hill. The manganese ore is bluish-black to brownish-black, massive to foliated, has a metallic luster and has nodular and botryoidal forms (Fig. 9.3a, b). The manganese minerals are mostly pyrolusite and psilomelane and the ore body is mainly bedded and fragmental type. The strike length of this ore body is 200 m having a width of 50–70 m on a hill slope at an elevation of 270 m which is trending in the N20°W–N40°E direction. At places, tight to isoclinals F_1 folds are also observed within it. The second band of manganese ore has been mapped towards the northeast of Amath having a dimension of 250 m and 40 m trending in the N10°W–N30°E direction. The manganese ore is mainly bluish-black, massive to foliated, has a metallic luster and has botryoidal and nodular forms (Fig. 9.3c, d).

Four bands of manganiferous quartzite have been mapped on the western flank of Sundari Dongri hill, southeast of Boringpadar with dimensions as follows: (i) 150 × 10 m trending NNE–SSW (ii) 230 × 10 m trending N40°E–S40°W, (iii) 150 × 5 m trending N40°E–S40°W and (iv) 100 × 5 m trending N–S to N45°E–S45° W. It is hard, white to black, highly jointed, fractured (Fig. 9.3e) and brecciated (Fig. 9.3f). A small band of ferruginous quartzite is observed at the contact of

Fig. 9.3 (**a, b**) Outcrop of massive manganese ore (bluish to brownish), which shows size variations of botryoidal form, southeast of Boringpadar. (**c, d**) Massive manganese ore in bluish-black displaying the botryoidal form, northeast of Amath. (**e**) Highly fractured and jointed manganiferous quartzite, southeast of Boringpadar. (**f**) Highly fragmented and brecciated pieces of manganiferous quartzite embedded in the matrix of manganese ore, Boringpadar. (**g**) Outcrop of ferruginous quartzite, southeast of Boringpadar. (**h**) Contact between manganiferous quartzite and ferruginous quartzite, southeast of Boringpadar

manganiferous quartzite and khondalite southeast of Boringpadar. It is hard, massive to foliated, chocolate, fine to medium-grained and consists of quartz as an essential mineral (Fig. 9.3g). The strike length of this band is 50 m and with a width of 10 m trending in the north-south direction. At places, sharp contact between manganiferous quartzite and ferruginous quartzite is also noticed (Fig. 9.3h).

Geophysical Study in Boringpadar and Amath Area

The Geophysical data was acquired on a 20 × 100 m grid and anomaly maps were prepared on a 1:12,500 scale in accordance with the Standard Operational Procedure (SOP) of the Geological Survey of India (GSI) Telford et al. (1976), Ramazi and Mostafaie (2013), Moreira et al. (2012) and Vieira et al. (2016). The base line in the Boringpadar area was laid along N40°E and the centre of the base line was marked as base station 0/0 (20°09′50″ N and 83°12′ 42″ E) for this investigation. In the Amath block the baseline was laid along N30°W and 0/0 was marked at the centre of the baseline (20°17′32″ N and 83°18′27″ E). The base line was 1000 m in the North and 400 m in the South to a base station, covering a total length of 1400 m. The Induced Polarization (IP) survey is considered a suitable method for disseminated sulphide-bearing zones while Self-Potential (SP) and Resistivity methods are utilized to supplement IP anomalies for the identification of mineralized zone. The interpretation methodology is largely based on qualitative analysis of anomaly maps. The Induced Polarization (chargeability) effect reflects the degree to which the subsurface is able to store electric charge, analogous to a leaky capacitor. It occurs when an electric current passes through a rock/soil. If the current is interrupted, a difference in potential, which decays with time, is observed. The rate of decay of this potential (induced polarization potential) depends on the lithology of the rock, its pore geometry and the degree of water saturation. Similar to the resistivity method, the IP survey measures variation in characteristics of subsurface materials. Since the measurement techniques of both methods are the same, IP and resistivity measurements were carried out simultaneously. The correlation of chargeability and resistivity values provides the subsurface structure of the target zones, and one of the most important objectives of this investigation is to delineate lateral as well as vertical extensions of mineralized zones. IP (time-domain) and resistivity methods are used to prepare pseudo sections for obtaining the sub surface information over the mineralized zone.

In Boringpadar, the apparent resistivity values in the survey area vary from 10 to 240 ohm-m. The resistivity survey has delineated significant low resistivity (conductive zone marked as zone C in Fig. 9.4a) over the exposed area of manganese over the traverses S4, S5, S6, S7 and S8. The NE-SE contact is represented by a black dotted line between the low and high resistivity zone. The significant low resistivity zone is indicative of a mineralized zone (Fig. 9.4a). The apparent chargeability contour map (Fig. 9.4a) shows a high anomaly which is marked as zone C along the lines S4, S5, S6, S7 & S8 with a magnitude of 20–42 mV/V over the background of 2–20 mV/V near the exposed area. Again, the high apparent chargeability zone almost follows the NE-SW trend. The zone with significantly high apparent chargeability indicates a mineralized zone.

In Amath, the apparent resistivity value in the survey area varies from 20 to 140 ohm-m. The resistivity survey has delineated significant low resistivity (conductive zone) over the exposed area of manganese over the traverses N3, N2 & N1. The low resistivity zone (marked as zone A) almost follows the NW-SE trend. The

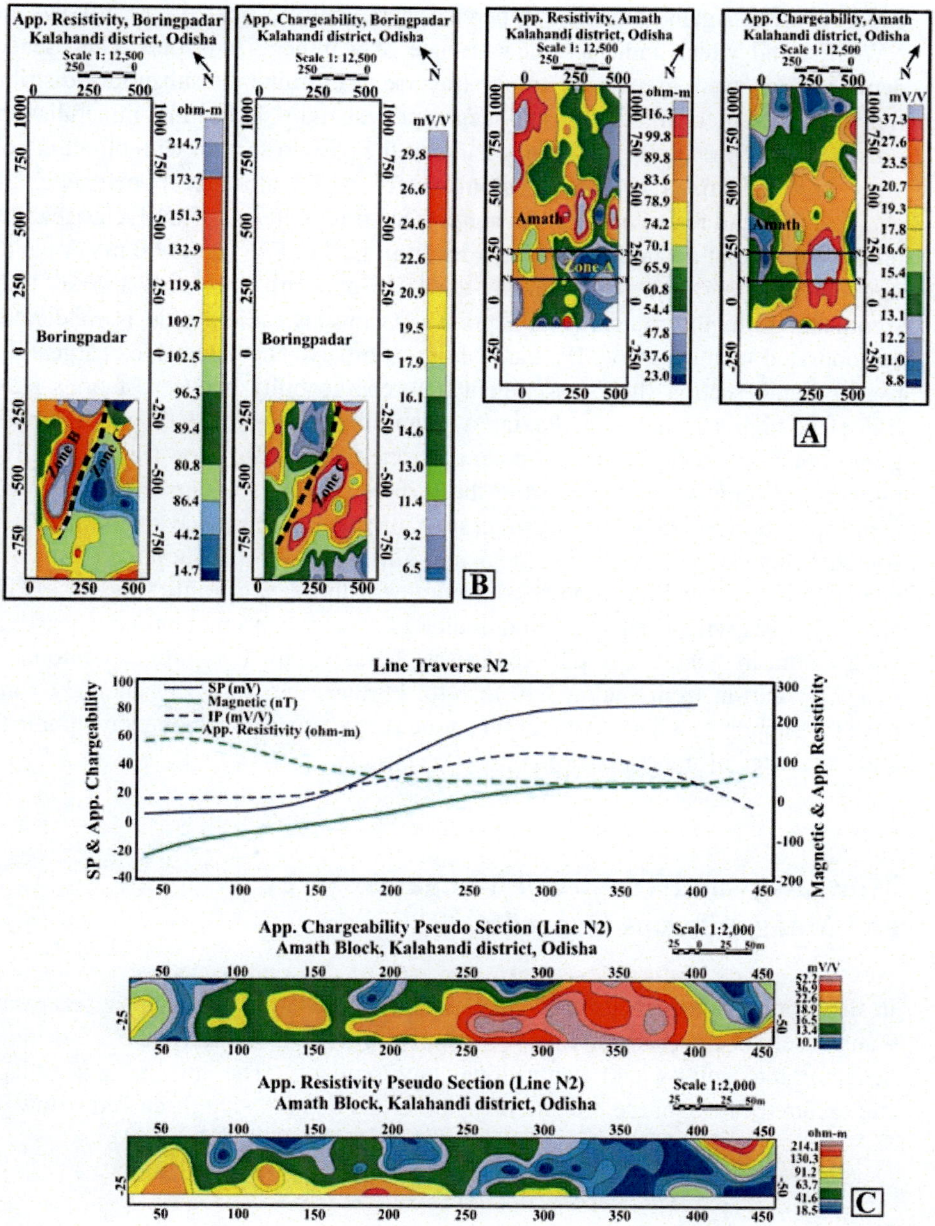

Fig. 9.4 (**a**, **b**) Apparent resistivity anomaly map and apparent chargeability anomaly map of Boringpadar and Amath area. (**c**) Magnetic, SP, IP and resistivity profile (top) along with pseudo-section of (middle) apparent chargeability and (bottom) apparent resistivity over traverse number N2 in Amath area

significant low resistivity zone is indicative of a mineralized zone (Fig. 9.4b). The apparent chargeability contour map shows a high anomaly along the lines N3, N2 & N1 having a magnitude of 35–75 mV/V over the background of 5–35 mV/V near the exposed area. Again, the high apparent chargeability zone almost follows the NW-SE trend which indicates the presence of a mineralized zone. The pseudo-depth sections were prepared over one traverse to examine the subsurface distribution of apparent chargeability and apparent resistivity indicating mineralization prospects (Telford et al. 1976). The results of the pseudo section with dipole-dipole array (a = 20 m, N = 1 to 5) along traverses N2 along with magnetic, SP, chargeability and resistivity profiles are presented in Fig. 9.4c. The IP/chargeability profile shows that an anomaly zone varies from E20 to E46 (up to 60 mV/V) and a well-matched low resistivity anomaly zone (20 ohm-m). This zone appears to be significant from the mineralization point of view. The SP profile is moderately corroborated with magnetic, IP/chargeability anomaly. The detectable chargeability pseudo-depth section shows prominent high chargeability between stations E2 to E46. This high anomaly zone having a magnitude of 40–68 mV/V over the background of 8–40 mV/V may be a potential zone of mineralization (Fig. 9.4c). This mineralized body starts almost from the surface and the contours of this anomaly zone open downward which is suggestive of depth persistence of the causative source below N = 5 level (i.e. >35 m). This zone is well supported by a low resistivity profile and moderately corroborates with the SP profile (Fig. 9.4c). The apparent resistivity pseudo section indicates that one prominent low resistivity 10–25 ohm-m zone from E30 to E40 is bounded by two higher contours of 175–225 ohm-m from station E40 to E46. Further, the resistivity anomaly zones have the support of a high chargeability zone suggesting a good zone, favourable for the occurrence of mineralization.

Mineralogy and Texture of Manganese Ore and Manganiferous Quartzite

In the Boringpadar-Amath area, the mineral assemblages of manganese ore and manganiferous quartzite and their textures suggest the occurrence of different metamorphic activity and depositional environments. The mineral assemblages and texture of manganese ore and manganiferous quartzite are discussed below.

Manganese Ore

Braunite, bixbyite, jacobsite and pyrolusite are primary minerals that are associated with host rocks like quartzite and these minerals occur as inclusions and anhedral in shape. The minerals of manganese ore are identified as pyrolusite, psilomelane, goethite and quartz gangue. Pyrolusite occurs as subhedral and relict pyrolusite and exhibits white, bluish grey, anisotropism with shades of grey and weak pleochroism, moderate reflectivity in the association of psilomelane and goethite. The contact is irregular, concave and contorted with secondary manganese ores and it has one set of cleavage. Coliform and spheroidal forms of pyrolusite, cryptomelane, psilomelane, goethite, quartz and these spherules sizes occur invariably at places. Spotted/porous textures are noticed in pyrolusite, cryptomelane-psilomelane. Psilomelane displays bluish grey, light grey to bluish-grey anisotropism along with fibrous cryptomelane (bluish-grey), and it subsequently altered to goethite with grey and dark brown internal reflection. A few secondary pyrolusite veins also occur along with the fractures in quartz. Manganese ore band exhibits replacement, colloform/spheroidal and vein textures (Fig. 9.5a). Braunite, manganite, pyrolusite, psilomelane-cryptomelane mineral assemblages are also noticed. Replacement and vein textures are common in it. Altered braunite is medium grained, grey and exhibits low reflectivity, weak pleochroism and anisotropism with one set of cleavage present. Subhedral braunite has greyish white to light grey with brownish-grey anisotropism and contains silicate gangue; it occurs as a vein and exhibits spotted/porous texture. Manganite, an alteration product of braunite, exhibits grey, weak pleochroism/anisotropism with low reflectivity in association with braunite and pyrolusite. Pyrolusite shows cream white to light grey, moderate reflectivity with an abundance of pores and one set of cleavage. Psilomelane-cryptomelane pyrolusite ore minerals (Fig. 9.5b) exhibit bluish-grey with strong anisotropism (bluish-grey to light grey) and fibrous. A few fibrous cryptomelane occur at the wall between silica and pyrolusite (Fig. 9.5c). In a few samples, manganese ore consists predominantly of todorokite and pyrolusite-psilomelane as an accessory along with quartz and garnet as gangues. Todorokite occurs as fibrous/fine bands (Fig. 9.5d), grey with dark brownish-grey anisotropism alternating with silica. Pyrolusite is light grey and occurs as non-crystalline or amorphous forms replacing todorokite. Sub-idioblastic garnets are replaced by psilomelane-goethite and their void spaces are occupied by quartz grains. Many garnets contain sub-rounded to rounded inclusion of braunite (Fig. 9.5e) which is light grey with a yellowish tint, strong pleochroic (brownish-grey to bluish grey) and moderate brown internal reflection. All these ores are cross-cut by ferruginous quartz veins (Fig. 9.5d). Idiomorphic manganite (Fig. 9.5f) occurs

Fig. 9.5 (continued) Photomicrographs of manganese ore and manganiferous quartzite under reflected light (RL): (**a**) Colloform and spheroidal textures of pyrolusite-psilomelane-cryptomelane in manganese ore. (**b**) Replacement of bixbyite to pyrolusite in manganese ore. (**c**) Fibrous cryptomelane at the wall between pyrolusite and silicate gangue in Mn ore. (**d**) Fibrous todorokite crosscutting by siliceous vein with garnets in Mn ore. (**e**) Inclusion of braunite in garnet where the garnet is completely replaced by manganese oxide in manganese ore. (**f**) Idiomorphic manganite and quartz gangue within manganese ore. (**g**) A thick vein of manganese ore (colloform texture) displays crosscutting relationship with the quartz. (**h**) Manganese ore showing colloform texture. (**i**) Brecciated braunite in manganiferous quartzite. (**j**) Manganese ore and quartz showing vein texture in manganiferous quartzite. (k) Veined manganese ore in brecciated quartzite. (**l**) Jacobsite and bixbyite occur within the secondary manganese ore seen in brecciated quartzite. (**m**) Concave and contorted contact of goethite and pyrolusite occur with secondary ore minerals. (**n**) Colloform texture well preserved along the manganese ore vein. (**o**) Zonal texture between manganite, pyrolusite and cryptomelane-psilomelane has been seen in manganiferous quartzite. (**p**) Sub-idiomorphic manganite with secondary manganite was noticed in manganiferous quartzite Abbreviations: *Bix* Bixbyite, *Jac* Jacobsite, *Brn* Braunite, *Mng* Manganite, *Pyr* Pyrolusite, *Psil* Psilomelane, *Crp* Cryptomelane, *Tod* Todorokite, *Goe* Goethite, *Grt* Garnet, *Qtz* Quartz

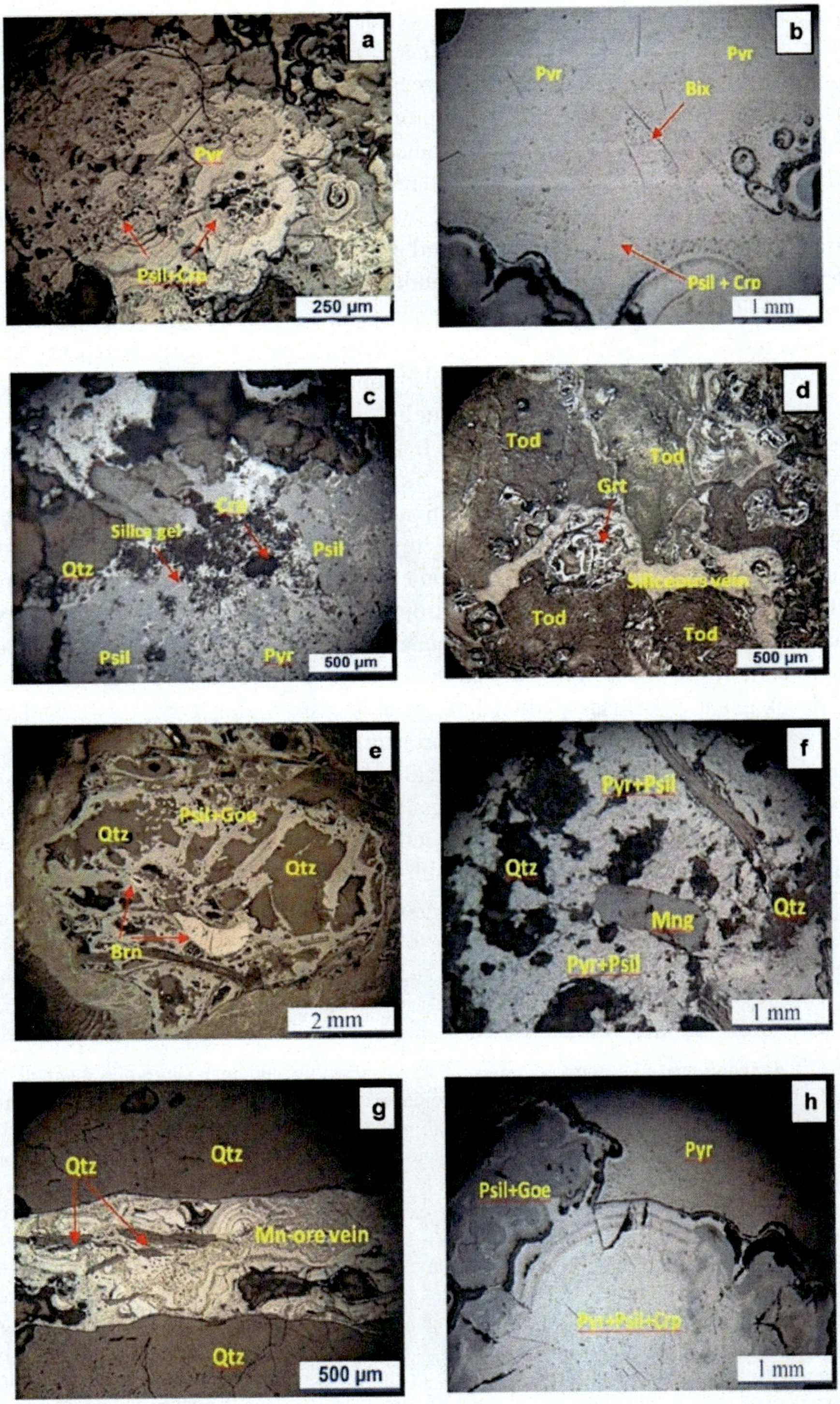

Fig. 9.5 (continued)

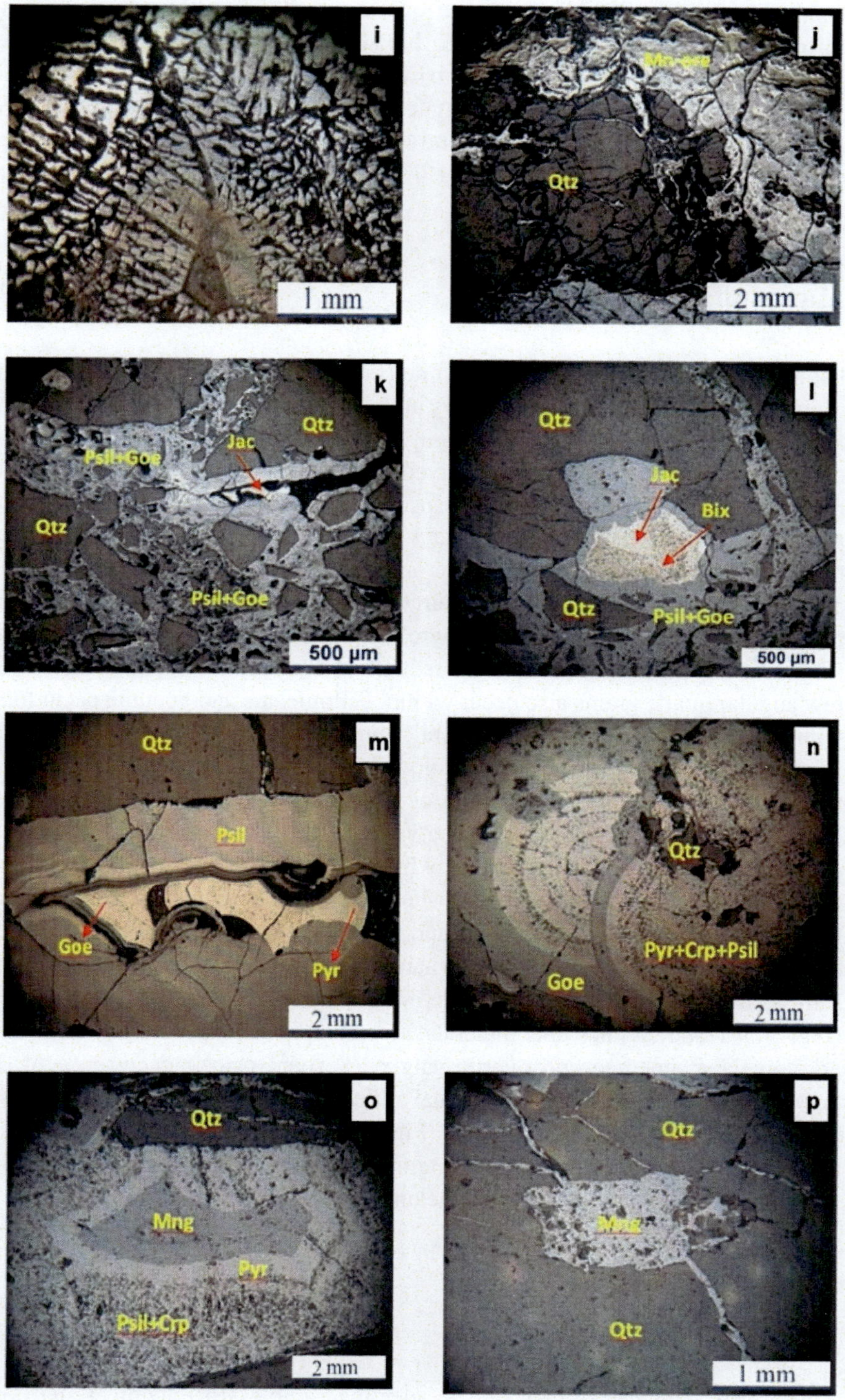

Fig. 9.5 (continued)

within secondary manganese ore and quartz which exhibits medium grey, weak anisotropic & pleochroism with strong blood-red internal reflection. In a few samples, it is mainly composed of bixbyite, braunite, pyrolusite, cryptomelane, psilomelane, goethite and quartz gangue. Bixbyite occurs as inclusion (Fig. 9.5b) within pyrolusite, and it is subsequently altered to psilomelane-cryptomelane and goethite and forms colloform texture (Fig. 9.5h).

Manganiferous Quartzite

Highly brecciated and elongated braunite (Fig. 9.5i) shows light grey with reddish-brown internal reflection, and it occurs in the manganiferous quartzite. Subhedral to anhedral braunite shows medium grey, brown internal reflection and irregular contact with pyrolusite is also observed within this unit at places. Cross-cutting relationship between the psilomelane-cryptomelane veins and the braunite-pyrolusite veins is also seen. The quartz grains occur as angular within the manganese ore matrix and a few manganese ore veins are also seen along the fracture of quartz (Fig. 9.5j). Manganiferous quartzite (sample BA/XRD/3 – Boringpadar) is made up of bixbyite, jacobsite, pyrolusite, psilomelane, goethite and quartz (Fig. 9.5k). These ore minerals interstitially occur between quartz grains, and a few angular quartz grains also occur within. Psilomelane and goethite occur together within the veins as the major minerals, and it has some primary minerals like jacobsite and bixbyite (Fig. 9.5l). Psilomelane shows bluish grey to cream white with strong brown internal reflection occurring with goethite and quartz. Goethite has a grey with bluish tint with a strong yellow-reddish brown internal reflection and is fibrous (colloform texture). Cream white with shades of grey-bluish grey anisotropism is seen in pyrolusite and jacobsite occurs as anhedral and is characterized by light grey, isotropic with concave and contorted contacts with psilomelane and quartz gangue (Fig. 9.5m). A thick manganese ore vein traversing in quartzite of sample BA/26 (Boringpadar) is seen which is represented by colloform texture (Fig. 9.5n) and defines the minerals assemblages like pyrolusite, psilomelane-cryptomelane, goethite and quartz gangue. A zonal patchy occurrence of darker grey manganite (bluish-brown internal reflection) and cream white pyrolusite is replaced by fibrous cryptomelane (Fig. 9.5o). Idiomorphic manganite shows medium grey with dark brown-red internal reflection, and it has remobilized along the fracture plane of quartz as psilomelane (Fig. 9.5p).

Analytical Techniques

Appropriate care has been taken in selecting samples for chemical analysis and 157 nos. of samples including bedrock samples (BRS), pitting/trenching samples (PTS) and soil samples (SS) from the manganese ore and manganiferous quartzite were collected for analysis of Mn, Fe, SiO_2, Al_2O_3, P, S and CaO. These samples were analysed using an X-Ray Fluorescence (XRF) instrument at the Geological Survey of India, Eastern Region, Kolkata, India. The analytical results are given in Tables 9.1 and 9.2.

Results and Discussion

Mode of Occurrence and Nature of Mineralization

Two bands of manganese ore and four bands of manganiferous quartzite are delineated southeast of Boringpadar and northeast of Amath. The manganese ore bodies occur as fragmental and discontinuous bands, pockets and stringers which are mainly restricted to the contact between calc-silicate granulite and garnetiferous quartzite. The ore is closely interbanded within calc-silicate granulite with diffused contact and very often shows pinch and swell structure along the strike. Manganese ore bands are generally noticed parallel to the prominent foliation of khondalite, calc-silicate granulite and garnetiferous quartzite. The schistosity and fracture planes probably represent small-scale paths for incompetent ductile flowage of manganese minerals to form cavity filling and replacement type of deposits. Lenticular bodies and discontinuous pockets of manganese ore are also recorded along joints and fractures planes. The first band of manganese ore is mapped to the southeast of Boringpadar which occurs on the lower flank of the hill. The Mn ore is bluish-black to chocolate (Fig. 9.3a), massive to foliated, has a metallic luster and has nodular and botryoidal forms (Fig. 9.3b). The manganese minerals are mostly cryptomelane, pyrolusite and psilomelane. The strike length of this ore body is 200 m having a width of 50-70 m which is trending in the N20°W-N40°E direction. The second band of manganese ore has been mapped towards the northeast of Amath having a dimension of 250 m and 40 m trending in the N10°W-N30°E direction. The manganese ore is mainly bluish black (Fig. 9.3c), massive to foliated, has a metallic luster and has botryoidal and nodular forms (Fig. 9.3d). At places, fragments of quartzite are embedded in a matrix of manganese ore. Four bands of manganiferous quartzite have been mapped to the southeast of Boringpadar which is hard, white to black, highly jointed, fractured (Fig. 9.3e) and brecciated (Fig. 9.3f). At places, the thin band of manganese ore occurs in the fracture plane of manganiferous quartzite. The geophysical investigation has been carried out in the Boringpadar and Amath areas to know the subsurface continuity of manganese mineralization. The apparent resistivity values vary from 10 to 240 ohm-m which displays low resistivity and the apparent chargeability contour map shows a high anomaly, indicative of a mineralized zone in the Boringpadar area. In Amath, the apparent resistivity value varies from 20 to 140 ohm-m and the apparent chargeability contour map shows a high anomaly. The pseudo-depth sections of Amath were prepared over geophysical traverses to examine the subsurface distribution of apparent chargeability and apparent resistivity which is suggestive of the depth persistence of the causative source below 35 m (Fig. 9.4c).

Controls of Mineralization and Types of Manganese Deposits

Manganese mineralization is usually controlled by stratigraphy, lithology and structure. In the studied area, it is mostly controlled by lithology and structure. The manganese ore bodies are strata bound occurring in between calc-silicate granulite and garnetiferous quartzite which define a narrow characteristics stratigraphic horizon within vast litho sequences of the Khondalite Group of rocks. Structurally, the ore-bodies are well preserved in F_1 synformal keels within the granitoid and gneisses, at F_2 fold closures and F_3 synforms. Richer concentrations are confined to schistosity, cleavage and fracture planes related to S_1 and S_2. On examination of trench and pits of Amath and Boringpadar areas, it is noted that the manganese ore and manganiferous quartzite bearing horizon is dominantly observed between calc-silicate granulite and garnetiferous quartzite of the Khondalite Group of rocks belonging to the Eastern Ghats Granulite Belt, thus indicating the stratigraphic control of manganese ore mineralization in the area.

Bands of manganese ore and manganiferous quartzite are delineated which show a predominantly syngenetic type of deposit, and these bodies occur in the form of fragmented bands and lenses at the contact of calc-silicate granulite and garnetiferous quartzite probably suggesting their syngenetic nature. Field evidence like sharp contact between these two litho-units and parallel alignment of manganese mineralization along these litho-units indicates that the manganese mineralization is of syngenetic type. In places, the epigenetic nature of mineralization is also noticed within these ore bodies which is mainly inferred from the replacement type of ore occurring as lenses, box-work, streaks, reticulate veins and discontinuous patches within the host rock along foliation/cleavage, fracture, joints and shear planes in the rocks as noted at some places towards the southeast of Boringpadar (Fig. 9.3e, f).

Paragenesis of Manganese Ore and Manganiferous Quartzite

The manganese ore minerals of Boringpadar-Amath areas show distinct textures due to the presence of different mineral associations. The texture of manganese ores is identified as a replacement, colloform/spheroidal, veined and brecciated textures. Based on textural evidence of manganese ore in the study area, it is concluded that pyrolusite, psilomelane-cryptomelane has a wide distribution in all the grades of metamorphism and all manganese ore horizon of different formation. Braunite, and jacobsite are recorded in many samples of manganese ore. Bixbyite, jacobsite and braunite are followed by manganite, pyrolusite, psilomelane-cryptomelane and goethite which are spotted in a few manganese ore. These are high-grade

Table 9.1 Analytical value (wt. %) of bedrock samples (BRS) and soil samples (SS) from the Boringpadar-Amath area

Sl. No	Sample No.	Rock name	SiO_2	Al_2O_3	Fe(T)	Mn	CaO	P_2O_5
1	BA/CH/1/1	Mn ore	10.59	2.84	20.88	27.65	2.37	2.37
2	BA/CH/1/2	Mn ore	14.22	3.16	18.57	28.18	2.45	1.77
3	BA/CH/2/1	Mn ore	27.60	5.30	36.88	2.26	0.52	1.89
4	BA/CH/2/2	Mn ore	32.94	5.20	31.09	3.22	0.42	2.52
5	BA/CH/3/1	Mn ore	20.10	4.69	42.39	0.58	0.27	2.36
6	BA/CH/3/2	Mn ore	28.78	5.04	32.53	5.81	0.27	2.1
7	BA/CH/3/3	Mn ore	40.78	2.78	31.89	1.21	0.49	1.7
8	BA/CH/3/4	Mn ore	30.41	4.72	29.51	7.50	0.75	2.10
9	BA/CH/3/5	Mn ore	15.31	5.46	39.68	5.26	1.06	2.81
10	BA/CH/3/6	Mn ore	18.28	13.39	28.17	9.17	0.62	2.19
11	BA/CH/3/7	Mn ore	18.72	12.72	28.28	9.14	0.63	2.25
12	BA/CH/3/8	Mn ore	17.77	15.97	26.88	9.28	0.46	1.47
13	BA/CH/4/1	Mn ore	31.79	2.04	16.21	20.04	3.29	2.73
14	BA/CH/4/2	Mn ore	29.99	3.12	13.08	23.78	2.67	1.95
15	BA/CH/4/3	Mn ore	54.50	1.79	10.60	14.82	1.09	0.77
16	BA/CH/4/4	Mn ore	34.95	2.95	14.56	19.33	2.79	1.86
17	BA/CH/4/5	Mn ore	49.58	1.97	11.98	16.46	1.26	0.99
18	BA/CH/4/6	Mn ore	30.44	3.29	17.08	20.33	1.95	1.71
19	BA/CH 5/1	Mn ore	24.68	4.47	39.49	1.93	0.55	–
20	BA/CH 5/2	Mn ore	20.92	2.91	45.32	1.13	0.43	–
21	BA/CH 5/3	Mn ore	23.02	4.14	41.87	1.51	0.67	–
22	BA/CH 5/4	Mn ore	37.73	6.33	30.60	0.77	0.41	–
23	BA/CH 5/5	Mn ore	29.95	6.17	34.54	1.14	0.78	–
24	BA/CH 5/6	Mn ore	31.33	4.77	36.73	1.36	0.38	–
25	BA/CH 5/7	Mn ore	30.84	7.6	32.72	0.47	1.35	–
26	BA/CH 5/8	Mn ore	34.73	5.52	32.59	1.02	0.81	–
27	BA/CH 5/9	Mn ore	33	2.35	37.30	1.53	0.54	–
28	BA/CH 6/1	Mn ore	35.88	3.34	33.23	1.36	0.50	–
29	BA/CH 6/2	Mn ore	37.22	1.22	37.62	0.78	0.30	–
30	BA/CH 6/3	Mn ore	32.34	2.74	35.31	1.53	0.72	–
31	BA/CH 6/4	Mn ore	23.95	5.83	39.47	1.64	0.53	–
32	BA/CH 6/5	Mn ore	16.55	4.90	46.56	0.53	0.18	–
33	BA/CH 6/6	Mn ore	20.76	3.32	42.57	1.22	1.02	–
34	BA/CH 6/7	Mn ore	34.28	6.32	32.52	1.41	0.34	–
35	BA/CH 6/8	Mn ore	31.83	6.05	33.56	0.80	1.00	–
36	BA/CH 6/9	Mn ore	14.51	12.25	38.48	0.19	0.8	–
37	BA/CH 6/10	Mn ore	30.8	3.15	36.84	1.50	0.47	–
38	BA/CH 6/11	Mn ore	8.62	1.47	56.31	1.16	0.63	–

(continued)

Table 9.1 (continued)

Sl. No	Sample No.	Rock name	SiO_2	Al_2O_3	Fe(T)	Mn	CaO	P_2O_5
39	BA/BRS/1	Manganiferous quartzite	41.95	2.19	24.49	9.60	0.09	–
40	BA/BRS/2	Manganiferous quartzite	58.46	1.43	25.04	0.26	0.05	–
41	BA/BRS/3	Manganiferous quartzite	61.07	2.56	20.16	0.88	0.09	–
42	BA/BRS/4	Manganiferous quartzite	92.12	2.29	1.87	1.23	0.11	–
43	BA/BRS/5	Manganiferous quartzite	31.21	2.81	23.04	16.57	0.20	–
44	BA/BRS/6	Manganiferous quartzite	77.64	1.58	11.67	0.19	0.12	–
45	BA/BRS/7	Manganiferous quartzite	37.09	1.01	39.09	0.12	0.03	–
46	BA/BRS/8	Manganiferous quartzite	77.70	2.79	10.92	0.09	0.17	–
47	BA/BRS/9	Manganiferous quartzite	33.98	1.06	43.93	0.05	0.01	–
48	BA/BRS/10	Manganiferous quartzite	72.03	3.75	14.27	0.26	0.17	–
49	BA/BRS/11	Manganiferous quartzite	72.03	10.11	4.82	2.22	0.23	–
50	BA/BRS/12	Manganiferous quartzite	72.3	7.31	5.76	4.40	0.4	–
51	BA/BRS/13	Mn ore	38.09	20.80	10.86	8.81	0.33	–
52	BA/BRS/14	Manganiferous quartzite	26.73	9.80	35.03	0.51	0.08	–
53	BA/BRS/15	Mn ore	18.72	9.67	38.92	0.95	0.08	–
54	BA/BRS/16	Manganiferous quartzite	28.39	12.59	26.00	6.25	0.15	–
55	BA/BRS/17	Mn ore	31.91	10.20	16.38	12.28	1.49	–
56	BA/BRS/18	Mn ore	34.42	7.34	26.57	0.48	4.73	–
57	BA/BRS/19	Mn ore	39.01	3.86	29.81	1.30	0.55	–
58	BA/BRS/20	Mn ore	21.4	2.25	42.84	1.85	0.59	–
59	BA/BRS/21	Mn ore	32.04	3.07	35.06	2.19	0.54	–
60	BA/BRS/22	Mn ore	8.45	3.05	30.49	20.09	2.86	–
61	BA/BRS/23	Ferruginous quartzite	74.78	4.50	8.30	0.64	0.01	–
62	BA/SS/1	In situ soil	51.93	20.21	6.25	0.23	0.82	0.20
63	BA/SS/2	In situ soil	53.35	18.92	5.98	0.20	0.57	0.16
64	BA/SS/3	In situ soil	54.46	17.71	5.59	0.24	0.75	0.20
65	BA/SS/4	In situ soil	54.04	18.08	5.17	0.22	0.98	0.26
66	BA/SS/5	In situ soil	51.13	19.43	6.27	0.27	1.18	0.25
67	BA/SS/6	In situ soil	52.97	19.83	6.17	0.23	0.80	0.38
68	BA/SS/7	In situ soil	52.77	20.03	6.04	0.24	1.15	0.34

(continued)

Table 9.1 (continued)

Sl. No	Sample No.	Rock name	SiO_2	Al_2O_3	Fe(T)	Mn	CaO	P_2O_5
69	BA/SS/8	In situ soil	51.82	19.32	5.84	0.21	0.94	0.28
70	BA/SS/9	In situ soil	54.02	20.39	5.51	0.26	0.69	0.35
71	BA/SS/10	In situ soil	51.85	20.67	9.85	0.30	0.66	0.11
72	BA/SS/11	In situ soil	55.05	17.99	9.11	0.27	0.64	0.13
73	BA/SS/12	In situ soil	55.96	18.29	8.87	0.25	0.56	0.12
74	BA/SS/13	In situ soil	57.41	16.93	8.11	0.28	0.91	0.14
75	BA/SS/14	In situ soil	55.92	16.75	8.20	0.28	0.89	0.14
76	BA/SS/15	In situ soil	57.41	17.96	7.49	0.25	0.69	0.20
77	BA/SS/16	In situ soil	53.21	20.55	7.60	0.21	2.52	0.12
78	BA/SS/17	In situ soil	58.54	18.59	7.62	0.25	0.68	0.20
79	BA/SS/18	In situ soil	56.36	18.05	7.85	0.62	0.75	0.14
80	BA/SS/19	In situ soil	56.52	17.99	7.98	0.55	0.81	0.16
81	BA/SS/20	In situ soil	56.96	18.39	7.96	0.45	0.60	0.14
82	BA/SS/21	In situ soil	55.28	19.22	8.39	0.30	0.61	0.11
83	BA/SS/22	In situ soil	56.92	18.83	7.80	0.29	0.88	0.12
84	BA/SS/23	In situ soil	54.66	21.12	9.94	0.21	0.40	0.07
85	BA/SS/24	In situ soil	57.34	18.82	7.93	0.30	0.79	0.12
86	BA/SS/25	In situ soil	57.23	18.91	7.79	0.29	0.63	0.09
87	BA/SS/26	In situ soil	58.82	19.15	7.49	0.22	0.34	0.07
88	BA/SS/27	In situ soil	57.86	20.12	7.91	0.31	0.59	0.08
89	BA/SS/28	In situ soil	56.55	19.47	7.97	0.37	1.38	0.12
90	BA/SS/29	In situ soil	56.75	18.44	7.72	0.45	1.59	0.12
91	BA/SS/30	In situ soil	55.01	19.09	8.87	0.27	0.65	0.10
92	BA/SS/31	In situ soil	56.08	19.37	8.39	0.25	0.49	0.09
93	BA/SS/32	In situ soil	55.75	19.72	8.71	0.27	0.41	0.09
94	BA/SS/33	In situ soil	57.67	20.75	7.82	0.17	0.23	0.05
95	BA/SS/34	In situ soil	56.69	19.92	7.55	0.18	0.28	0.06
96	BA/SS/35	In situ soil	56.52	18.99	7.73	0.23	0.42	0.08
97	BA/SS/36	In situ soil	57.62	20.96	6.46	0.17	0.27	0.06
98	BA/SS/37	In situ soil	60.86	19.49	5.98	0.17	0.19	0.05
99	BA/SS/38	In situ soil	57.09	19.91	7.77	0.20	0.31	0.06
100	BA/SS/39	In situ soil	55.70	19.77	8.34	0.22	0.38	0.08
101	BA/SS/40	In situ soil	56.58	18.97	7.95	0.23	0.79	0.09
102	BA/SS/41	In situ soil	56.97	19.32	7.93	0.25	0.46	0.08
103	BA/SS/42	In situ soil	54.93	20.15	8.92	0.20	0.83	0.08
104	BA/SS/43	In situ soil	52.02	20.02	8.66	0.18	0.61	0.07
105	BA/SS/44	In situ soil	52.65	20.90	9.13	0.21	0.72	0.07
106	BA/SS/45	In situ soil	51.23	21.14	9.25	0.29	0.97	0.08
107	BA/SS/46	In situ soil	51.90	20.48	9.41	0.25	0.82	0.09

Table 9.2 Analytical value (wt. %) of pitting/trenching samples from Boringpadar-Amath area

Sl. No	Sample No.	Rock name	SiO_2	Al_2O_3	Fe(T)	Mn	P_2O_5
1	BA/T1/1	Mn-Graphite bearing khondalite	34.01	11.04	28.43	0.47	1.12
2	BA/T1/2	Mn-Graphite bearing khondalite	41.01	10.56	20.96	0.40	2.22
3	BA/T1/3	Mn-Graphite bearing khondalite	38.45	9.32	22.66	0.36	3.19
4	BA/T1/4	Mn-Graphite bearing khondalite	38.99	6.35	23.07	0.43	5.43
5	BA/T1/5	Mn Ore	27.58	10.26	6.41	24.48	1.18
6	BA/T1/6	Mn Ore	33.90	6.55	28.95	3.40	2.24
7	BA/T1/7	Mn Ore	31.00	9.09	34.07	0.48	1.39
8	BA/T1/8	Mn Ore	15.53	5.00	46.85	1.58	0.91
9	BA/T1/9	Mn Ore	36.16	6.60	21.93	11.18	1.53
10	BA/T1/10	Mn Ore	29.09	7.13	26.00	5.60	1.83
11	BA/T1/11	Mn Ore	28.18	4.69	38.19	0.29	2.29
12	BA/T2/1	Mn Ore	18.19	7.79	44.70	0.30	0.77
13	BA/T2/2	Mn Ore	30.13	7.17	36.75	0.79	0.94
14	BA/T2/3	Mn Ore	19.30	6.36	45.25	0.33	0.80
15	BA/T2/4	Mn Ore	35.83	14.11	18.73	10.21	0.86
16	BA/T2/5	Mn Ore	35.94	9.01	29.02	2.72	0.75
17	BA/T2/6	Mn Ore	29.77	11.36	33.02	0.16	0.53
18	BA/T2/7	Mn Ore	35.73	14.00	18.21	10.44	0.85
19	BA/T3/1	Mn Ore	30.77	9.04	28.98	5.01	0.95
20	BA/T3/2	Mn Ore	41.93	7.58	22.18	6.69	0.54
21	BA/T3/3	Manganiferous quartzite	52.79	25.08	2.56	0.12	0.36
22	BA/T3/4	Mn Ore	23.80	5.82	40.84	1.84	0.72
23	BA/T3/5	Mn Ore	44.71	10.10	20.99	4.21	0.62
24	BA/T4/1	Mn bearing khondalite	47.83	25.40	7.55	0.12	0.08
25	BA/T4/2	Mn bearing khondalite	36.38	22.12	20.33	1.94	0.47
26	BA/T5/1	Mn Ore	61.91	1.41	16.98	6.08	0.28
27	BA/T5/2	Mn Ore	73.02	2.47	8.93	2.19	0.77
28	BA/T5/3	Mn Ore	43.60	1.66	29.98	2.60	0.68
29	BA/T5/4	Mn Ore	38.71	1.33	35.20	1.22	0.29
30	BA/T5/5	Mn Ore	46.28	1.45	29.58	1.78	0.21
31	BA/T5/6	Mn Ore	59.95	2.67	18.55	2.67	0.31
32	BA/T5/7	Mn Ore	64.73	1.74	18.12	1.07	0.18
33	BA/T5/8	Manganiferous quartzite	73.48	3.15	10.06	2.76	0.10
34	BA/T5/9	Mn Ore	51.95	2.30	23.15	1.78	0.73
35	BA/T5/10	Mn Ore	37.87	2.12	31.57	3.78	0.92
36	BA/T5/11	Mn Ore	58.85	1.40	17.93	5.70	0.40
37	BA/T5/12	Manganiferous quartzite	52.92	1.78	23.80	4.97	0.37
38	BA/T6/1	Mn bearing khondalite	36.70	12.12	29.27	0.09	0.16

(continued)

Table 9.2 (continued)

Sl. No	Sample No.	Rock name	SiO_2	Al_2O_3	Fe(T)	Mn	P_2O_5
39	BA/T6/2	Mn bearing khondalite	44.09	25.14	15.25	0.10	0.18
40	BA/T6/3	Mn bearing quartzite	80.67	10.21	5.47	0.09	0.08
41	BA/T6/4	Mn bearing khondalite	47.06	25.60	11.06	0.05	0.03
42	BA/T6/5	Mn bearing quartzite	87.54	6.72	2.11	0.10	0.06
43	BA/T6/6	Mn bearing khondalite	26.66	7.14	37.94	1.97	0.76
44	BA/T7/1	Mn Ore	13.70	4.68	30.09	16.08	2.75
45	BA/T7/2	Mn Ore	19.24	11.64	19.17	18.84	1.29
46	BA/T7/3	Mn Ore	29.59	4.17	16.07	22.53	0.55
47	BA/T7/4	Mn Ore	59.89	4.55	10.01	8.20	0.90
48	BA/T7/5	Mn Ore	9.75	3.28	21.41	27.77	2.42
49	BA/T7/6	Mn Ore	22.92	11.78	17.47	17.10	1.98
50	BA/PT/1	Manganiferous quartzite	60.06	20.55	5.01	0.73	0.33

metamorphic minerals. Pyrolusite, psilomelane-cryptomelane and manganite are identified as low-grade metamorphic minerals. Exsolution of manganese oxides in jacobsite, braunite and pyrolusite also occur in manganese ore, and they have been introduced by higher pressure-temperature conditions. Second generation braunite has replaced only primary braunite and secondary pyrolusite and psilomelane have also invaded and replaced all the primary minerals abundantly. Pseudo-colloform and spheroidal textures are formed by the alteration of jacobsite and pyrolusite to psilomelane-goethite. Replacement texture and relict texture are common between braunite-manganite-pyrolusite, jacobsite-braunite-pyrolusite, and pyrolusite-psilomelane-cryptomelane-goethite. Along the margin of these pyrolusite-jacobsite veins, psilomelane-cryptomelane also occurs. Tiny fibrous/needle-shaped cryptomelane is spotted. Goethite, quartz and garnets are associated with manganese ore minerals which are metamorphic minerals. In the study area, two types of minerals are identified. These identified minerals are categorized as primary minerals and secondary minerals. Braunite, bixbyite, jacobsite and pyrolusite are primary minerals that are associated with host rock like quartzite, and these minerals occur as inclusions and anhedral in shape. Primary manganese ore minerals are considered as most probably initial sedimentary manganese ore or metasediment which are laterally metamorphosed. The influx of hydrothermal to colloidal solution might be derivatives of the dissolution of the metamorphic manganese and the supergene enrichment might be due to the mobilization both of metamorphic and colloidal manganese. Secondary minerals are formed by the process of colloidal influx or metasomatism psilomelane-cryptomelane, and pyrolusite and goethite are secondary minerals which occur in higher quantities.

Manganese Mineralization in Manganese Ore and Manganiferous Quartzite

The chapter has brought to light an array of manganese mineralization with promising values, hosted in manganese ore and manganiferous quartzite. The analysed samples show high values of Mn, Fe, SiO_2, Al_2O_3, P, S and CaO (Table 9.1). SiO_2 values in bedrock samples (BRS) vary from 8.45 to 92.12 wt. % and Al_2O_3 values range from 1.01 to 20.80 wt. %. The Fe(T) value ranges from 1.87 to 56.31 wt. %. The Mn value ranges from 0.05 to 23.78 wt. %. The high values of Fe(T) were recorded from manganese ore of the Boringpadar area and the maximum value of Mn comes from manganese ore of the Amath area. CaO and P_2O_5 were also analysed. CaO values recorded in manganese ore vary from 0.01 to 3.29 wt. %. The P_2O_5 value varies from 0.77 to 2.81 wt. % in manganese ore. Analytical results of 50 nos. of pitting/trenching samples (PTS) also show encouraging results of Mn, Fe, SiO_2, Al_2O_3, P, S and CaO (Table 9.2). SiO_2 and Al_2O_3 values range from 9.75 to 87.54 wt. % and 1.3–25.60 wt. %, respectively. The Fe(T) value ranges from 2.11 to 46.85 wt. %. The Mn value ranges from 0.05 to 27.77 wt. %. P_2O_5 was also analysed, and the value varies from 0.03 to 5.43 wt. %. In addition, 46 nos. of soil samples have also been collected in the zone of manganese mineralization, and the analytical results of manganese and other elements are presented in Table 9.1. SiO_2 values vary from 51.1 to 60.86 wt. % and Al_2O_3 values vary from 16.75 to 21.14 wt. %. The Fe(T) and Mn values vary from 5.51 to 9.85 wt.% and 0.17–0.62 wt.%, respectively. Apart from these, CaO and P_2O_5 were also analysed, and the value varies from 0.19 to 2.52 wt. % and 0.05–0.38 wt. %, respectively.

Source, Deposition and Responsible Climatic Environment for the Origin of Manganese

The manganese accumulations predominantly occur in the Archean and Proterozoic periods and a few occurrences were also noticed in the Phanerozoic Era due to the continental crustal growth, decreasing mantle heat flux, tectonic evolution, sedimentary environment, and biosphere–atmosphere–hydrosphere interaction (Acharya et al. 1994; Roy 2000). The rapid burial of organic matter, increased weathering on tectonically amalgamated supercontinents, reduction of reduced gases from a gradually oxygenated mantle, and the net increase in photosynthetic O_2 was also noticed at the transition of the Archaean-Proterozoic. However, manganese rocks and ore were also derived from volcanogenic and/or, hydrothermal and freshwater sources which are also reported from different parts of the world (Acharya et al. 1994; Roy 2000, 2006; Mishra 2015). In the study area, two bands of manganese ore and four bands of manganiferous quartzite are delineated towards the southeast of Boringpadar and northeast of Amath which are mainly strata bound occurring in between calc-silicate granulite and garnetiferous quartzite. The occurrence of a well-defined manganese band having the same strike and dip as that of the country rocks (khondalite) having signatures of deformation on both, strongly suggests a meta-sedimentary parentage and continental erosion appears to be the most probable source for this deposit. Manganese ore bodies appear to be confined to definite stratigraphic horizons in the Khondalite suite of rocks in the part of the Eastern Ghats Mobile Belt. Based on field observations, ore petrographic, and available geochemical inputs indicate that the manganese ores and manganiferous quartzite of the study area might be formed in the freshwater environment. The study is also supported by the non-existence of rocks derived from volcanic activity in the near vicinity of the area under investigation, and the absence of wall rock alteration and manganese sulphide minerals. In addition, further detailed studies are required for identifying the link between manganese mineralization and the climatic changes in the study area and the same area which can be correlated with worldwide deposits, especially in the Proterozoic Era.

Conclusions

Two bands of manganese ore and four bands of manganiferous quartzite are reported southeast of Boringpadar and northeast of Amath area belonging to the Eastern Ghats Mobile Belt, India, and have been described in terms of their field distribution, ore-petrology and chemistry to find out their genetic evolution. The manganese ores are represented by bluish-black to brownish-black, massive to foliated, metallic luster, nodular and botryoidal forms and manganiferous quartzite is hard, white to black, highly jointed, fractured and brecciated, respectively. The geophysical study indicated that the mineralized body starts almost from the surface and the contours of this anomaly zone open downward which is suggestive of depth persistence of the causative source below 35 m. The values of Mn and Fe (T) ranged from 0.05 to 23.78 wt. % and 1.87–56.31 wt. %, respectively. In pitting/trenching samples, the Mn value ranges from 0.05 to 27.77 wt. % and the Fe(T) value ranges from 2.11 to 46.85 wt. %. The highest values of Mn come from the manganese ore of the Amath area, and maximum values of Fe(T) were recorded from manganese ore of the Boringpadar area. Based on the analytical results of the study area, it appears to be a resource potential for manganese and iron mineralization. The field observations, ore petrographic and available geochemical inputs indicated that the manganese ores and manganiferous quartzite of the study area might be formed in the freshwater environment.

Acknowledgements The authors convey sincere thanks to the Director-General, Geological Survey of India for permitting publication of the paper and are also thankful to the Deputy Director-General, SU: Odisha, Bhubaneswar for supporting necessary logistics while carrying out field work and laboratory studies. The authors would like to express their heartfelt thanks and gratitude to the Deputy Director-General, SU: Bihar, Patna for his cooperation, constant encouragement, guidance and valuable suggestions during the work.

References

Acharya BC, Rao DS, Sahoo RK (1994) Mineralogy and genesis of Kutinga manganese deposit, South Orissa, India. J Min Petr Econ Geol 89:317–328

Ajmal M (1990) Trace element study of manganese ores of the Barbil area, Keonjhar district, Orissa, India. Mineral Deposita 25:160–162

Anhaeusser CR, Wilson JF (1981) Southern Africa: the granitic-gneiss greenstone shield. In: Hunter DR (ed) Precambrian of the Southern Hemisphere. Elsevier, Amsterdam, pp 423–499

Basu NK (1969) Origin of the lateritoid manganese deposits of Keonjhar district, Orissa. Q J Geol Min Metall Soc India 41:183–187

Bhattacharya A, Raith M, Hoernes S (1998) Geochemical evolution of the massif-type Anorthosite complex at Bolangir in the eastern Ghats Belt of India. J Petrol 39(6):1169–1195

Chadwick B, Ramakrishnan M, Viswanatha MN (1981a) Structural and metamorphic relations between Sargur and Dharwar supracrustal rocks and Peninsular Gneiss in Central Karnataka. J Geol Soc India 22:557–569

Chadwick B, Ramakrishnan M, Viswanatha MN (1981b) The stratigraphy and structure of the Chitradurga region: an illustration of cover-basement interaction in the late Archaean evolution of the Karnataka craton, southern India. Precambrian Res 16(1–2):31–54

Chetty TRK, Murthy DSN (1994) Collision tectonics in the late Precambrian Eastern Ghats Mobile Belt: mesoscopic in satellite scale structural observations. Terra Nova 6. https://doi.org/10.1111/j.1365-3121.1994.tb00635.x

Condie KC (1981) Archean greenstone belts. Elsevier, Amsterdam, 434p

Dasgupta HC, Sambasiva Rao VV, Krishna C (1999) Geology, geochemistry and genesis of the freshwater, Precambrian manganese deposits of the Iron Ore Group from Noamundi basin, Eastern India. Indian J Geol 71(4):247–264

Dash C, Behera SN (2009) Search for potential manganese ore bands/bodies within the manganiferous horizon of the eastern part of the Eastern Ghats granulite belt in Orissa covering parts of Angul and Boudh districts, (P-I). Unpublished report of Geological Survey of India, FS, pp 2005–2006

Dash C, Behera SN, Patel SN (2005) Delineation of potential manganese ore bands/bodies within the Manganiferous horizon of the central sector of the Eastern Ghats Granulite Belt in Orissa covering parts of Bolangir, Kalahandi and Rayagada districts (P-II). Unpublished report of Geological Survey of India, FS: 2003–05

DeWit M, Hart R, Stern C, Barton CM (1980) Metallogenesis related to seawater interaction with 3.5 b.y. oceanic crust. EOS 61:386

Dimroth E, Imrch L, Rocheleau M, Goulet N (1982) Evolution of the south-central of the Archean Abitibi Belt, Quebec. Part I: stratigraphy and paleogeographic model. Can J Earth Sci 19:1729–1758

Engineer BB (1956) Geology and economic aspects of the manganese ore deposits of Jamda Koira valley; Keonjhar, Bonai, Orissa, India. In: 20th international geological congress & symposium on Manganese (Mexico) 4, pp 25–39

Former LL (1909) The manganese ore deposits of India. Mem Geol Surv India 37:610p

Gandhi A (2010) Manganese Report. Ideas 1st Information Services Pvt. Ltd, Mumbai, pp 5–7

Ghosh R, Chakraborty D, Halder M, Baidya TK (2015) Manganese mineralization in Archean greenstone belt, Joda-Noamundi sector, Noamundi basin, East Indian Shield. Ore Geol Rev 70: 96–109

Goodwin AM (1973) Archean iron-formation and tectonic basins of the Canadian Shield. Econ Geol 68:915–933

Gross GAL (1986) The metallogenetic significance of iron-formation and related stratafer rocks. J Geol Soc India 28:92–108

Hallberg JA, Glikson AY (1981) Archean granite-greenstone terrains of W. Australia. In: Hunter DR (ed) Precambrian of the Southern Hemisphere. Elsevier, Amsterdam, pp 33–103

Jayaram BN (1956) Report on the investigation of reported occurrences of manganese ore in the Boudh subdivision of Phulbani district, Orissa. Unpublished report of Geological Survey of India, FS: 1954–55

Jena SK, Devdas V (1994) Interim progress report on the exploration for manganese in Kutinga-Devjhola block and adjoining areas of Koraput and Rayagada districts, Orissa. Unpublished report of Geological Survey of India, FS: 1992–93

Krishna Rao JSR (1963) Microscopic Examination of Manganese Ores of Srikakulam and Visakhapatnam Districts, A.P., India. Econ Geol 58:434–440. https://doi.org/10.2113/gsecongeo.58.3.434

Kuleshov V (2017) The major epochs and phases of manganese accumulation in the Earth's history. Isotope Geochem:353–383. https://doi.org/10.1016/B978-0-12-803165-0.00004-5

Machado N, Carnerio M (1992) U–Pb evidence of Late Archean tectono-thermal activity in the southern Sao Francisco Shield, Brazil. Can J Earth Sci 29(11):2341–2346

Martin H, Peucat JJ, Sabate P, Cunha JC (1997) Crustal evolution in the early Archean of South America: example of the Sete Voltas massif, Bahia state, Brazil. Precambrian Res 82(1–2): 35–62

Mishra RN (1994) Geological environment of manganese ores of North Orissa — exercise in modelling of deposit. Indian Mining and Engineering Journal, (October and November), 23–26

Mishra B (2015) Precambrian metallic mineralization in India. Geol Soc Lond Mem 43:327–337. https://doi.org/10.1144/M43.22

Mishra P, Mishra SK, Singh PP, Mohapatra BK (2016) Reworked manganese ore bodies in Bonai-Keonjhar belt, Singhbhum Craton, India: petrology and genetic study. Ore Geol Rev 78:361–370

Mohapatra MN, Bagchi TC (1961) Mineralogy and paragenesis of the lateritic manganese ores around Joda, Keonjhar district, Orissa. Q J Geol Min Metall Soc India 23(1):5–14

Mohapatra BK, Paul AK, Sahoo RK (1989) Characterisation of manganese ores of a part of Western Koira valley, Keonjhar dist. Orissa. J Geol Soc India 34(6):632–646

Mohapatra BK, Mishra PP, Singh PP, Rajeev (2009) Manganese ore deposits in Koira- Noamundi province of Iron Ore Group, North Orissa, India: in the light of geochemical signature. Chem Erde 69:377–394

Mohapatra BK, Singh SK, Rao DS, Martha RK, Mishra S (2010) Characterisation of bog manganese ore associated with BIF in Iron Ore Group, North Orissa for its optimum utilization. International Seminar on Mineral Processing MPT-2010, Jamshedpur, pp 31–38

Mookherjee D (1966) Mode of occurrence of manganese ores in different rock types around Roida in Keonjhar and proposed classification of the associated ores. J Mines Metals Fuels 14:388–391

Moreira CA, Borges MR, Vieira GL, Malagutti FW, Fernandes MMA (2012) Geological and geophysical data integration for delimitation of mineralized areas in a supergene manganese deposits. Geofis Int 53(2):199–210

Murty VN, Ghosh BK (1971) Manganese ore deposit of the Bonai-Keonjhar belt, Orissa. Indian Minerals 25(3):201–210

Myers JS, Kröner A (1994) Archean tectonics: continental deformation. In: Hancock PL (ed) . Pergamon Press, Oxford, pp 355–369

Narayanswami S (1975) Proposal for the charnockite-khondalite system in the Archean shield of Peninsular India. Geol Surv India Misc Publ 23(1):1–16

Pascoe EH (1950) A manual of the geology of India and Burma, vol 1. Government of India Publication. Geological Survey of India, 485p

Prasad Rao GHSV, Murty YGK (1956) Manganese ore deposits of Orissa & Bihar India. In: 20th International geological congress & symposium on Manganese (Mexico) 4, pp 115–131

Ramakrishna M, Nanda JK, Augustine PF (1998) Geological evolution of the Proterozoic Eastern Ghats mobile belt. In: Proceedings of workshop on Eastern Ghats Mobile BeltGeological Survey of India Special Publication, 44, pp 84–92

Ramakrishnan M, Viswanatha MN, Swami Nath J (1976) Basement-cover relationships of Peninsular gneiss with high-grade schists and greenstone belts of southern Karnataka. J Geol Soc India 17:97–111

Ramazi H, Mostafaie K (2013) Application of integrated geoelectrical methods in Marand (Iran) manganese deposit exploration. Arab J Geosci 6(8):2961–2970

Rao GV (1969) The geology and manganese ore deposits of parts of Vizakhapatnam (Visakhampatnam) Manganese Belt, Srikakulam District, A.P., India. Bull Geol Surv India 35 (A):129

Ray BC (1954) The manganese ore deposit of Orissa. Indian Miner 8(4)

Ray S (1955) Ore microscopic studies of the manganese minerals of Jamda valley, Keonjhar. Proc Indian Acad Sci 42:90–97

Roy S (1981) Manganese deposits. Academic, London

Roy S (2000) Late Archean initiation of manganese metallogenesis: its significance and environmental controls. Ore Geol Rev 17:179–198

Roy S (2006) Sedimentary manganese metallogenesis in response to the evolution of the Earth system. Earth Sci Rev 77:273–305

Saha AK (1994) Crustal evolution of Singhbhum-North Orissa, eastern India. Mem Geol Soc India 27:341p

Schidlowski M (1988) A 3800-million-year isotopic record of life from carbon in sedimentary rocks. Nature 333:313–318

Schidlowski M (1993) The beginning of life on Earth: evidence from the geological record. In: Greenburg JM et al (eds) The chemistry of life's origins. Kulwar, Dordrecht, pp 389–414

Sen B (1951) Manganese ores of Keonjhar. Trade Min Geol Metall Soc India 25(2):51–60

Siddiquie FN, Raza M (2008) Geochemical environment & metallogenesis of manganese ores of Srikakulam-Visakhapatnam Manganese Belt (A.P.). In: Proceedings of the 24th convention and national seminar on sedimentary basin of India and Their Economic Resources, Aligarh, pp 86–87

Siddiquie FN, Shaif M (2015) Geochemistry of major oxides in host rocks in Vizianagarm manganese Ores Belt (A.P.), India. Int J Geosci 6:350–372. https://doi.org/10.4236/ijg.2015.64028

Siddiquie FN, Alam J, Shaif M (2015) Occurrence of manganese ore deposits and their mineralogy in Vizianagaram-Visakhapatnam manganese Ores Belt (Andhra Pradesh) India. Int J Geosci 6: 549–566. https://doi.org/10.4236/ijg.2015.66043

Singh V, Biswas A (2017) Physicochemical processing of low-grade ferruginous manganese ores. Int J Miner Process 158:35–44

Spencer E (1948) Manganese ore deposits of Jamda-Koira valley. Trade Min Geol Metall Soc India 44(2):61–77

Straczek JA, Krishanswamy S (1956) Manganese Ore Deposits of the Vizagapatnam (Visakhapatnam) District, A.P., India. In: Proceedings of the 20th International Geological Congress, Mexico, D. F., Symposium on Manganese, 4, pp 141–157

Teixeira W, Carneiro MA, Noce CM, Machado N, Sato K, Taylor PN (1996) Pb, Sr and Nd isotope constrains on the Archean evolution of gneissic–granitoid complexes in the southern Sao Francisco craton, Brazil. Precambrian Res 78(1–3):151–164

Telford W, Geldart L, Sheriff R, Keys D (1976) Applied geophysics. Cambridge University Press, New York

Vieira LB, Moreira CA, Côrtes ARP, Luvizotto GL (2016) Geophysical modelling of the manganese deposit for induced polarization method in Itapira (Brazil). Geofis Int 55(2):107–117

Windly BF (1982) The evolving continents, 2nd edn. Wiley, pp 28–65

Yadav PK, Subhadarshini S, Das M, Singh A, Shyam R, Gokul GS (2017) Final report on the reconnoitory survey for manganese ore in Boringpadar-Amath block in the Eastern Ghats Granulite belt, Kalahandi district, Odisha (G4). Unpublished report of Geological Survey of India, FS: 2016–17

Chapter 10
Archaeological Evidence of Climate Change: Potential Source to Understand the Past Climate Trend

P. D. Sabale

Abstract Each river develops a typical type of topography according to her reaches because each flowing channel shapes the Earth's surface through the process of erosion, transportation and deposition of sediments from source areas and deposition of sediment in low-lying areas and streams. The old deposited sediments during the Quaternary era are immensely important from an archaeological point of view, because human development has taken place contemporary to this period. Therefore, most of the contemporary cultural deposit is deposited in these sediments, and such sediment is rich in flora and fauna, which help to reconstruct the palaeoenvironment of those particular areas concerning human development. In the present context, the details of the palaeoenvironment along with the man–land observed in the Upper Reaches of the Sina are studied by taking the Quaternary trenches or excavations at suitable cultural settlement localities, in parts of the south Ahmednagar district of Maharashtra.

Keywords Paleontological · Palaeoclimate · Palaeoenvironment · Fertile floodplains · Flora-fauna

Introduction

The sediment, plant, paleontological and animal records of natural climate archives and archaeological records offer unique opportunities to observe, measure and understand the cultural activities and how humans have responded to a wide range of climatic conditions in the past. This forms a basis for the prediction of how climate change could transform our lives in the future and offers a range of possible solutions (Hambrecht et al. 2020). The record of cultural archives from archaeological sites offers opportunities to understand the complex interaction between man and the environment under different climatic zones. To study the past environments and to identify the factors that promoted humans in the past and apply the knowledge

P. D. Sabale (✉)
Department of Archaeology, Deccan College Deemed University, Pune, India

N. Khare (ed.), *Science, Policies and Conflicts of Climate Change*, Springer Climate,
https://doi.org/10.1007/978-3-031-16254-1_10

gained to the present, contributing a much-needed, long-term perspective to climate research. The important strengths of such a record are the cultural diversity it encompasses. Humans have successfully faced similar challenges in the past. Such archaeological record is a highly valuable source of information that has been used on a wide-scale in climate research. Therefore, climate science has had an impact on past research. Hence, arising from the importance of the same, the multidisciplinary, intersectorial research teams are no longer exceptions and many teams operating today have mature working partnerships.

Natural Resources and Cultural Sites

The river basins especially in severe, chronic, drought prone semi-arid regions of India acted as lifelines for the development of ancient civilizations, but they are also important to open the continental archive in the reconstruction of human history within the Quaternary palaeoclimate and palaeoenvironment (Gupta 2007). Major rivers which flow through a variety of geological and geomorphological features are important; therefore, we need to study such basins to understand the sediment transport characteristics, stratigraphic development, flood episodes, hazards and their effects on human culture. This chapter presents an overview of the research on the settlements in semiarid zones in India, especially in Maharashtra and identifies major evidence of climatic changes and unanswered research questions.

In many places, these rivers have sustained human civilizations for more than 5000 years and the deposition of rivers' alluvium provided fertile floodplains for agriculture. Examples of such ancient river valley civilizations are seen on the Indus, Nile and Ganga, where quite advanced flood protection measures were taken (Gupta 2007).

Need of Work

The records for the Quaternary era, especially Pleistocene, are rather scarce from most parts of the Indian peninsular. This may be due to lack of exposed sections, depositional and erosional fluvial activities and resulting changes in the morphology of the river, intensive agricultural practices resulting in changing the land surface, large-scale rural and urban development, etc. The Quaternary landform is hidden/ buried under other recent deposits. Therefore, there is an urgent need to study such Quaternary stratigraphy down to the bottom level (i.e. up to the hard rock). During the study of the natural and cultural landscape, the chronological framework of the stratigraphic records also needs to be strengthened and a stronger database is needed for developing high-resolution stratigraphy for the Quaternary period to help paleo-climatic reconstruction and test climate modelling for future projections (Burke et al. 2021). To understand the details of the palaeoenvironment along with man–land relationships of the particular area, we need to study the litho, bio and chronostratigraphy in context with the cultural stratigraphy of the region.

Archaeology as an Interdisciplinary Science

Archaeology of climate change emerges seamlessly from the long-standing collaboration between archaeology and the natural sciences that has provided climatological and environmental data by studying the Quaternary archaeological site from litho, bio and chronostratigraphical points of view and comparing this data critically for cultural stratigraphy. Because the archaeological record captures the breadth of past human adaptations, the archaeology of climate change is well situated to highlight alternative strategies that have worked in the past and address the social and economic ramifications of global warming for the diverse global community. Since the nineteenth century, such techniques have been widely used to study the palaeoenvironment against which past human activities were undertaken.

Therefore, the mutual benefits to be gained from working together fostered new collaborations between, earth scientists, palaeoclimatologites and archaeologists to revive interest in human-environment interactions. Sometimes, due to the lack of deposits, it is difficult to correlate cultural and natural stratigraphy. Therefore, most of the time, detailed field investigations will be helpful to correlate climate and cultural stratigraphy.

Effects of Climate Change on Human Society

Environmental changes and effects of warming result in a complex series of biological feedbacks which affect the economic and social challenges for human populations. Agriculture – an important occupation which caters to the need for food – sometimes leads to fertile lands dramatically transformed into dry and deserted lands where, it is difficult to cultivate any crop. There are ample examples of such landscapes which are already observed in some regions today and are likely to accelerate in the near future (Berteaux et al. 2018). Such slow or rapid transformations will affect food security for the next generations and have far-reaching consequences for the physical and psychological well-being of human populations in these regions (Ford et al. 2018). In this manner, the poor farmers, who rely on their relations to the land and access to its natural resources, lose their farm production and its results are felt in the overall economy.

Such local effects of production can affect the global economy and played an important economic role in many developmental activities (Brody 1981). Therefore, in this way, in economic development, agricultural production played a very important role. In this process of agriculture-based human development, climate change poses a fundamental threat to this group (Anonymous 2018). Evidence of climate change in Quaternary strata are recorded in the form of different evidence, like deposition of sediments layers, nature of the size of sediments, thickness of sedimentary strata, nature of cementation and material and mode of preservation and deposition of fossils in the bed.

Background

Evolutionary theory and the standard geological time scale have provided a rich, conceptual framework of events with activities that help to study the dynamics of past flora–fauna, environment and human interactions. The Quaternary period witnessed various stages of human evolution. The natural and cultural remains are very well preserved in Quaternary deposits. Thus, to understand the exploitation of the natural resources in each part of the river such as rocky terrain in the source region, while undulating topography partly covered with soil and rocky patches in the middle reaches and fully fertile, thick soil zone in the lower reaches, a detailed geoarchaeological exploration was carried out in field season 2010–11 and reported a number of Quaternary sites in each sector.

In the present context, the aim was to understand the details of geological formations and geomorphic features development of certain landscape resources. In what way do these landscapes attract humans and how do they exploit the natural resources. To understand the same, three localities in each reaches of Sina basin were systematically studied with the help of proper Quaternary trenching/excavation at each locality for better study of cultural and natural evidence found in the stratas. The three localities are Math-Pimpri, Pimpri and Hatvalan in the Upper; Panch-Pimpla, Khairav and Virvade in the Middle; while Rajur, Hattarsang and Kudal in the lower reaches (Fig.10.1). In the present context, paleoclimatic inferences observed in the Upper Reaches of the basin are discussed.

Methodology Adopted

The area under study was quite vast and forms the unexplored complete basin of the Sina, a left-bank tributary of Bhima. Therefore, it was decided to carry out a detailed geoarchaeological exploration and excavations to reconstruct the palaeoenvironmental conditions, geomorphology, geology and the landscape of the whole river basin.

The Quaternary is characterized by repeated climatic changes of considerable amplitude; analysis of these fluctuations reveals the ubiquity of change and the fundamental dynamism of earth systems. Change is normal and, even though southern Africa was not subject to Quaternary glaciations per se, the influence of variations in, say, amount and seasonality of rainfall, has been very marked indeed. The Quaternary also represents the time period during which people have become a

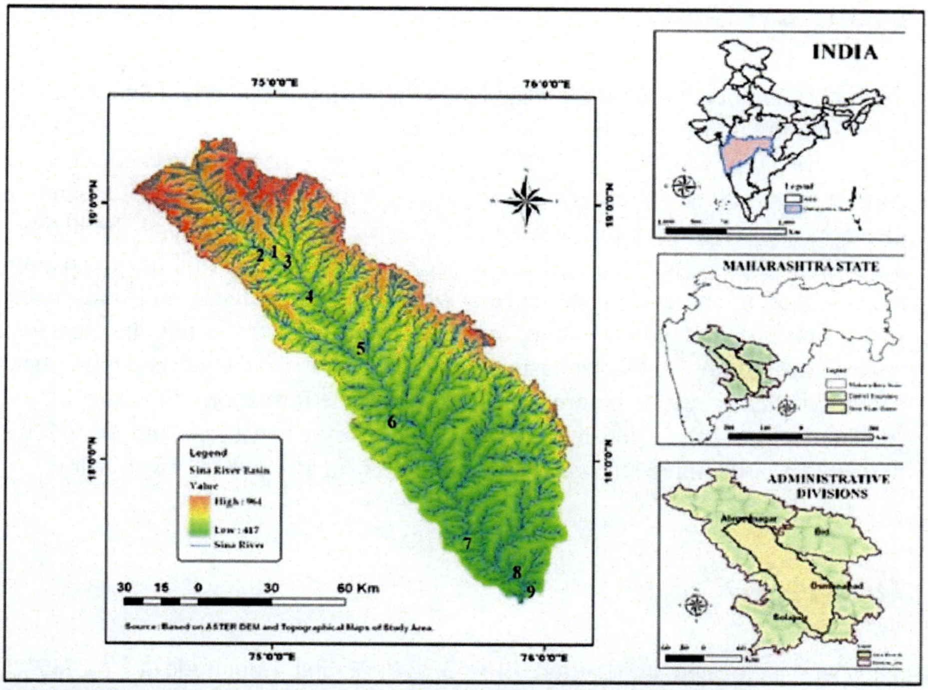

Index: 1.Math Pimpri 2.Pimpri 3.Hatvalan 4.Panch-Pimpla 5. khairav 6.Virwade Bk 7.Rajur 8.Hattarsang 9.Kudal

Fig. 10.1 Location map of the study area. (Modified after, Kadam 2018)

dominant environmental agent in the subcontinent; southern Africa has played a crucial role in the evolution of hominids and their development has been influenced strongly by accompanying environmental change.

The most important kind of specific environmental interpretation one can try to make is to figure out the nature of the depositing fluid flow. Most of the sediments but not all are deposited by flowing fluids. This presents some possibilities for making paleoflood interpretations by examining old clastic sedimentary sequences. When such interpretations can be made, they serve as guides or constraints in framing a broader picture of the depositional environment. Therefore, the possibilities for making such interpretations are numerous and varied, but they are still limited. Further work is going to reveal a lot of useful interpretive approaches based on features of beds we still don't know how to interpret.

Therefore, at most of the interesting localities, to study the Quaternary (Pleistocene and Holocene/flood) and habitation deposit, separate trenches in each deposit were taken. Wherever possible, the available different lithological sections of the river bank, dug well, road, etc., are checked and these yielded highly useful interpretations. Especially in the river cutting section, the stratigraphy of the particular area with the thickness of each bed sequence, thickness, its nature, type and texture of the material is studied. In each layer, fossils of plants and animals are recorded properly. Different types of samples of sediments, fossils and cultural materials were collected for the lab study, as per their norms.

Quaternary Sites

To understand details about the man–land relationship of the driest Sina basin, two sites from the upper, middle and lower reaches were excavated. These excavations were carried out from 2012 to 2020 season with some specific aims and objectives: (i) to reconstruct the cultural development and settlement patterns; (ii) to study the Quaternary environment of the Sina river basin; (iii) to understand the nature of deposition of different shape and size of sedimentary material with plant and animal assemblages in more or fewer layers with variable thickness in stratigraphical succession from the base level, i.e. from the hard rock; (iv) to study the Quaternary environment of the Sina River basin; and (v) to reconstruct the palaeoenvironmental conditions that prevailed before, during and after the formation of the site.

The details of these localities, where Quaternary excavations and studies were undertaken in the three reaches of the Sina River (of Bhima) are given below.

Math-Pimpri

This site is located at (Latitude 18°53.334′ N and Longitude74° 52.712′ E), 24 kilometres in the South Ahmednagar district, along the Solapur–Ahmednagar National Highway. This cultural settlement is present on the left bank of an obtuse meander of the Sina River, in between the present and palaeochannel of the Sina River (Figs. 10.1 and 10.2).

To study the cultural development and settlement patterns as well palaeoenvironment in the upper part of the basin, the site was excavated in two seasons, i.e. 2013–14 and 2014–15 (Sabale 2012). With the help of three trenches, three different types of material were excavated up to a depth of 17.5 m.

The work carried out at this site has produced a fourfold cultural sequence, namely

Locality-I: Holocene and Pleistocene deposits that rest on hard rock, i.e. Deccan Trap basalt. This trench was taken along the left bank of the Sina River towards the habitation site. This trench was made down to hard rock basalt (bottom most) rock. The section consists of various grades of sediments rich in animal remains and Mesolithic tools (Plates 10.1 and 10.2).

Locality-II: This trench is present just upstream of the previous locality-I in between habitation and the Quaternary trench. A huge thick, sediment strata is observed. It consists of concretized, humus-rich, agricultural-virgin soil (black cotton) layer of flood deposit (in the middle) (Plate 10.3).

Locality-III: Cultural deposits of Early Historic (*Satvahana*) and Early Medieval period (on the top of the mound), grinding stones, sling ball, blade and chip of agate and quartz, etc., were collected. A huge red slip ware from *Satvahana* and grey coarse ware of the medieval period were collected from the excavation trench and

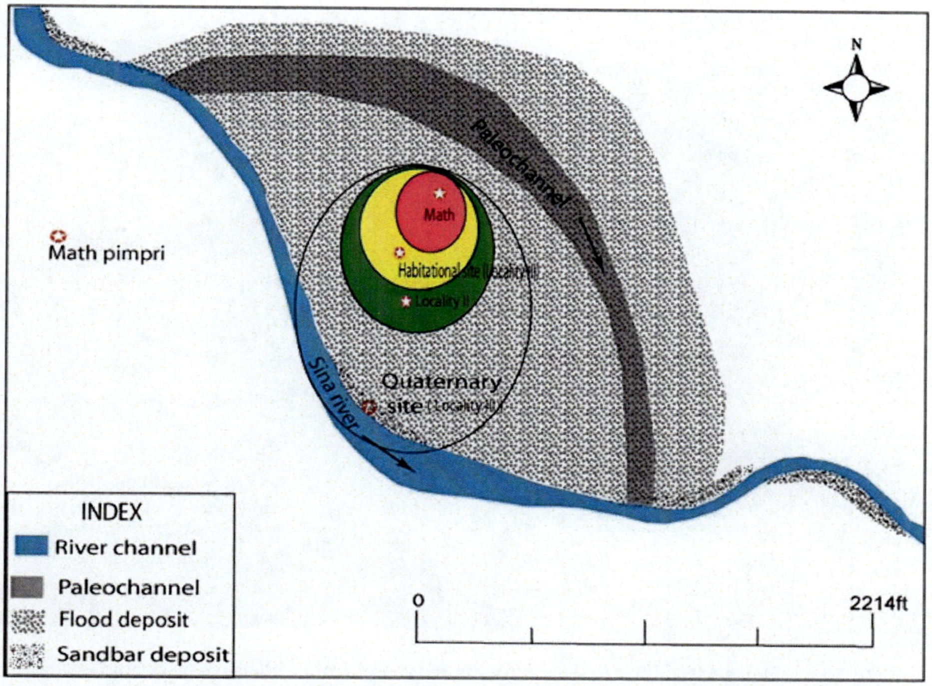

Fig. 10.2 Schematic map of the paleochannel of the Sina River at Math-Pimpri site

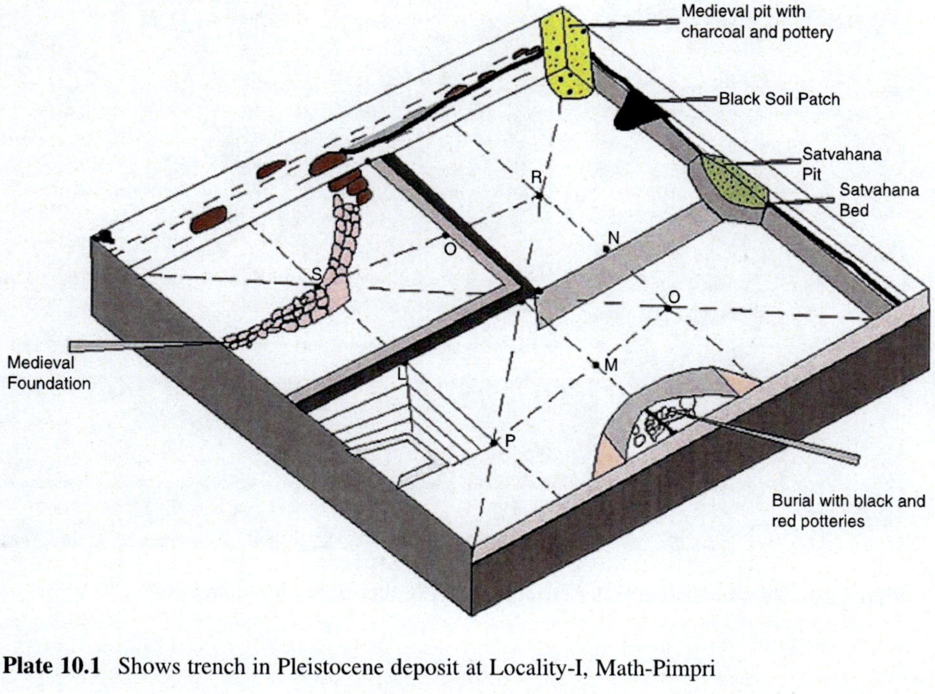

Plate 10.1 Shows trench in Pleistocene deposit at Locality-I, Math-Pimpri

Plate 10.2 Shows Mesolithic tools and fossils recovered from a Pleistocene deposit from Math-Pimpri

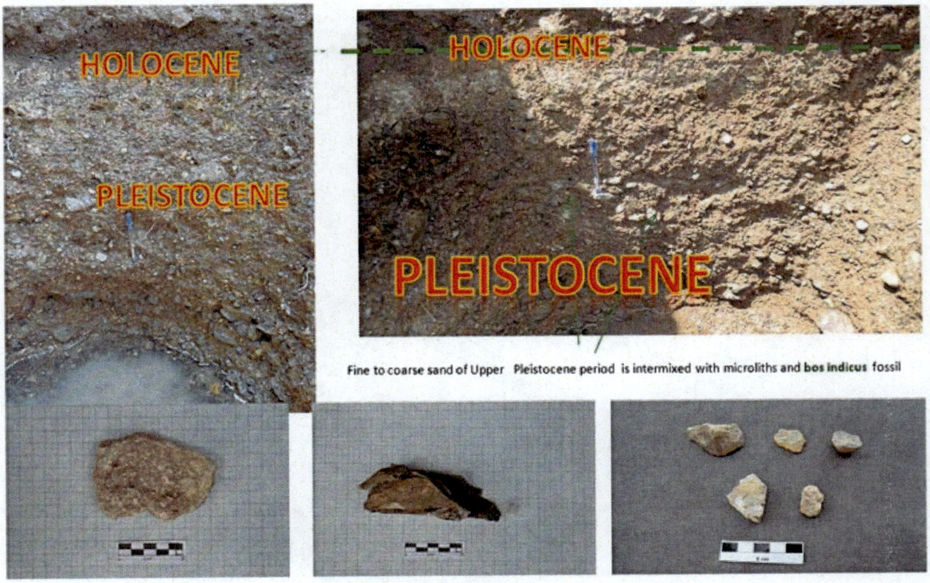

Plate 10.3 Shows trench in Holocene deposit at Locality-II, Math-Pimpri

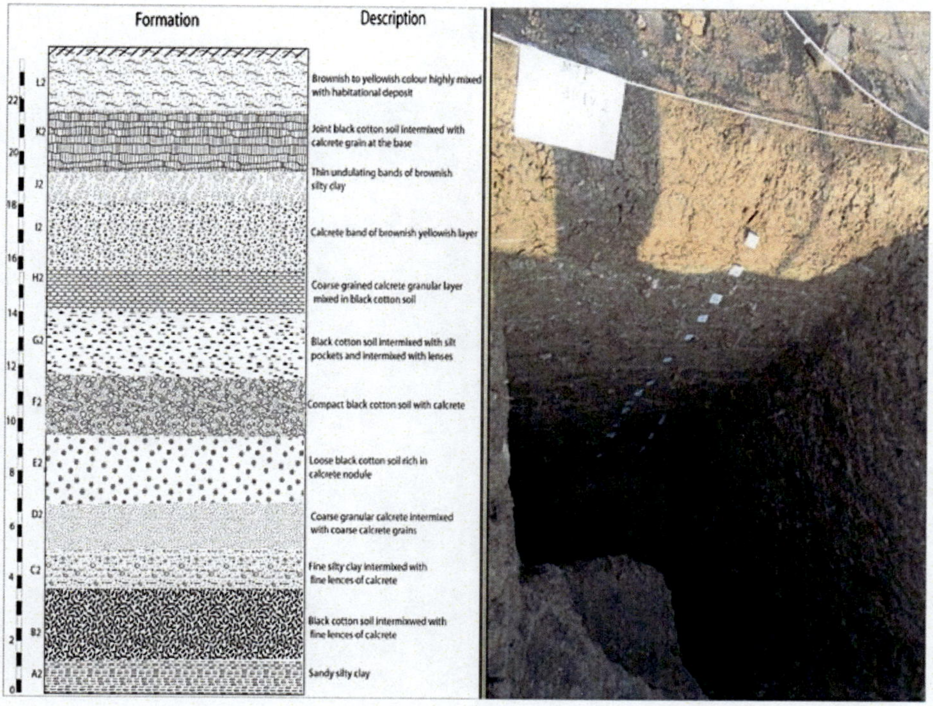

Plate 10.4 Trench in habitation deposit at Locality-III

surface exposures (Plates 10.4 and 10.5). In addition to this, a red ware bowl and one black ware plate of the medieval period were collected from secondary burial (Fig. 10.3).

Pimpri

This site is located on the right bank of the Sina River, in the vicinity of the present Math-Pimpri village in South Ahmednagar. A very good section of Quaternary is exposed in the form of a dug well, which shows all the stratigraphy of the Upper reaches of Sina (Fig. 10.1, Plate 10.6). This is a very long, detailed, complete and rare section that starts with agricultural soil at the top and extends to the hard rock – compact basalt of the Deccan Trap formation. This Quaternary deposit had prehistoric tools. A palaeolithic stone tool of the flake category was recovered from the dug well section (Kshirsagar et al. 2014). The different types of ceramic and historic period pottery wares were collected around the dug well and the Pimpri site was also recovered during the progress of works. It includes early historic (*Satvahana*) to medieval pottery.

Plate 10.5 Shows trench in habitation deposit at Locality-III, Math-Pimpri. (**a**) and (**b**) pottery recovered from the deposit and, (**c**) mesolithic tools

Hatvalan

It is present 4 km downstream of Math-Pimpri, in the South Ahmednagar district. The site section Quaternary (Middle-Lower Pleistocene) deposit is located on the right bank of the channel, where the diversion of the river channel takes place and a small island is developed. A huge section of pebble-cobble-rubble-coarse sand and silt, cemented with calcrete cement layers overlain by clay and fine silt sediment on the top surface, was observed along the channel. Out of these, the middle layers are rich in Mesolithic tools, especially agate, Jasper, milky, rosy, and banded haematite quartzite. In addition, the section exposed at the foundation of the present settlement shows early historic pottery (Fig.10.1, Plate 10.7).

Results and Discussion

In the dry land regions, drought and flood periods are most frequently reported (Endfield et al. 2004). Therefore, to quantify the long series of palaeoenvironmental records, a systematic study of each Quaternary site was taken up in each reach of the river with landform characters that provide unparalleled information.

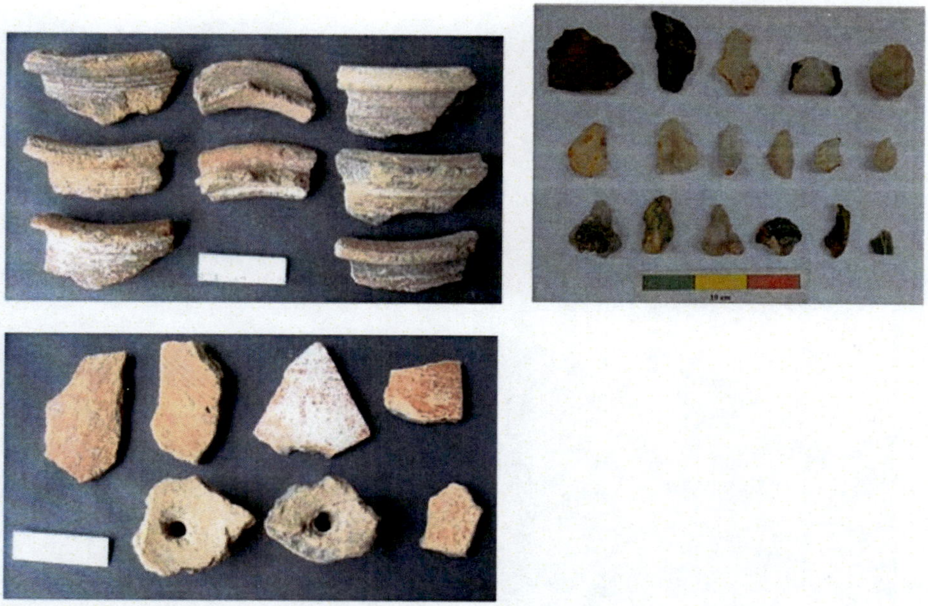

a. b. c.

Fig. 10.3 Three dimensional models of cultural deposit at Locality-III

In the present context, the palaeoenvironmental investigations to understand the geological formations, geomorphic features and development of certain landscape resources, in short *'Evolution of landscape and Man–land relationship'* in the driest river – Sina, in the Bhima basin, since 2009 (Sabale and Kshirsagar 2014). To understand in what way these landscapes attracted the humans and how they exploited the natural resources, three localities in each reach of the Sina basin were systematically studied with the help of proper Quaternary trenching/excavation at each locality to better study the cultural and natural evidence found in the stratas. These localities are Math-Pimpri, Pimpri and Hatvalan in Upper, Panch Pimpla, Khairav and Virvade in Middle, while Rajur, Hattarsang and Kudal are in the Lower reach.

From the upper reaches, Math-Pimpri is a unique site which is situated in between the two channels of the Sina River, due to highly suitable landscape conditions. Catchment analysis of Math-Pimpri has been carried out up to a 10 km radius from the site, which also covers other sites in the vicinity like Pimpri and Hatvalan. Qualitative and quantitative geomorphic analyses were undertaken to understand the landscape and topographical characters. The details of natural features in and around the study area were also recorded. Morphometric analyses of the catchment areas were also carried out to understand the landscape details about their suitability for the setup of the cultural settlement. An extensive, two-season excavation was carried out (Sabale et al. 2015). Field and lab studies of various cultural, sediment, fossil materials, etc., were carried out to understand the source of sediments and the

Plate 10.6 View of Quaternary section exposed in the dug well at Pimpri

different depositional environments (Jyotibala et al. 2019). Pollen and archaeobotanical samples were collected by the flotation method and were brought to the lab for various analyses. The faunal assemblages Bos-Indicus collected from the middle-lower Pleistocene layer of locality-I were handed over for fluorine dating analysis. The assemblage of the cattle burial was handed over to the Anthropological lab for further analysis.

Conclusions

The earth archives and cultural records offer opportunities to study the deposits showing complex interactions between the environment and humans under different climate regimes and at different spatial and temporal scales. Environmental archaeology offers opportunities to identify the variations in palaeoclimate that promoted human resilience in the past and apply the knowledge thus gained to the present, contributing a much-needed, long-term perspective to climate research (Haywood et al. 2019).

Plate 10.7 Shows the Quaternary section exposed at Hatvalan

Such effects in climatic variations result in disrupting social, economic or political systems and culturally appropriate strategies (Hussain and Riede 2020). Such changes one can systematically study by using the natural climate archives which include sediment records, pollens, ice core and paleontological and archaeological records that offer unique opportunities for observing, measuring and understanding how humans have responded to a wide range of climate events in the past. Such data is found to be very useful in the prediction of climate change and understanding how it will transform our lives in the future and offering a range of possible solutions with reference to the cultural periods (Hambrecht et al. 2020). Here, the observed cultural evidence serves as a very valuable source of information that has been largely overlooked in climate research (Ford et al. 2018; Berteaux et al. 2018; Brody 1981; Hambrecht et al. 2020; Anonymous 2018; Jackson et al. 2018; Hudson et al. 2012; Sandweiss and Kelley 2012; Cooper and Peros 2010). Therefore, the reconstruction of the palaeoenvironment with the help of litho-chrono and biostratigraphical parameters is a multidisciplinary aspect of the past that provides a solid foundation for assessing the implications of climate change across cultures and helps design sustainable development strategies.

Continental and oceanic sediments from tropical Africa, Australia and South America have recorded analogous changes in climate and stream behaviour during the late Quaternary. Recently, data from north-central India have provided evidence for late Quaternary changes in river regime, in response to global changes in climate in the intertropical zone (Kale and Rajaguru 1987).

In this study, an effort has been taken to collect the archaeological and natural evidence based on Quaternary excavation to understand climate change and adoption of human culture under different changes in the Sina River, part of the Bhima basin in Western Maharashtra. It is clearly observed that according to the youth, mature and old stages in the upper, middle and lower reaches of the Sina, the rate of weathering, resulting in the dissection of surface, topography, landform and drainage characters are changed accordingly. The hilly and rocky topography in the upper source region is altered into an undulating plain in middle reaches and in the lower reaches it has turned into the plain type. Owing to the rocky surface in the upper reaches, it is very difficult to use the land for agriculture. While in the middle reaches, rocky masses are surrounded by depressed areas, which are filled with soil, which is widely used for agricultural activities. A very thick and fertile alluvial soil/sediment is deposited in the lower plain surface, so it is extensively used for agriculture activities. The archaeological and geological evidence can reveal how societies have responded to climate variability in the past. It clearly shows that the societal responses could not have been predicted in advance and that the successful adaptations of people to the new climatic conditions depended on their good or bad choices.

In the present context, the archaeological excavation based on observations and recovered documents provides valuable information about their past conditions, economy, culture, as well as climatic conditions. In most regions, due to anthropogenic activity, the evidence thins out and often becomes discontinuous (National Research Council 2006). To understand the differences in landscape and distributions of natural resources accordingly, and how humans captured the river bank area in its Upper reaches, three Quaternary excavated sections at Math-Pimpri, Pimpri and Hatvalan areas were studied systematically. It is observed that in present drought prone regions of Deccan land, in which the Sina flows, such climate was more or less the same from its origin. However, in the meantime, the dry period was altered into wet, resulting in terminal Pleistocene, a huge flash flood developed which resulted in the deposition of thick layers/terraces with coarse size sediments, especially rubble-cobble-pebble sediments. At the same time, animal and plants also wash out and are deposited in these deposits. Therefore, numerous floral and faunal remains are also found in the Quaternary deposit.

The geomorphological studies have indicated that gradational and erosional phenomena of the rivers have been controlled to some extent by large-scale tectonic activity (Rajaguru and Pappu 1970). The geomorphic features such as cascades, entrenched meander and palaeochannel observed at Math-Pimpri, Hatvalan areas support the tectonic stability.

Rajaguru (1973) and Pappu (1970, 1973) gave some clues about the deposition of sediments in a stream flowing in Peninsular India. They suggested that broad river basins have higher discharge and more sedimentary load at the time of deposition of older alluvium, suggesting a somewhat humid climatic condition results in slightly more precipitation and flood frequency than the present. The resulting deposition of sediments takes place along the channel. The same scenario is also observed in Sina at several places.

The sedimentological record observed in the upper reaches of Sina, at the Math-Pimpri site, represents a vertical succession of several facies that belong to the channel and floodplain deposits in the lower–middle and upper parts of a fluvial depositional system. The lithostratigraphic section shows the evidence for both fluvial and dry environments.

The primary observation of the sections and various tests performed in the laboratory indicate that the area has undergone a fluctuation of dry and wet periods of climate, i.e. fluctuation in the environment. To overcome especially the drought situation in this area, people exploited the available natural resources which are present along the bank of the river. Since the river is dry compared to the other rivers in the Bhima basin, good agricultural practices are observed only along the bank, i.e. on the fertile track, humus-rich soil and of course, hence the big-sized, rich settlements of different cultural periods are settled on both banks of this river. Still, further study is required to reconstruct the palaeoenvironment and man–land relation of the basin.

There are 37 sedimentological sequences recorded at Locality-I at Math-Pimpri which are dominated by over bank deposits, indicating the successive flooding episodes. The upper part of each layer shows a slightly erosive surface, while the middle and lower portions of sequences are characterized by depositional episodes. The relatively fine-grained sediments and amalgamation of channel fill suggest a river system surrounded by a floodplain. However, at places, the deposition of calcrete layers in between the flood layers indicate a drought situation due to very little rainfall in the upper catchment areas. Such calcium carbonates in calcrete layers cement, spread and capture the lose sediments in the upper and lower contact layers. Based on the results of the present work, it is clear that the river flowing in the severe, chronic, drought-prone, semi-arid region of Maharashtra is also rich in faunal records of a variety of species, which occurs in other comparatively perennial rivers like Bhima. Therefore, one can conclude that the paleoenvironmental conditions and man–land relationship in Sina were not that much different than those in Bhima, even though it is extremely dry today and other rivers are flowing in severe, chronic drought prone regions. Detailed investigations are still required to understand the biodiversity of the basin in the Quaternary Era (Sabale et al. 2021).

Acknowledgements I thank the authorities of Deccan College Post Graduate and Research Institute, Pune, for their guidance and encouragement during the work. I am grateful to the colleagues and students who visited the site and helped in the work. I am also thankful to the villagers for their help during the progress of the excavation.

References

Anonymous (2018) Utqiagvik declaration. Inuit Circumpolar Council

Berteaux D et al (2018) Northern protected areas will become important refuges for biodiversity tracking suitable climates. Sci Rep 8:4623

Brody H (1981) Maps and dreams. Pantheon, New York, p 297

Burke A, Peros MC, Wren CD, Pausata FSR, Riel-Salvatore J, Moine O, de Vernal A, Kageyama M, Boisard S (2021) The archaeology of climate change: the case study for cultural diversity. PNAS July 27, 2021 118(30):e2108537118. Ed. R. Piperno

Cooper J, Peros M (2010) The archaeology of climate change in the Caribbean. J Archaeol Sci 37: 1226–1232

Endfield GH, Tejedo IF, O'Hara SL (2004) Drought and disputes, deluge and dearth: climatic variability and human response in colonial Oaxaca, Mexico. J Hist Geogr 30:249–276

Ford JD, Couture N, Bell T, Clark DG (2018) Climate change and Canada's north coast: research trends, progress, and future directions. Environ Rev 26:82–92

Gupta A (2007) Geology of large river systems. In: Gupta A (ed) Large rivers: geomorphology and management. Wiley, pp 1–5

Hambrecht G et al (2020) Archaeological sites as distributed long-term observing networks of the past (DONOP). Quat Int 549:218–226

Haywood AM et al (2019) What can palaeoclimate modelling do for you? Earth Syst Environ 3:1–18

Hudson MJ, Aoyama M, Hoover KC, Uchiyama J (2012) Prospects and challenges for an archaeology of global climate change. Wiley Interdiscip Rev Clim Chang 3:313–328

Hussain ST, Riede F (2020) Paleoenvironmental humanities: challenges and prospects of writing deep environmental histories. Wiley Interdiscip Rev Clim Chang 11:e667

Jackson RC, Dugmore AJ, Riede F (2018) Rediscovering lessons of adaptation from the past. Glob Environ Change 52:58–65

Jyotibala S, Sangode SJ, Sabale PD (2019) Mineral magnetic characterization of archeological potsherds: an example from the Deccan Province, Western Maharashtra, India. Curr Sci 108

Kadam (2018) Crop combination regions in Sina River Basin: Maharashtra

Kale VS, Rajaguru SN (1987) Late Quaternary alluvial history of the northwestern Deccan upland region. Nature 325:612–614

Kshirsagar S, Mishra S, Sabale PD (2014) Some Acheulian localities in the Sina basin, vol 74. Bulletin of the Deccan College Post-Graduate and Research Institute, pp 21–26

National Research Council (2006) Surface temperature reconstructions for the last 2,000 years. The National Academies Press, Washington, DC. https://doi.org/10.17226/11676

Pappu (1970) Pleistocene geomorphology of the Upper Krishna Basin. Indian Antiquary- IV 1–4: 191–204

Pappu (1973) Quaternary studies in peninsular India. Bulletin of the Deccan College XXXIII(1–4)

Rajaguru (1973) On the problems of stone age chronology at Chirki Nevasa, District Ahmednagar, Maharashtra. Bulletin Deccan College, Pune, pp 360–369

Rajaguru SN, Pappu RS (1970) On the problem of Stone Age Chronology at Chriki-Nevasa, District Ahmednagar. Bulletin of Deccan College Post Graduate and Research Institute 31:360–369

Sabale PD (2012) Geoarchaeological investigations in the middle reach of Bhima. An unpublished report of major Research Project submitted to University Grant Commission, New Delhi

Sabale PD, Kshirsagar SD (2014) A preliminary report on geoarchaeological explorations in upper reaches of Sina river basin: a case study of Ahmednagar, Bid, Osmanabad and Solapur districts of M.S. Bulletin of the Deccan College 7:127–138

Sabale PD, Deshmukh JD, Kshirsagar SD, Pokharkar DM, Sarwade KG, Kamble PV, Rao N (2015) Preliminary report on quaternary geoarchaeological excavation at Math-Pimpri, District Ahmednagar. Bulletin of the Deccan College Research Institute 75:25–36

Sabale PD, Badam GL, Kshirsagar SD (2021) Quaternary faunal records from upper reaches of Sina basin: a case study of Math-Pimpri. Quaternary Climate Change over the Indian Subcontinent Springer, pp 345–256

Sandweiss DH, Kelley AR (2012) Archaeological contributions to climate change research: the archaeological record as a paleoclimatic and paleoenvironmental archive. Annu Rev Anthropol 41:371–391

Chapter 11
A Review Study on Youngest Toba Tuff Ash, Artifacts and Vertebrate Remains Preserved in the Quaternary Alluvial Sediments of India and Their Significances in Climate and Paleoenvironment Interpretations

Ajab Singh, Ashok K. Srivastava, Nura Abdulmumini Yelwa, and Neloy Khare

Abstract A review study of Youngest Toba Tuff (YTT) ash, artifacts, and vertebrate remains uncovered in the alluvial deposits of various river basins of India has been attempted in order to comprehend paleoclimate and paleoenvironmental conditions throughout the Quaternary period. The YTT eruption is regarded as the biggest eruption on the Earth's surface during the Pleistocene period, and hence it is considered a catastrophe for ecosystems and mass populations. The ash of the same eruption has been found in laminated, sandy to clay alluvial columns in the form of light gray to yellowish brown, massive to structured, pockets, lenticles, and continuous to discontinues beds ranging in thickness from 0.2 to >3 m thick. Based on bedding geometries, preservational characteristics and physical appearances, the ash is classified as primary and secondary. Geochemically and geochronologically, it overlaps elemental composition and isotopic age of the Oldest Toba Tuff (OTT) ash. Artifacts belonging to the Middle Paleolithic to Microlithic period are retained in the fluvial profiles that form pre- and post-YTT columns. Based on the typological and carving similarities, it has been linked to the Middle Paleolithic artifact industry of Africa, indicating migration of hominids in India, whereas the majority of vertebrate remains reported mainly from both pre- and post-tephra successions of various

Ashok K. Srivastava has died before the publication of this book.

A. Singh (✉)
Department of Geology, Faculty of Science, University of Malaya, Kuala Lumpur, Malaysia

N. A. Yelwa
Department of Geology, Faculty of Science, University of Malaya, Kuala Lumpur, Malaysia

Department of Geology, Usmanu Danfodiyo University, Sokoto, Nigeria

N. Khare
Ministry of Earth Science, New Delhi, India

N. Khare (ed.), *Science, Policies and Conflicts of Climate Change*, Springer Climate,
https://doi.org/10.1007/978-3-031-16254-1_11

alluviums reveal the existence of herbaceous animal communities, implying the presence of small to tall grass vegetation and water pools to swamps environments under humid to temperate climates. Based on preservations of artifacts and vertebrate remains in both pre- and post-YTT ash strata, the study suggests the existence of hominids and animals in the subcontinent prior to and after the eruption, indicating negligible effects of the eruption on the living media.

Keywords Youngest Toba Tuff · Artifacts · Vertebrate remains · Vegetation · Quaternary

Introduction

Light colored, unconsolidated to slightly compact tephra, uncovered in the stratified, sandy to silty-clay lithounits of the various river basins of India has been regarded as the remains of the Youngest Toba Tuff (YTT, 75 ka) eruption of Toba Caldera, Sumatra, Indonesia, based on the geochemical and geochronological similarities (Acharya and Basu 1993; Westgate et al. 1998; Petraglia et al. 2007; Smith et al. 2011; Srivastava and Singh, 2019a, b, 2020a, b) (Fig. 11.1). Prior to the YTT eruption, the Toba Caldera had generated three other low magnitude eruptions in chronostratigraphic order as Haranggoal Dacite Tuff (HDT, 1.2 Ma) (Nishimura et al. 1977), Oldest Toba Tuff (OTT, 840 ka) (Diehl et al. 1987) and Middle Toba Tuff (MTT, 504 ka) (Chesner et al. 1991). In these aforesaid eruptions, the YTT erupted with an 8.8 magnitude, forming a 100 km long and 30 km wide caldera and generated about 2800 km^3 of rhyolitic magma, including ignimbrite and co-ignimbrite materials covering about 20,000–30,000 km^3 in Southeast Asia (Rose and Chesner 1987). According to Rampino and Self (1992), ignimbrite material containing sulfur gases blasted from the caldera moved toward the atmosphere and formed umbrella shaped volcanic clouds. These clouds prevented sun radiation from passing through the earth surface, resulting in cooling for almost 6 years, whereas the sulphur gases generated sulphur di-oxides combined with prevailing hydrogen ions in the atmosphere and dropped on the surface in the form of acid rains (Rampino and Self 1992). It has been hypothesized that the combination of cooling and acid rain decimated the living mass populations and vegetation across Southeast Asia, including East Africa (Ambrose 1998; Rampino and Amrose 2000; Williams et al. 2009); however, a few researchers have argued a minimal impact of the YTT eruption on the mass population and vegetation (Schulz et al. 2002; Oppenheimer 2002; Gathorne-Hardy and Harcourt-Smith 2003), as a result, the magnitude of impact of the said eruption is still debatable among the scientific community.

The YTT exposures are reported from both marine and terrestrial deposits, including lake sediments. The marine includes cores of Indian Ocean (Pattan et al. 2002), South China Sea (Liu et al. 2006), Bay of Bengal (Gasparrotto et al. 2000)

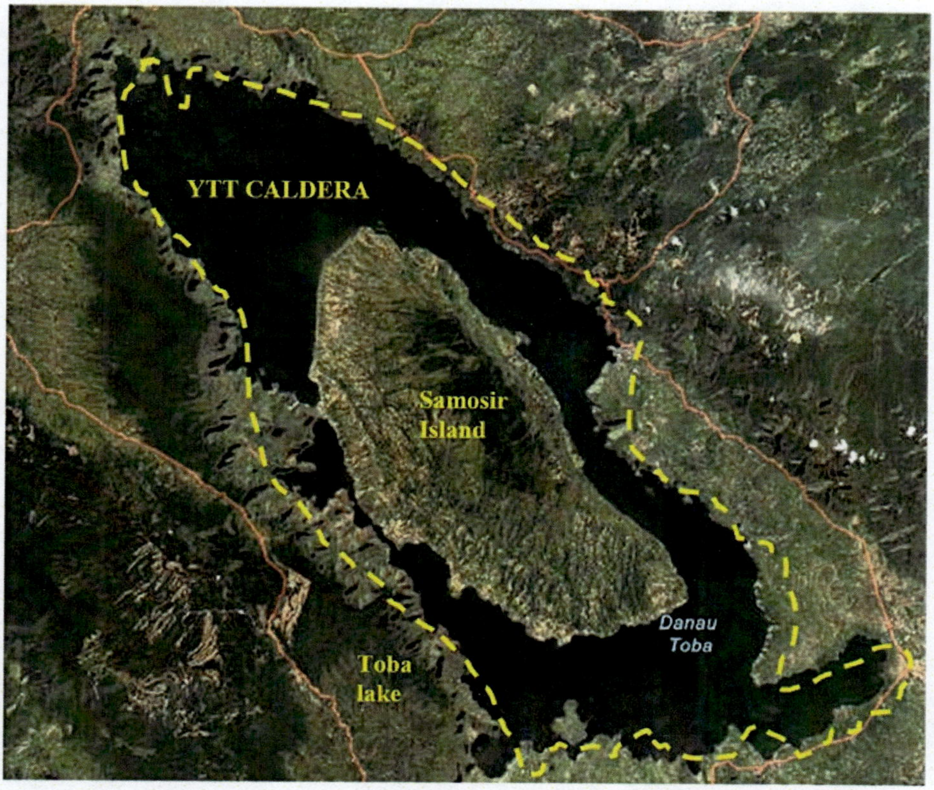

Fig. 11.1 Map showing location of the Toba Caldera and the YTT. (Image source – Earthstar Geographics)

and Arabian Sea (Jayaprakash et al. 2009), whereas the terrestrial represents stratified, fluvio-lacustrine, Quaternary sediments of various Indian river basins, viz., Krishna, Godavari, Mahanadi, Vasandhara, Kukadi valley, Karha valley, Middle Son valley, Narmada valley (Acharya and Basu 1993), Wardha river basin (Westway et al. 2011) and Purna valley (Srivastava and Singh, 2019a, b, 2020a, b); in addition to, Lenggong valley of Malaysia (Gatti et al. 2013; Singh et al. 2021a, b). Lane et al. (2013) also reported the same ash from the sediments of lake Malawi and extended hypothesis of global distribution of the YTT ejecta. In addition, a sulphate rich layer of ash is reported from the Greenland ice core GISP2 of Antarctica that has been equated with YTT based on its geochronological and geochemical similarities (Rampino and Self 1992). The extensive literature survey reveals the global dispersion of YTT ash and its climatological significances. Given the climatological implications of the same and its wide geographical spread, the present study aimed to carry out a comprehensive evaluation of the ash in India and its significance in understanding paleoclimate and paleoenvironment. It also adds to the body of knowledge on the climatic conditions in India prior to, during, and after the YTT eruption, as well as its impacts on humans and vegetation.

Exposures of YTT Ash in India

Figure 11.2 reveals the locations of the YTT-bearing river basins of India. The ash under consideration has been investigated in India over the past four decades after the first report from the Son valley during archeological explorations headed by Williams and Royce (1982). After that, Korisetter et al. (1988) reported preservations of the same from a stratified lithounit exposed at the Bori site lying along the Kukadi River and claimed the same as the remains of the YTT based on the

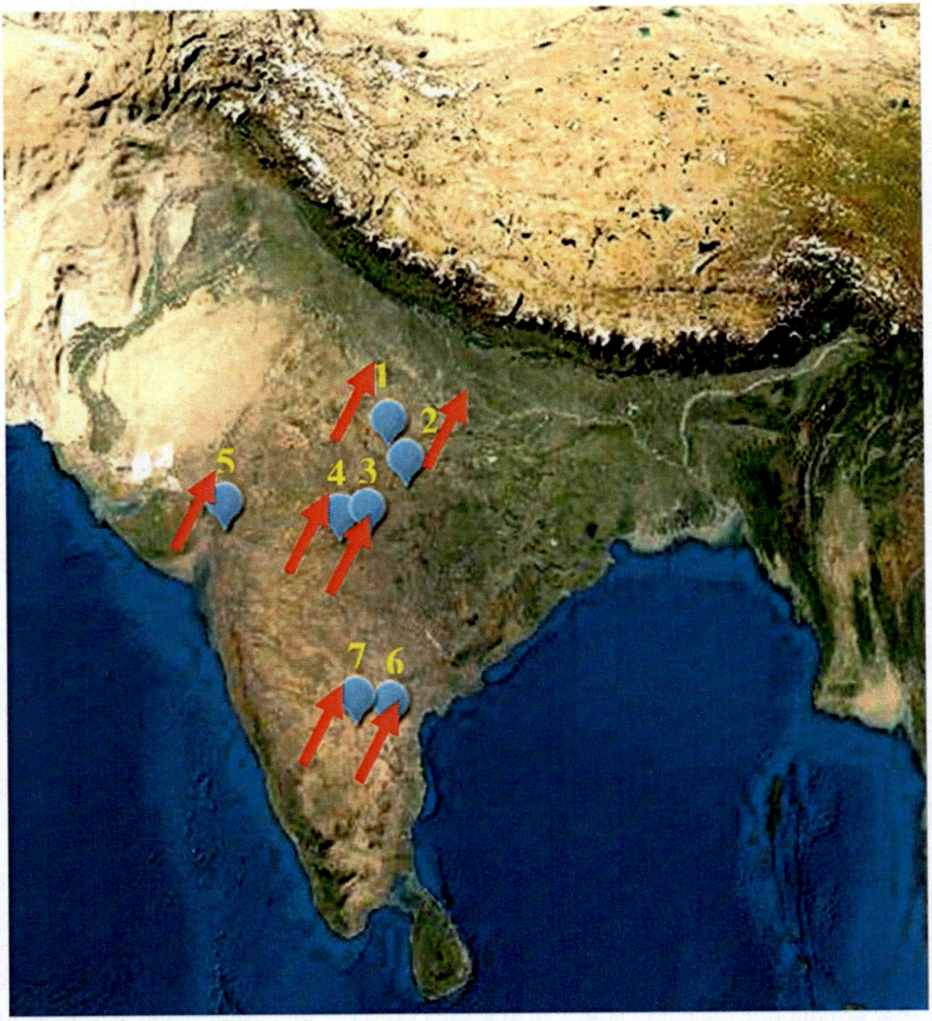

Fig. 11.2 Map revealing locations of the various investigated localities such as (1) Ghoghara area, Son valley, (2) Pawalaghat locality of the Narmada basin, (3) Simbhora site, Wardha valley, (4) Masod area, Purna basin, (5) Tejpur locality of the Madhumati River, (6) Sagileru bridge site, Sagileru valley and (7) Jwalapuram area of the Jerreru valley. (Image source – Earthstar Geographics)

geochemical signatures matching the proximal deposits in Indonesia. Furthermore, it had also been recorded from a number of other river basins, viz., Godavari, Mahanadi, Jurerru, Sagileru, Kukadi, Karha, Narmada, Madhumati (tributry of Narmada), Son, Purna and Wardha and investigated for a variety of reasons (Acharya and Basu 1993; Westgate et al. 1998; Raj 2008; Westway et al. 2011). Recently, the authors carried out field work in the YTT-bearing basins to learn more about the ash preservation and collected samples for various analyses. A brief idea about various ash-bearing basins and their localities are as follows:

(i) **Son valley** – The Son, longest river (784 km) in India having E-W flow along the direction of the tectonic lineament of Son-Narmada Fault, as of Narmada River, merges into the Ganga River. It originates at Amarkantak, Madhya Pradesh, just east of the Narmada River and flows over the Middle Proterozoic limestone and shales of the Vindhyan Super Group and Middle Pleistocene to Holocene aged alluvial plains, before changing its course toward eastward to incised Middle Proterozoic sandstones of the Kaimur Range of Vindhyan Super Group (Singh 1980). The distal ash in the valley has been encountered at Ghoghara (24°30′096″ N, 82°01′172″ E) in the form of light gray to yellowish-brown, powdery, massive to structured, 0.5–3 m thick beds, preserved in the yellowish-brown, stratified silty-clay lithounits (Williams and Royce 1982; Acharya and Basu 1993; Gatti et al. 2011) (Fig. 11.3a). It has been regarded as YTT based geochemical and geochronological similarities with the Toba Caldera (Smith et al. 2011; Mark et al. 2014).

(ii) **Narmada valley** – The Narmada valley drained by one of the major rivers in India, viz., Narmada, originates in the Satpura Hill range at Amarkantak and flows westward, opposite to the regional slope, due to Narmada-Son Fault and debouches into the Gulf of Combay after covering the catchment area of about 98,796 km^2 (Chamyal et al. 1997). The Pawlaghat locality (23°06′803″ N, 79°23′048″ E) of this valley is comparatively well documented for preservation of light gray, powdery, massive to structured, 0.8 m thick bed of tephra retained in the reddish to yellowish-brown, silty-clay lithounit (Acharya and Basu 1993) (Fig. 11.3b). The ash has been correlated with the YTT of Toba Caldera based on geochemical and geochronological fingerprintings (Westgate et al. 1998).

(iii) **Wardh valley** – The Wardh River, a major branch of the Godavari River, originates from Satpura hill range at Khairwani locality near Multai town of Betul district, Madhya Pradesh and flows from west to east direction before debouching in the Bay of Bengal. The light gray to yellowish-brown, powdery, reworked and ~ 1 m thick bed of ash is exposed at the Simbhora locality (21°26′50″ N, 78°57′16″ E) (Fig. 11.4a) near Morshi town of Amravati district, Maharashtra (Tiwari et al. 2010). Westway et al. (2011) reported 827 ± 39 ka age of the same using $^{40}Ar/^{39}Ar$ technique on the glass shards and suggested that it belongs to the Oldest Toba Tuff (OTT, 840 ka). However, Westway's hypothesis has been rejected by several subsequent researchers

Fig. 11.3 Field photographs demonstrating preservation of ash in the alluvial profiles exposed at (**a**) Ghoghara locality of Son valley and (**b**) Pawalaghat area, Narmada valley (arrows)

using geochemical and geochronological comparabilities with the other YTT ash of India, including Toba Caldera (Westgate et al. 2013; Pearce et al. 2014).

(iv) **Purna valley** – The Purna River draining the valley is a major tributary of Tapti River and originates from the Satpura Hills at Chincholi village of Betul district, Madhya Pradesh and flow roughly SSE for approximate 75 km and further takes a westerly direction for 120 km and finally merges into the Tapti River (Srivastava and Kale 2010). The ash in this basin are preserved at eight different localities, i.e., (i) Masod, (ii) Parad, (iii) Bhog Nala, (iv) Hudki, (v) Sukali, (vi) Gandhigram, (vii) Kapileshwar and (viii) Andura. In this basi area, the ash is uncovered in the form of light gray to yellowish-brown,

Fig. 11.4 Field views revealing exposures of the tephra preserved at (**a**) Simbhora area of Wardha valley, (**b**) Masod locality of Purna basin and (**c**) Tejpur site, Madhumati valley (arrows)

powdery, massive to reworked and constitute 0.15–3 m thick beds hosted in the yellowish-brown, laminated silty-clay lithounits (Srivastava and Singh 2019a, b, 2020a, b, 2021a, b; Srivastava et al. 2020a, b; Singh and Srivastava, 2021a, b, 2022). Although, in the present attempt, we have explored Masod site (20°14′07″N, 77°39′07″E) of the same basin area, where the tephra is attributed with yellowish-brown color, thin elongated bed of massive nature and 0.4–0.5 cm thickness (Fig. 11.4b)

(v) **Madhumati valley** – The Madhumati valley of the Lower Narmada basin is drained by the Madhumati River that flows over the Late Cretaceous to Eocene Deccan basaltic lava flows and overlying Tertiary sediments in the further western part. A light gray colored, powdery, 0.4 m thick bed of volcanic ash, exposed at Tejpur locality (21°46′414″ N, 73°17′876″ E) (Fig. 11.4c) near the Rajpardi town of Gujrat state, has been reported and correlated with the YTT tephra based on its geochemical and glass shards morphological similarities (Raj 2008).

(vi) **Sagileru valley** – The Sagileru River, draining the valley, is a tributary of the Penner River that originates at Cumbum town of Kurnool District, Andhra Pradesh and flows between the Nallamalai and Velikonda hill ranges of the Eastern Ghats (Blinkhorn et al. 2014). The Proterozoic Cuddappah rocks with almost flat beds of sandstone, limestone, and shale forms the basement for Quaternary alluvial deposits that are reddish-brown to yellowish-brown, stratified, calcareous, hard to semi-consolidated, sandy, silty-clayey lithounits. These deposits exhibit abundant preservations of termite structures, calcretes, and root casts of different sizes and shapes and Middle Paleolithic tools. It also preserves light gray to yellowish-brown, powdery to hard, 2 m thick bed of volcanic ash showing diverse sedimentary structures near Sagileru Bridge, Porumamilla area (14°57′002″ N, 78°56′789″ E) (Fig. 11.5a). The ash has been reported from Vankamarri and Jammulamadugu localities and studied for geochemical and geochronological aspects as well as correlated with other YTT tephra of India (Blinkhorn et al. 2014; Geetanjali and Achyuthan 2019; Geetanjali et al. 2019; Krishnan and Achyuthan 2019).

(vii) **Jurreru valley** – The Jurreu valley, a major river basin in the central part of Andhra Pradesh state, is drained by an easterly flowing tributary of the Penner River marked with steep sides and relatively wide bottom that flows over the landforms of Cuddaph rocks (Clarkson et al. 2012). The ash, exposed at the Jwalapuram locality (15°19′525″ N, 78°07′990″ E) of this valley, is light gray to yellowish-brown, >3 m thick and divisible into primary and secondary based on the preservational and depositional characteristics. It is highly intermixed with the reddish to yellowish-brown river sediments along with calcretes of various morphotypes, including rhizoliths (Fig. 11.5b). The same has been investigated comprehensively for geochemistry, geochronology, and paleoclimatic reconstructions (Petraglia et al. 2007; Mark et al. 2014; Srivastava et al. 2019). The Jwalapurma locality is considered as a significant site because of rich occurrences of artifacts below and above the ash strata, suggesting occupation of humans in the area prior to and after the eruption under reference (Petraglia et al. 2007, 2012; Haslama et al. 2012).

Preservational Attributes of YTT Ash in India

Figures 11.3, 11.4, and 11.5 plainly reveal that the YTT ash in the Indian subcontinent is preserved in the form of discontinued beds, lenticles, and pockets of variable thickness primarily in laminated, fine to medium sand mixed, yellowish-brown silty-clay lithounits, although, at places, disturbed and evenly scattered on the surface. It is principally whitish, light gray to yellowish-brown in color, fine to medium grained, powdery to semi-consolidated and massive to structured with a thickness ranging from 0.05 to 3 m. Based on their colors, thicknesses, bedding geometries and retention of sedimentary structures, the ash is divided into two distinct sorts as

Fig. 11.5 Field views showing ash deposit at (**a**) Sagileru bridge, Porumamilla area of Sagileru valley and tephra excavation for commercial purposes at (**b**) Jwalapuram locality of Jurreru valley

primary and reworked. The primary ash is whitish to light gray, fine-grained and massive to structureless with a thickness of 0.05 to 0.20 m. It usually serves as a basement for the deposition of reworked tephra units and at places, is deposited as a single elongated bed attaining a thickness of 0.20 m. The reworked tephra, on the other hand, is mostly yellowish-brown, relatively coarse grained, structured with a thickness up to ~3 m and reveals preservations of low energy produced sedimentary structures, i.e., parallel to cross laminations, trough cross beddings, sand laminae, grove casts, and soft sediment deformational features (Acharya and Basu 1993; Gatti et al. 2011; Jones 2010; Srivastava and Singh 2019a, b, 2020a, b). Fieldwork carried out by the authors in all the tephra bearing river basins mentioned above suggests that the ash in the south Indian river basins, viz., Jurreru and Sagileru is intermixed with host lithounits having low thicknesses and high amounts of calcium carbonate precipitation in the form of calcretes, whereas the tephra in the river basins of central and western India is preserved as elongated beds, pockets, and lenticles. However, substantial precipitation of calcium carbonates has been noticed mainly in the pre-tephra successions classified as paleosols and at places within tephra beds.

YTT Impacts and Human Occupations

The impacts of YTT eruption on the vegetation and modern humans have been well documented with the latter experiencing bottleneck and dispersing across Africa to Southeast Asia, including India. Several studies have confirmed changes in demography of the ancient hominids after the eruption in the form of artifact transformation from lithic to microlithic (Petraglia et al., 2012; Clarkson et al. 2020), although certain authors favored partial effects of the YTT in places where hominids were concentrated prior to and after the said eruption (Oppenheimer 2002; Timmereck et al. 2012). However, according to Ambrose (1998), the hominids had witnessed climatic havoc caused by the YTT for a long time as it posed delicate climatic difficulties to humans in the form of volcanic cooling and poisonous acid rain (Ambrose 1998). Both cooling and rains combined to be a disaster that forced them to scatter eastward out from the Africa continent across the globe in order to avoid the hostile conditions (Ambrose 1998). The hypotheses about catastrophic nature of the YTT toward the lives and demography of humans are exemplified by the shift in technologies from Levallois to Microlithic, which is retained in close association with the lower and upper horizons of the YTT ash, exposed in the river basins of India. These archaeological tools are thought to have been produced by the modern humans who existed prior to and after the eruption, indicating YTT had no impact on humans (Oppenheimer 2002; Petraglia et al. 2007, 2012; Jones 2010; Haslama et al. 2012; Clarkson et al. 2012, 2020).

Debate Among Researchers

The YTT in India is a topic of discussion among researchers because of differing viewpoints and assertions based on contradictory results of $^{40}Ar/^{39}Ar$, geochemistry, luminescence dating, and effects on the living masses and their migration from a native place to distal sites. Certain studies based on the $^{40}Ar/^{39}Ar$ dating of the glass shards have predicted that the ash in the Kukadi, Karha, Wardha, and Purna river basins has an age equivalent to the Oldest Toba Tuff (OTT, 840 ka) (Mishra et al. 1995; Westway et al. 2011). However, using the $^{40}Ar/^{39}Ar$ technique on biotite grains of the ash from Jurreru and Son valleys, Mark et al. (2014) postulated YTT affinity and eliminated the theories of OTT in India. Simultaneously, certain geochemical studies of glass shards have been made to shed light on these debates. However, the geochemical composition of glass shards proved to be supporting tool instead of decisive factor to identify YTT or OTT as Smith et al. (2011) found similar geochemical concentrations of both YTT glass shards from Jurreru and Son valley and OTT of Toba Caldera, and opined that the geochemistry of glass shards is unreliable and an unrealistic tool in correlation. Based on minor and trace element concentrations of the glass shards collected from entire Indian Peninsula and source Toba Caldera, Westgate et al. (2013) and Pearce et al. (2014) argued that the ash in India belongs to the YTT. Srivastava and Singh (2020b) used $^{40}Ar/^{39}Ar$ dating on the glass shards of two different localities, viz., Gandhigram and Hudki of Purna alluvial basin, Central India and reported 790 and 770 ka ages, although discarded the same based on Ar and K isotopes incompatibility in the glass shards during magmatic and post-magmatic process. According to them, this incompatibility in the isotopes of glass shards plays an essential role in generating spurious or older age while using the Ar/Ar technique on the same material.

Several optically stimulated luminescence (OSL) dating have also been carried out in order to evaluate the bracketing time of the tephra within the fluvio-lacustrine sediments exposed as lower and upper horizons of the same. Petraglia et al. (2007) used OSL dating to determine the age of the lower and upper horizons of YTT-bearing exposures at Jwalapuram locality, Jurreru valley, South India and came up with 70 ka and 71 ka ages, respectively. They also hypothesized that the ash was capped within the sediments after thousands of years of YTT eruption. In the queue of the work, Biswas et al. (2013) applied the same technique on 11 samples collected from pre-, within, and post-tephra sediments of Ghoghara and Khuntheli localities of Son basin, Morgaon of Karha basin, Bori of Kukadi basin and Tejpur of Madhumati river basin and reported the age as 70 ± 9 & 40 ± 5 and 61 ± 7, 41 ± 5, 27 ± 3 and 74 ± 7 & 60 ± 9 ka ages, respectively. Neudorf et al. (2014) sampled the sediments from pre- and post-tephra horizons at the Ghoghara area of Son valley and did the same with the OSL technique. They generated single quartz ages for the samples GHO3 and GHO2 representing above and below sections as 44 ± 3 and 36 ± 3 ka and claimed that tephra would have been bracketed within the sediments after tens of thousands of years of YTT eruption. Singh et al. (2022) extended the investigation on the same aspect for the samples taken from below and above strata of the Hudki and Sukali localities, Purna alluvial basin and reported 70 ± 04 & 57 ± 5 and 66 ± 5-- 67 ± 4 ka ages for the same, respectively. They suggested that the ash settled on the surface from the atmosphere was likely deposited at the present sites tens of thousands of years after the YTT eruption. They also stated that the ages of the pre- and post-tephra strata differ due to reshuffling and reworking of older and youngest sediments inherited with the ash during high energy condition of depositing medium. However, an overview of various research demonstrates that the bulk of investigations, whether Ar/Ar dating, geochemistry, and luminescence contradict each other most likely due to flaws in sample collection, material calculation and analyses procedures. In light of these considerations, the author proposed doing time assessments of all the ash exposed in various basins using high-precision instruments which will provide a clear picture for regional correlation and other interpretations such as paleoclimate and paleoenvironment.

Existence of Vegetation and Its Deterioration in India

Several studies have documented the existence of rainforest vegetation in Southeast Asia, particularly in India and African prior to the eruption as well as their explicit destruction by post-YTT produced volcanic winter and H_2SO_4 rains as interpreted based on the stable isotope analyses of calcretized nodules occurred in close affinity with YTT deposits (Rampino and Ambrose 2000; Williams et al. 2009; Jones 2010; Haslama et al. 2012; Timmreck et al. 2012; Srivastava and Singh 2019a, b, 2020a, b). Based on the botanical evidence and model simulations prepared to comprehend the effects of YTT eruption, Rampino and Ambrose (2000) asserted that the post-Toba cooling may have destroyed the local and regional vegetation of Southeast Asia and also suggested that the vegetation in the tropical regions with cold sensitive characteristics would have been completely destroyed. They went on to say that the most temperate and sub-Arctic vegetation at higher latitudes would have been ravaged and taken a long period to recover because the temperature had been reduced up to ~10 °C for a long time. Following the global climatic model in relation to the YTT eruption, Rampino and Ambrose (2000) extended their hypothesis for atmospheric perturbations and asserted that the eruption had led to severe droughts specifically in the tropical rainforests and monsoonal regions for a considerable period that could have created hostile conditions for global ecology by reducing vegetation and mass populations. In the context of the Indian Subcontinent, only a few studies have demarcated the severe effects of Toba ash on the vegetation and its rebirth after thousands of years (Oppenheimer 2002; Gathorne-Hardy and Harcourt-Smith 2003). In order to validate this, Williams et al. (2009) concentrated on pollens from laminated Toba ash preserved in the marine cores of Bay of Bengal and soil carbonates from sediments above and below the tephra units from the Son and Belan river basins. On the basis of pollen analyses, i.e., herbaceous taxa, Amaranthaceae, Artemisia, Cyperaceae, Poaceae and Arboreal texa, they revealed that the YTT eruption resulted in prolonged cooling and aridity, which reduced vegetation and growths in India and adjacent areas. On the basis of carbon isotopic composition of soil calcrete nodules, they also hypothesized that C_3 vegetation was changed into C4 woodlands and grasslands in Indian Subcontinent. However, Williams et al. (2009) stated that the ash had only modest effects on the forests and humans. Jones (2010) discovered fossilized plants and insect remains from the ash-bearing lithounit of Jurreru valley, implying the presence of vegetation and insects prior to the fallout of the YTT eruption. Srivastava and Singh (2019a, b, 2020a, b) reported calcretized spherical to rounded bodies of rhizospheres and rhizoliths preserved in paleosol units making a pre-tephra column in the Purna, indicating the existence of vegetation previous to and during the YTT eruption. According to them, the preservation of abundant rhizospheres and rhizoliths in pre-tephra lithounits and their sparse occurrence in post-tephra horizons, suggest possible minor effects of the winter and acid rains under the reference. Recently, Singh and Srivastava (2021b) revealed preservation of four unique types of rhizospheres, i.e., concentric, rimmed, nodular and mottled preserved in the pre-tephra lithounits of the Purna basins and claimed that these four sorts are the products of both C_3 and C_4 vegetation. Furthermore, based on the distinct morphologies of the rhizospheres, they projected mixed plant and grassland floras in the basin area prior to the YTT eruption as well as suggested modest effects of the said eruption on the vegetation.

According to Gathorne-Hardy and Harcourt-Smith (2003), the impact of the Toba eruption would be limited to a radius of 350 km. This conclusion is based on the study of nine endemic species, i.e., four primates, four squirrels, and one mouse and termite structures from the Mentawai islands, located 350 km southeast of Toba and 130 km Sumatran mainland. They further claimed that this island emerged from the sea during Late Pliocene and was separated from the mainland for approximately three million years. Gathorne-Hardy and Harcourt-Smith (2003) postulated the existence of evergreen and everfruiting rainforest flora associated with the abovementioned species on the island after its separation from the mainland. They, additionally, hypothesized that the termite structures indicate that the island was formerly a forested refuge.

Artifacts

Preservation in Quaternary Sediments and Correlation with the African Industry

The artifacts belonging to the Middle Paleolithic to Microlithic ages are primarily reported from both the pre- and post-YTT tephra-bearing successions exposed in various river basins in the form of stratified, sandy to silty-clay lithounits. It has also been recovered from boulder-conglomerate beds that make up pre-tephra strata. These artifacts are assigned as the product of the modern humans (Petraglia et al. 2007; 2012; Haslama et al. 2012; Clarkson et al. 2020; Joshi 2020). Based on the morphological and typological studies of the artifacts, several researchers have revealed that hominids existed on the Indian Subcontinent and migrated in groups from Africa prior to the YTT eruption, implying continuity in population even during the climatic repercussions caused by the said eruption (Petraglia et al. 2007, 2012; Haslama et al. 2012; Clarkson, 2020). Petraglia et al. (2007) reported occurrences of Middle Paleolithic to Microlithic industries in the pre- and post-YTT bearing horizons at the Jwalapuram area of the Jurreru valley, southern India as well as asserted continuity in human populations even during the prevailing of volcanic cooling led by the eruption under study. Based on the similarities in carvings on the lithic, they established a superior correlation between the Jwalapuram industry and the Middle Paleolithic technology of Africa and suggested hominids were migrated from the latter continent. Furthermore, based on the presence of artifacts in the lower and upper horizons of the YTT-bearing sediments, Petraglia et al. (2012) asserted that Indian Subcontinent was the home of modern humans between ~100,000 to

200,000 years ago, and opined that the number of humans may have reduced significantly during climatic changes induced by the YTT eruption. They went on to say that the mass population would have likely been diminished owing to the climate fluctuations, as a result, humans might have altered their skills to produce more advanced artifacts. Haslama et al. (2012) excavated a site at Dhaba locality about a kilometer from the YTT-bearing Ghoghara locality, Middle Son valley and found Acheulean, Middle Paleolithic and Microlithic industries within the sedimentary profiles of the river. They attempted to distinguish the industries based on surfaces structures and source of their material and concluded that Acheulean artifacts have bifaces and cleavers, whereas Middle Paleolithic tools lack the same. They further asserted that the Acheulean artifacts could be between 125 and 140 ka age and are correlatable with the earlier findings in India and outside. The Microlithic technology could not be placed in a chronological order, although it can be linked to other similar findings dating from MIS 3 to Holocene period. Clarkson et al. (2020) excavated the Dhaba site in order to examine more assemblages for superior correlations with African industries to build a broad picture of human habitation in India. Based on luminescence dating of lithic containing sediments, they predicted that humans occupied the area around 80 ka even before the YTT catastrophic event, but experienced behavioral development after the eruption exemplified by introducing new technology, i.e., Microlithic. Joshi (2020) discovered occurrences of the Acheulean artifacts at two distinct sites, viz., Bhatala and Temburda of the Wainganga River, eastern Maharashtra state of India and suggested wide distribution of humans in India.

Migration of Early Humans and Their Occupation on the Indian Peninsula

The YTT-led fine grained material produced cooling and acid rains that destroyed vegetation and hominids survival and compelled them to migrate from indigenous places to other continents via the Indian subcontinent. As a result, it is heavily viewed as a catastrophic event during the Quaternary period (Ambrose 1998; Oppenheimer, 2002). Ambrose (1998) made a descriptive study on the existence, bottleneck, growth, genetic diversity, and dispersal of the modern humans after the eruption based on Weak Garden of Eden model proposed by Harpending et al. (1993). He suggested that the modern humans migrated from native places in Africa to different regions about 100 ka years ago and had witnessed bottleneck after the YTT, although experienced dramatic growth and technological advancement around 50 ka (Ambrose 1998). On the basis of geological and climatic evidence as well as discovery of the Upper Paleolithic technology in the Equatorial Africa, Ambrose (1998) went on to state that humans first saw genetic diversity, technological improvement, and population growth in Africa and afterward, in Eurasia. He further concluded that the volcanic winter lowered the population, caused genetic alterations, and forced them to adapt local needs in order to reproduce more quickly. On the basis of genetic study, Rampino and Ambrose (2000) claimed that prior to ~60,000 ka ago, approximately 3000–10,000 hominids had seen severe destructions, and subsequently, their population, technological developments, and migration for suitable settings had occurred. They emphasized that the bottleneck and migration of humans would have taken place as a result of climatic effects of the Toba eruption and concluded that environmental and ecological effects of the eruption may support the connection between the Toba event and mass population bottleneck as well as destruction of other living species at the same timeframe. Following the hypotheses of climatic effects, Oppenheimer (2002) delineated a route, passing through the Indian subcontinent, was followed by the Anatomically Modern Humans (AMH) for spreading out from Africa to other continents, as well as highlighted their locations and probable dates in these continents. He also claimed that humans travelled from their native land to Sahara to Levant during the Eamian interglacial period (ca. 125 ka), but could not have made it to Europe which was later prominently Neanderthals. Oppenheimer (2002) further stated that AMH left Africa to India in groups via a southern route and disseminated across the subcontinent. Petraglia et al. (2007, 2012) carried out an extensive study of artifacts found in close proximity to YTT ash in the Jurreru valley, concluding that the archeological industries, above and below the tephra beds, are more closely related to the Middle African Middle Stone Ages tradition, i.e., Howieson's Poort than contemporaneous Eurasian Middle Paleolithic industries, i.e., discoidal and Levallois techniques. Several investigations based on technological parallels have been carried out to interpret human existence prior and after YTT eruption. Haslama et al. (2012) conducted a similar investigation at the Locality 22 of the Jwalapuram area, Jurreru valley and claimed that humans were present in the Indian subcontinent before and after the YTT eruption. They have also noted changes in hominid technological industries based on the preservation of artifacts in pre- and post-tephra sediments in the basin area. On the basis of similarities in lithic industry with the stone tools of the native place in Africa, Haslama et al. (2012) went on to say that hominids would have migrated to India from Africa through a southern route, categorizing the markings on the artifacts as discoidal to Levallois techniques that are very similar to the technology of the African human community. Recently, Clarkson et al. (2020) excavated the Dhaba site near YTT deposit at Ghoghara locality situated at Son-Rehi confluence and recorded breakthrough evidence for the existence of human populations dating back ~80 ka based on the similarities with the lithic industries of the basin area to Africa, including the Jurreru valley. They pointed out an unchanged lithic industry buried under the sediments of 79.6 ± 3.2 & 65.2 ± 3.1 ka that resembled the African Middle Stone Age (MSA), Arabia, and Australia and suggested that the same would have likely been formed by the modern humans as they moved eastward from Africa.

The literature review reveals the route modern humans used to move from Africa to India on to other continents, along with findings of different types of lithic industries, and report of the first human skull from Narmada basin are confined specifically in the south-west Indian river basins. Interestingly, the river basins of north-central India, e.g., Indo-Gangetic and Yamuna river basins lacks such findings and none such kind of industry have been reported yet; nonetheless, a large number of faunal discoveries have been made. This indicates that the climatic impacts of the YTT eruption were most likely limited to central and southern India. The authors working on the Purna alluvial basin for the past few years for a variety of reasons, including past climate conditions in relation to the YTT, have not yet found any industry either close to the ash exposures or even in the entire river basin. Even though Mishra (2016) reported two different types of industries from the river course at the Kapileshwar locality of the same basin area, the present authors disagree based on the quantity and authenticity of material. Considering the route delineated by Oppenheimer (2002), we believe that the humans would not have travelled to central and northern India when travelling to the other continents as the high hills of the Satpura Mountain may have created a barrier to crossing central India and heading further north.

Vertebrate Remains

Distribution in the Alluvial Deposits of India

The vertebrate remains have been found in almost all major river basins of India including caves sediments of Late Pleistocene to Holocene period and correlated with contemporaneous assemblages of Africa and Europe fauna. These remains are recorded in groups or scattered occurrences in river courses and sometimes, in vertical to lateral sedimentary profiles developed as banks of the rivers (Badam 2005; Chauhan 2008; Patnaik et al. 2009; Deshpande-Mukherjee 2014). The alluvial basins reviewed to compile the reports of fossil remains are as follows.

Manjara Valley, Maharashtra

The Manjara valley, a major tributary of the Godavari river basin, occupies the western part of the Maharashtra State, is considered a house of distinct types of animal communities existed during Late Pleistocene to Recent (Badam 2000). The vertebrate remains from this valley have been recorded from four different localities, viz., Dhanegaon, Wangdari, Tadula and Ganjur as well as identified as *Elephus hysudricus*, *Stegodon insignis ganesa*, *Equus namadicus*, *Axis axis*, *Antilope cervicapra*, *Cervus unicolor*, *Cervus duvaueeli*, *Hexaprotodon palaeindicus*, *crocodylus sp.,* and Chelonia (Badam 2000). Based on the variety of animal communities in the basin area, Badam (2000) opined that the valley was concentrated majorly by elephant populations, indicating the existence of a high vegetated forest in the valley. She further went on to assert that the presence of cattle remains indicates savannah type landscape and deep water pools under humid climate during the period of animal existence in the valley.

Mahanadi Valley

The Mahanadi valley is widely documented for various aspects related to Quaternary set-ups and preserves good assemblages of paleoclimatic intermediaries, including vertebrate remains (Tripathi 1972). The vertebrate remains are reported from four different localities, viz., Nandghat, Sigma, Somnath, and Rajnandgaon and have been identified as *Bos nomadicus*, *Bos Sp.*, *B. indicus*, *Bubalus bubalis*, *Cervus sp.*, *Equus nomadicus,* etc. (Badam 1979, 2000; Joshi et al. 1980). On the basis of diversity found in remains, Badam (2000) asserted that the valley was dominated by savannah vegetation and water pools to support the life of these animals during the Late Pleistocene period.

Narmada Valley

The Narmada valley has been a matter of concern among researchers from the eighteenth century to date with respect to preservation of vertebrate remains. The vertebrate remains are primarily reported from the calcretized gravel bed interbedded with sand and clay horizons. The gravel bed also possesses human remnants and Paleolithic artifacts (Theobold 1860; Khatri 1966; Sonakia 1984; Badam and Sathe 1995). Based on the different types of fauna preservation in the Narmada basin, Theobold (1860) divided the same into two separate groups, i.e., (i) Lower Narmada Group and (ii) Upper Naramda Group. The former contains the sediments of the Middle Pleistocene age and comprises fossil assemblage of *Equus nomadicus*, *Bos nomadicus*, *Bubalus palaeindicus*, *Hexaprotodon nomadicus*, *Sus nomadicus*, *Elephas hysudricus*, and *Stegodon insignis-ganesa*, whereas the latter comprises the deposits of Early Upper Pleistocene period and exhibits vertebrate remains of *Equus nomadicus*, *Bos nomadicus*, *Hexaprotodon palaeindicus*, *Elephas hysudricus*, *Stegodon insignis-ganesa* (Joshi et al. 1978), *Cervus sp.*, *Cervus duvauceli*, and *Rhinocerosunicornis* (Sahu 1988).

Son Valley

The Son valley is well documented for having preservations of diverse paleoclimatic proxies including fauna and artifacts in the gravel bed and loose and stratified sandy to silty clay lithounits. The reports of both lithic and vertebrate remains have been made from various locations lying along the river course and tributaries of the Son valley, of which, 7 vertebrate remains bearing sites are Kharabara, Ramnagar, Baghor, Tariha Dhaba, Odara, Baliar, and Pawariah (Blumenschine and Chattopadhyaya 1983). Initially, Blumenschine and Chattopadhyaya (1983) reported the bones from coarse member of the Baghor Formation developed between Patpara and Baghor localities; however, identification was very difficult even in both field and laboratory due to calcified gravel material on the remains. As a result, they have been unable to report 70% of specimens; however, asserted their ages to be terminal Pleistocene. On the basis of a large collection of fauna and their analyses, Blumenschine and Chattopadhyaya (1983) opined arid climatic conditions during terminal Pleistocene glacial maximum that supports mosaic vegetation of grass to woodland over the floodplains. Later on, Badam (1989) added more fossiliferous localities, including Ramdiha, Doraon, Kunjhar, Baghor, and Khunteli and agreed with the paleoenvironmental conditions proposed by Blumenschine and Chattopadhyaya (1983) with the assertion of the presence of a swampy environment in the Son valley. Based on the occurrence of ostriches' eggshell fossils in the basin area, Badam (2005) further supported the hypothesis of Blumenschine and Chattopadhyaya regarding arid climatic conditions in Central India during terminal Pleistocene.

Sahu (1988) reported certain vertebrate remains from the Belan and Son valleys, i.e., *Elephas indicus, Hippopotamus palaeoindicus, Bos sp., Bubalus sp., Cervus, and Ovidae/Capridae* and added his contribution toward diversity among the animal masses during Upper Pleistocene period based on findings of remains of diverse animals. In the successive years, Badam (1989) carried out extensive studies in the Son river basin and reported vertebrate remains from the interface of Baghor and Patapra formations, including remains of *Bos namadicus*, *Bibos gaurus*, *Hippopotamus sp.*, *Equus sp.*, *Cervus sp.*, and *Axis axis/porcinus*. Again, Badam (2005) collected the rich assemblage of fauna from Baghor Formation and opined remains of *Bos namadicus, Bos/Bibosgaurus, Bubalus sp., Antilope cervicapra, Bos elaphus tragocamelus, Tetracero squadricornis, Gazella gazella, Axis axis, Cervus unicolor, Cervus duvauceli, Muntiacus muntjak, Sus sp., Hippopotamus sp., Rhinoceros sp. Equus namadicus, Equus asinus/hemionus, Elephas sp., Panthera sp. andsomeother carnivores.*

Godavari Valley

The Godavari valley occupies the largest area of the Godavari basin and is the longest river in India, with a 1400 km long river course that stretches from its origin at Trimbek place in Western Ghat of Nasik District, Maharashtra to the Bay of Bengal, where it debouches in it (Chauhan 2008). The Godavari river basin principally consists of sediments transported by more than ten major and minor tributaries running off over the flat floodplains of the Daccam Trap and surrounded by Balaghat ranges and Ajanta hills (Saraf and Regulwar 2016). Certain localities, viz., Gangapur, Nandur and Paithan, Chirki, Hathiwell, Mula dam, and Yeldari dam in respect of vertebrate remains have been investigated by Badam (1989) and thoroughly reviewed by Chauhan (2008). Badam (1989) reported the faunal remains of *Bos nomadicus, Bos sp., Elephas, Elephas sp., Equus, Bubalus bubalis, Elephas hysudricus, Stegodon insignis, Elephas nomadicus, Elephas antiquus,*

Hippopotamus palaeoindicus, and *Equus nomadicus* and suggested the existence of various diverse animal communities in the basin area during Late Pleistocene.

Ganga Valley

The Ganga valley, a pious river in India, forms the Ganga river basin believed to be produced by the India-Asia collision. It is joined by several major tributaries originating from Himalaya, viz., Yamuna, Ramganga, Ghaghra, Gandak, Kosi, and Tista and merges into the Bay of Bengal after occupying the world's largest alluvial plains (Srivastava and Shukla 2009). Apart from these tributaries, the several rivers, viz., Chambal, Son, Yen Betwa, and Punpun flowing over the Indian cratons and Deccan Basalt terrain also join the same and make a significant junction of Quaternary deposits (Sinha et al. 2005; Srivastava and Shukla 2009). It is highly documented for having a thick pile of Quaternary sediments preserving various climate controlled intermediaries represented primarily by calcretes, artifacts and vertebrate remains (Singh et al. 1999). The vertebrate remains reported from the basin area comprise *hippopotamus, camel, Stegodon sp., Stegodon insignis*, and *Palaeoloxodon nomadicus* (Verma et al. 1998). Singh et al. (1999) traced out faunal remains at Kalpi locality of the Yemuna River and identified them as shoulder blade and tusk of elephant, molars of *equus, bovids, bos,* etc., in close vicinity of human occupation at the same locality.

Krishna Valley

The Krishna valley, a less attempted river basin for a thorough paleontological investigation, originates from the Western Ghat at Mahabaleshwar plateau in Maharashtra state and drains southeastern for 1400 km before debouching into the Bay of Bengal. The Krishna basin named after the Krishna river comprises approximately 8 sub basins and 13 tributaries, in which major rivers include Ghatprabha, Malprabha, Bhima, and Tungbhadra (Harini et al. 2018). The credit to carry out first paleontological studies in the same basin area goes to Foote's (1876) who reported vertebrate assemblage from Chikdauli area and identified them as *Bos namadicus and Rhinoceros deccanensis.* After finding a 35 ka year-old elephant tusk near Dhom Dam, Badam (1985, 1989) concluded that the fauna assemblages in this basin area are relatively less explored probably due to lack of effort of a bias survey and taphonomic investigation.

Tarafeni Valley, West Bengal

The Tarafeni valley, a lesser known paleontological area of interest, is the tributary of the Kansavati River that starts flowing from Midnapur District of West Bengal. It is surrounded by the hills of Bankura in the northwestern part and flows over the

Precambrian terrain (Basak et al. 1998). The vertebrate remains found from this river basin area go back to the nineteenth century, even though minor assemblages were discovered by Ghosh and Majumdar (1981, 1991). Later on, Badam (2000) carried out systematic investigations in the same valley near the Dhuliapur area and found preserved remnants of *Equus sp., Bos nomadicus, Axia/axia, Antilope, cervicapra* along with fragments of *turtule carapaces*. She dated the same and opined their period of existence in the basin area during late/terminal Pleistocene vis-a-vis in relation to the geology and archeology. Based on fossil studies, Badam (2000) concluded that the area had witnessed savannah vegetation, water pools, grasslands over the foothills and hard ground in the water sheet under the sub-humid climate.

Kurnool Caves

The Kurnool caves are very famous for deposition of Late Pleistocene-Holocene sediments and preserving faunal remains of various animal species and lithic artifacts therein. They are situated in the Betamcherla town of Kurnool district of Andhra Pradesh (Patnaik et al. 2008). The first discovery of the fossil remains was made by Newbold (1844) at Billa Surgam cave followed by Foote's and his son. Later on, various studies were made in reference to searching for vertebrate remains and their systematic investigations. Murty (1985) and Ansari (1999-2000) pointed out charred bodies of mammal remains in the cave and asserted that the small animals *Hystrix indica* and *Lepus nigiricollis* would have been slaughtered for consumption. Based on the presence of fossil remains belonging to *Antilope cervicapra, Gazella gazella, Cervus unicolor, Boselaphus tragocamelus*, ursus, *Bos* and *Bubalus*, Badam (1979) went on to state that the area had witnessed vegetation represented exceptionally by scrub to tall grasses and forests with gullies of grasslands during Upper Pleistocene period. The report of large fossil assemblage after foot's (1984) was carried out by Raju and Venkatasubbaiah (2001a, b) and identified them as *Semnopithecus entellus, Felis tigris/leo, Felis chaus, Felis rubiginosa, Hyaena crocuta, Vivera karnulensis, Rhinoceros karnulensis, Bos or Bubalus, Equus asinus, Boselaphus tragocamelus, Antilope cervipara, Gazella bennetti, Cervus axis or Axis axis, Sus cristatus, Ursus, Presbytisentellus, and Capra/Ovis sp.* Deshpandey-Mukherjee (2005) reported the vertebrate fauna of *Rhinoceros* from the same area and asserted that the same basin might have witnessed a swampy environment during Upper Pleistocene period.

Vegetational Scenario and Mortality of Animals During the Quaternary Period

The literature survey published on the animal remains in the Quaternary deposits and their mortalities in the above mentioned river basins suggest the existence of diversified fauna, vegetation, i.e., grasslands to deep forests and paleoenvironment, i.e., water pools to swamp during the Late Pleistocene to Holocene period. Certain researchers, viz., Badam, Patanik, Murthy, Ansari, Raju, and Venkatasubbaiah, etc., have pointed out the extinction of animal communities due to extensive hunting by the ancient humans and climatic fluctuations in the Indian Subcontinent. It has been noted that almost all of the river basins preserve the distal volcanic ash of Youngest Toba Tuff eruption, Indonesia, in close associations of the vertebrate remains. The same eruption is considered to have destroyed the vegetation and life of animals and forced them to migrate from one place to another, by inducing a long volcanic winter and acid rains (Ambrose 1998; Rampino and Ambrose 2000). The assertion of Ambrose about the complete destruction of living media in Southeast Asia was countered by Harcouth-Smith (2003), Oppenhemeir (2002) and Roberst et al. (2014) based on the findings of fossil, insect, and archeological remains from the post-tephra successions and asserted partial impact of the YTT. Similarly, our study also suggests partial or negligible impacts of the YTT ash on the animal community as evident by continuity in the population in India throughout from Late Pleistocene to Holocene period.

Conclusions

1. The YTT tephra is uncovered in the form of lenticles, elongated beds, and thick pockets in several river basins of south, western and central India, primarily in stratified, sandy to silty clay lithounits. Given its bedding geometries, depositional and sedimentological attributes, it has been viewed as primary and secondary ash and suggests calm and low energy fluvial environments in the subcontinent during tephra accumulations in the basins.
2. It is well recorded that the ash caused volcanic cooling and acid rains for a considerable period that wreaked havoc on vegetation, animals, and hominids life and forced them to migrate from Africa to other continents via a southern route. However, it is disputed among the scientific community, for example, Rampino and Self (1992) and Ambrose (1998) believe that the impact of the eruption was extremely powerful enough to create hostile conditions for living masses and vegetation, whereas Gathorny-Hardy and Hardcourt-Smith (2003), Oppenheimer (2003) and Singh and Srivastava (2021, 2022) argue against aforementioned hypotheses and suggest that the YTT had no effects on the mass populations and vegetation. Furthermore, based on the study of rhizospheres from the Purna alluvial basin, Singh and Srivastava (2020a, b, 2022) reported the existence of

mixed C3 and C4 vegetation in the Indian subcontinent prior, during, and after the YTT eruption.

3. Despite the above controversies, the ash in India is thought to be YTT based on its geochemical and chronological examinations; however, certain studies disagree. The geochemical investigation of the glass shards, collected from Son and Jurreru valley by Smith et al. (2011), show good compatibility with the OTT. Similarly, the geochronological data, obtained by Westway et al. (2011) and Srivastava and Singh (2021) using $^{40}Ar/^{39}Ar$ technique on the glass shards, reveal good correlations with the age of the OTT. Nevertheless, recent studies carried out by Westgate et al. (2013), Pearce et al. (2014), and Srivastava and Singh (2020a, b), made a clear distinction between YTT and OTT, implying that the ash in the Indian Peninsula belongs to the YTT rather than OTT.
4. The artifacts belonging to the Middle Paleolithic to Microlithic period have been retrieved from the pre- and post-YTT ash strata exposed in various river basins of India indicating continuity in the hominid populations and changes in the lithic technologies after the eruption. These lithic industries are consistent with the Middle Paleolithic artifacts of Africa suggesting migration of humans in India from the distal places.
5. The vertebrate remains found in the calcretized gravel beds exposed as pre-YTT column in several river basins have been identified as the remnants of the majority of herbaceous animal populations that existed during Late Pleistocene to Holocene, indicating a continuity in the mass community and the existence of small to tall vegetation, swamp, and water pool environments in India during the aforesaid period.

Acknowledgements This work has been carried out under the financial support provided by Science and Engineering Research Board (SERB), New Delhi, India, in the form of major research project (SB/S4/ES-692/2013) awarded to AKS.

References

Acharya SK, Basu PK (1993) Toba ash on the Indian subcontinent and its implications for correlations of late Pleistocene alluvium. Quat Res 40(1):10–19

Ambrose SH (1998) Late Pleistocene human population, bottlenecks, volcanic winter, and differentiation of modern humans. J Hum Evol 34:623–651

Ansari S (1999-2000) Small game hunting Musahars: an ethnoarchaeological approach. Puratattva 30:142–150

Badam GL (1979) Pleistocene Fauna of India. Deccan College Postgraduate and Research Institute, Pune

Badam GL (1985) Palaeontology of the Krishna Basin with special reference to Karnataka. Archaeology in Karnataka, papers presented at the National Seminar in archaeology. Directorate of Archaeology and Museums, Mysore, pp. 58–66

Badam GL (1989) Observations on the fossil hominid site at Hathnora, Madhya Pradesh, India. In: Sahni A, Gaur R (eds) Perspectives in human evolution. Renaissance Publishing House, Delhi, pp 153–172

Badam GL (2000) Pleistocene vertebrate palaeontology in India at the threshold of the millennium. J Palaeontol Soc India 45:1–24

Badam GL (2005) A note on the ostrich in India since the Miocene. Man and Environment 30(2): 97–104

Badam GL, Sathe V (1995) Palaeontological research in India: retrospect and prospect. In: Wadia S, Korisettar R, Kale VS (eds) Quaternary environments and geoarchaeology of India, vol 32. Essays in Honour of Professor S.N. Rajaguru. Memoirs of the Geological Society of India, Bangalore, pp 473–495

Basak B, Badam GL, Kshirsagar A, Rajaguru SN (1998) Late quaternary environment, palaeontology and culture of Tarafeni valley, Midnapur District, West Bengal—a preliminary study. J Geol Soc India 51:731–740

Biswas RH, Williams MAJ, Raj R, Juyal N, Singhvi AK (2013) Methodological studies on luminescence dating of volcanic ash. Quat Geochronol 17:14–25

Blinkhorn J, Smith VC, Achyuthan H, Shipton C, Jones S, Ditchfield P, Petraglia MD (2014) Discovery of youngest Toba tuff localities in the Sagileru valley, South India, in association with Palaeolithic industries. Quat Sci Rev 105:239–243

Blumenschine RJ, Chattopadhyaya UC (1983) A preliminary report on the terminal Pleistocene Fauna of the middle Son Valley. In: Sharma GR, Clark JD (eds) Palaeoenvironments and prehistory of the middle Son Valley, Madhya Pradesh, North Central India. Abinash Prakashan, Allahabad, pp 281–284

Chamyal LS, Khadkikar AS, Malik JN, Maurya DM (1997) Sedimentology of the Narmada alluvial fan, Western India. Sediment Geol 107:263–279

Chauhan PR (2008) Large mammal fossil occurrences and associated archaeological evidence in Pleistocene contexts of peninsular India and Sri Lanka. Quat Int 192:20–42

Chesner CA, Rose WI, Deino A, Drake R, Westgate JA (1991) Eruptive history of the earth's largest quaternary caldera (Toba, Indonesia) clarified. Geology 19:200–203

Clarkson C, Jones S, Harris C (2012) Continuity and change in the lithic industries of the Jurreru Valley, India, before and after the Toba eruption. Quat Int 258:165–179

Clarkson C, Harris C, Li B, Neudorf C, Roberts M, Richard G, Lane C, Norman K, Pal JN, Jones S, Shipton C, Koshy J, Gupta MC, Mishra DP, Dubey AK, Boivin N, Petraglia M (2020) Human occupation of northern India spans the Toba super-eruption ~74,000 years ago. Nat Commun 11:961. https://doi.org/10.1038/s41467-020-14668-4

Deshpande-Mukherjee A, Murty MLK, Badam GL (2005) Terrestrial gastropod shell assemblage from the Mesolithic cave site of Muchchatla Chintamanu Gavi in the Kurnool District, Andhra Pradesh. Man and Environment 30(1):86–93

Deshpande-Mukherjee A, Prova KK, Meshram P, Bhoyar A, Thakur A, Sontakke V (2014) A study of the animal remains from the early historic settlement of Chandankhera, district Chandrapur, Maharashtra. Bulletin of the Deccan College Research Institute 74:63–72

Diehl JF, Onstott TC, Chesner CA, Knight MD (1987) No short reversals of Brunhes age recorded in the Toba tuffs, North Sumatra, Indonesia. Geophysical Research Letter 14:753–756

Gasparotto G, Spadafora E, Summa V, Tateo F (2000) Contribution of grain size and compositional data from the Bengal Fan sediment to the understanding of Toba volcanic event. Mar Geol 162(2–4):561–572

Gathorne-Hardy FJ, Harcourt-Smith WEH (2003) The super-eruption of Toba, did it cause a human bottleneck? J Hum Evol 45:227–230

Gatti E, Durand AJ, Gibbard PL, Oppenheimer C (2011) Youngest Toba tuff in the son valley, India: a weak and discontinuous stratigraphic marker. Quaternary Science Reviews 30:3925–3934

Gatti E, Saidin M, Talib K, Rashidi N, Gibbard P, Oppenheimer C (2013) Depositional processes of reworked tephra from the late Pleistocene youngest Toba tuff deposits in the Lenggong Valley, Malaysia. Quat Res 79(2):228–241. https://doi.org/10.1016/j.yqres.2012.11.006

Geethanjali K, Achyuthan H, Jaiswal MK (2019) The Toba tephra as a late quaternary stratigraphic marker: investigations in the Sagileru river basin, Andhra Pradesh, India. Quat Int 513:107–123

Ghosh RN, Majumder S (1981) Neogene-Quaternary sequence of Kasai Basin, West Bengal, India, Proc. Field Conf Neogene-Quaternary Boundary, India, 1979, (Ed.) Sastri et al., Geol Surv India pp. 63–73

Ghosh RN, Majumdar S (1991) Geology and morphostratigraphy of West Bengal: a database for archaeological exploration. In: Datta A (ed) Studies in archaeology: papers presented in memory of P.C Dasgupta. Books and Books, New Delhi, pp 7–37

Harini P, Sahadevan DK, Das IC, Manikyamba C, Durgaprasad MB, Nandan MJ (2018) Regional groundwater assessment of Krishna River basin using integrated GIS approach. Journal of the Indian Society of Remote Sensing 46:1365–1377

Harpending HC, Sherry ST, Rogers AR, Stoneking M (1993) The genetic structure of ancient human populations. Curr Anthropol 34:483–496

Haslama M, Harris C, Clarkson C, Pal JN, Shipton C, Crowthera A, Koshy J, Bora J, Ditchfield P, Rame HP, Price K, Dubey AK, Petraglia M (2012) Dhaba: an initial report on an Acheulean, middle Palaeolithic and microlithic locality in the middle Son Valley, north-Central India. Quat Int 258:191–199

Jayaprakash C, Jayakumar R, Dinesh AC, Maran N, Shareef NM (2009) Tephra from quaternary sediments of western slope of Lakshadweep ridge, Arabian Sea. Curr Sci 97(11):1670–1674

Jones SC (2010) Palaeoenvironment response to the ~74 Toba ash-fall in the Jurreru and Middle Son valleys in southern and north-central India. Quat Res 73:336–350

Joshi PM (2020) Palaeolithic occupation and cultural transition in the Wainganga River basin, India. Antiquity 94:300–322

Joshi RV, Badam GL, Pandey RP (1978) Fresh data on the quaternary animal fossils and stone age cultures from the Central Narmada Valley, India. Asian Perspect 21(2):164–181

Joshi RV, Badam GL, Pandey RP (1980) Late Pleistocene animal fossils and stone age cultures in upper Mahanadi Valley, M.P. The Eastern Anthropologist 33(2):169–175

Khatri AP (1966) The Pleistocene mammalian fossils of the Narmada river valley and their horizons. Asian Perspect 9:113–133

Korisettar R, Mishra S, Rajaguru SN, Gogte VD, Ganjoo RK, Venkatesan TR, Tandon SK, Somayajulu BLK, Kale VS (1988) Age of the Bori volcanic ash and lower Palaeolithic culture of the Kukdi valley, Maharashtra. Bulletin Deccan College and Research Institute 48:135–138

Krishnan G, Achyuthan H (2019) Incipient weathering of the ∼74ka young Toba tuff (YTT) biotite grains, Sagileru Valley, Andhra Pradesh. J Geol Soc India 94, 573–578 https://doi.org/10.1007/s12594-019-1362-y

Lane CS, Chorn BT, Johnson TC (2013) Ash from the Toba supereruption in Lake Malawi shows no volcanic winter in East Africa at 75 ka. Proceeding of the National Academy of Science of the Unites States of America 110(20):8025–8029

Liu Z, Colin C, Trentesaux A (2006) Major elements geochemistry of glass shards and minerals of the youngest Toba Tephra in the southwestern South China Sea. J Asian Earth Sci 27:99–107

Mark DF, Petraglia M, Smith VC, Morgan LE, Barfod D, Ellis B, Pearce NJ, Pal JN, Korisettar R (2014) A high precision $^{40}Ar/^{39}Ar$ age for the young Toba Tuff and dating of ultra-distal tephra: forcing of quaternary climate and implications for hominin occupations if India. Quat Geochronol 21(1):90–103

Mishra S (2016) The Narmada River in Indian prehistory. In: Dr. a.P. Khatri memorial lecture, delivered on 17th December 2015, Hyderabad

Mishra S, Venkatesan TR, Rajaguru SN, Somayajulu BLK (1995) Earliest Acheulian industry from peninsular India. Curr Anthropol 36:847–851

Murthy MLK (1985) Late Pleistocene Fauna of Kurnool caves, South India. In: Clason AT (ed) Archaezoological studies. Special publication, North Holland Publishing Company, Amsterdam and New York, pp 132–138

Neudorf CM, Roberts RG, Jacobs Z (2014) Testing a model of alluvial deposition in the middle Son Valley, Madhya Pradesh, India—IRSL dating of terraced alluvial sediments and implications for archaeological surveys and palaeoclimatic reconstructions. Quat Sci Rev 89:56–69

Newbold TJ (1844) Notes on the osseous breccia and deposit in the caves of Billasurgam, southern India. Journal Asiatic Society of Bengal 13(2):610–615

Nishimura S, Yojoyama T, Wirasantosa S, Dharma A (1977) Danau Toba – the outline of Lake Toba, North Sumatra, Indonesia: Paleolimnology of Lake Biwa. Japan Pleistocene 5:313–332

Oppenheimer C (2002) Limited global change due to the largest known quaternary eruption, Toba ≈74 kyr BP? Quat Sci Rev 21:1593–1609

Oppenheimer C (2003) Climatic, environmental and human consequences of the largest known historic eruption: Tambora volcano (Indonesia) 1815. Prog Phys Geogra 27(2):230–259

Patnaik R, Badam GL, Murty MK (2008) Additional vertebrate remains from one of the late Pleistocene—Holocene Kurnool caves (Muchchatla Chintamanu Gavi) of South India. Quat Int 192:43–51

Patnaik R, Chauhan PR, Rao MR, Blackwell BA, Skinner AR, Sahni A, Chauhan MS, Khan H (2009) New geochronological, paleoclimatological, and archaeological data from the Narmada Valley hominin locality, central India. J Hum Evol 56(2):114–133

Pattan JN, Pearce NJG, Banakar VK, Parthiban G (2002) Origin of ash in the central Indian Ocean Basin and its implications for the volume estimate of the 74,000 year BP youngest Toba eruption. Curr Sci 83(7):889–893

Pearce NJ, Westgate JA, Gatti E, Pattan JN, Parthiban G, Achyuthan H (2014) Individual glass shard trace element analyses confirm that all known Toba tephra reported from India is from the c. 75-ka youngest Toba eruption. J Quat Sci 29:729–734

Petraglia M, Korisetter R, Boivin N, Clarkson C, Ditchfield P, Jones S, Koshy J, Lahr M, Oppnheimer C, Pyle D, Roberts R, Schwenninger J, Arnold L, White K (2007) Middle Paleolithic assemblage from the Indian subcontinent before and after the Toba super-eruption. Science 317:114–116

Petrglia M, Ditchfield PS, Jones SC, Korisetter R, Pal JN (2012) The Toba volcanic super-eruption, environment change, and hominin occupation history in India over the last 140,000 years. Quat Int 258:119–134

Raj R (2008) Occurrence of volcanic ash in the quaternary alluvial deposits, lower Narmada basin, western India. J Earth Sys Sci 117(1):41–48

Raju DR, Venkatasubbaiah PC (2001a) Late Pleistocene Fauna from Kurnool caves. In: Settar S, Korisettar R (eds) Indian archeology in retrospect-prehistory, vol 1. Manohar and Indian Council of Historical Research, New Delhi Appendix I

Raju DR, Venkatasubbaiah PC (2001b) Late Pleistocene Fauna from Kurnool caves. In: Settar S, Korisettar R (eds) Indian archeology in retrospect-prehistory volume 1. Manohar and Indian Council of Historical Research, New Delhi Appendix I

Rampino MR, Ambrose SH (2000) In: McCoy FW, Heiken G (eds) Volcanic winter in the Garden of Eden: the Toba supereruption and the late Pleistocene human population crash, vol 345. Volcanic Hazards and Disasters in Human Antiquity: Geological Society of America, Special Paper, pp 71–82

Rampino MR, Self S (1992) Volcanic winter and accelerated glaciation following the Toba supereruption. Nature 359:50–52

Roberts P, Delson E, Miracle P, Ditchfield P, Roberts RG, Jacobs Z, Blinkhorn J, Ciochon RL, Fleagle JG, Frost SR (2014) Continuity of mammalian over the last 200,000 y in the Indian subcontinent. Proceeding National Academy Science of the United States of America 111(15): 5848–5853

Rose WI, Chesner CA (1987) Dispersal of ash in the great Toba eruption. Geology 15:913–917

Sahu BP (1988) From hunters to breeders. Anamika Prakashan, Delhi

Saraf VR, Regulwar DG (2016) Assessment of climate change for precipitation and temperature using statistical downscaling methods in upper Godavari River Basin, India. Journal of Water Resource and Protection 08:31–45

Schulz H, Emeis KC, Erlenkeuser H, Rad U, Rolf C (2002) The Toba volcanic event and interstadial/stadial climates at the marine isotopic stage 5 to 4 transition in the Northern Indian Ocean. Quat Res 57:22–31

Singh IB (1980) Precambrian sedimentary sequences of India; their peculiarities and comparison with modern sediments. Precambrian Res 12(1–4):411–436

Singh IB, Srivastava P, Sharma S, Sharma M, Singh DS, Rajagopalan G, Shukla U (1999) Upland interfluve (Doab) deposition: alternative model to muddy overbank deposits. Facies 40:197–210

Singh A, Muhammad RFBH, Taib NI, Jha D, Srivastava AK (2021a) Surface texture, mineralogy and stable isotope studies of nodular calcretes preserved in the YTT ash of Padang Terap river basin and lenggong valley, peninsular Malaysia: implications in its origin and paleoclimatic reconstruction. Rhizosphere 19:100380

Singh A, Srivastava AK (2021b). Rhizosphere: a fascinating paleovegetational and paleoclimatic new intermediary in the quaternary fluvio-lacustrine set-up of the Purna alluvial basin, Central India. Rhizosphere 20(2), 100430

Singh A, Srivastava AK (2022) Had youngest Toba tuff (YTT, ca. 75 ka) eruption really destroyed living media explicitly in entire Southeast Asia or just a theoretical debate? An extensive review of its catastrophic event. J Asian Earth Sci X. 7:100083

Singh A, Srivastava AK, Chauhan N (2022) Luminescence dating and bracketing time of the youngest Toba tuff deposits in the quaternary sediments of Purna Alluvial Basin, Central India. J Earth Sci, 33, 1007–1016.

Sinha R, Jain V, Babu GP, Ghosh S (2005) Geomorphic characterization and diversity of the fluvial systems of the Gangetic Plains. Geomorphology 70:207–225

Smith VC, Pearce NJG, Matthews NE, Westgate JA, Petraglia M, Haslam M, Lane CS, Korisettar R, Pal JN (2011) Geochemical fingerprinting of the widespread Toba tephra using biotite compositions. Quat Int 246:97–104

Sonakia A (1984) The skull-cap of early man and associated mammalian fauna from Narmada Valley alluvium, Hoshnagabad area, Madhya Pradesh (India). Geological Survey of India 113(6):159–172

Srivastava AK, Kale VM (2010) Sediment characteristics and depositional environment of Purna alluvial basin, Maharashtra. Gondwana Geological Magazine, Special Publication 12(1): 293–302

Srivastava P, Shukla UK (2009) Quaternary evolution of the Ganga River system: new quartz ages and a review of luminescence chronology. Himal Geol 30:85–94

Srivastava AK, Singh A (2019a) YTT ash from quaternary sediments of Kapileshwar area, Purna alluvial basin, Central India. Quat Int 500:96–107

Srivastava AK, Singh A (2019b) Nature, occurrence and lithological set-up of the Youngest Toba Tuff volcanic ash, Purna alluvial basin, Central India. The Journal of Geol 127:593–610

Srivastava AK, Singh A (2020a) Lithological, physical and chemical attributes of primary volcanic ash of YTT, Purna alluvial basin, Central India. Geol J 55(10):7011–7023

Srivastava AK, Singh A (2020b) Weathering pattern of youngest Toba Tuff, Purna alluvial basin, Central India. Arab J Geosci 13:607. https://doi.org/10.1007/s12517-020-05641-y

Srivastava AK, Singh A (2021a) Geochemistry and Constrained 40Ar/39Ar dating of Youngest Toba Tuff glass shards, Purna alluvial basin, Central India. Journal Earth System Science 130. https://doi.org/10.1007/s12040-020-01513-x

Srivastava AK, Singh A (2021b) Rhizosphere: A fascinating paleovegetational and paleoclimatic new intermediary in the Quaternary fluvio-lacustrine set-up of the Purna alluvial basin, central India. Rhizosphere 20(2), 100430.

Srivastava AK, Singh A, Muhammad RF, Pattan J, Sharma N (2019) Geochemical characterization and regional correlation of youngest Toba tuff (YTT, 75 ka) glass shards, Purna alluvial basin, Central India. Arabian Journal of Geoscience 14(19) https://doi.org/10.1007/s12517-021-07848-z.

Theobold W (1860) On tertiary and alluvial deposits of the central portion of the Narmada valley. Memoirs of the Geological Survey of India 14(1):66–125

Timmereck C, Graf H-F, Zanchettin D, Hagemann S, Kleinen T, Krüger K (2012) Climate response to the Toba super-eruption: regional changes. Quat Int 258:30–44

Tiwari MP, Bhai HY, Varade AM (2010) Stratigraphy and tephra beds of the Purna quaternary basin, Maharashtra, India. Gondwana Geological Magazine 12:283–292

Tripathy KC (1972) South Orissa prehistory-the first record of stone age tools. Asian Perspect 15: 47–59

Verma BC, Mishra VP, Kumar U (1998) Discovery of early quaternary vertebrate fossils beneath the Ganga River bed at Bhagalpur, Bihar: their age and biostratigraphic implications. J Palaeontol Soc India 43:35–40

Westgate JA, Shane PAR, Pearce NJG, Perkins WT, Korisetter R, Chesner CA, Williams MAJ, Acharyya SK (1998) All Toba tephra occurrences across peninsular India belong to the 75,000 yr B.P. eruption. Quat Res 50:107–112

Westgate JA, Pearce NJ, Perkins WT, Preece SJ, Chesner CA, Muhammad RF (2013) Tephrochronology of the Toba tuffs: four primary glass populations define the 75-ka youngest Toba Tuff, northern Sumatra, Indonesia. J Quat Sci 28:772–776

Westway R, Mishra S, Deo S, Bridgland DR (2011) Methods for determination of the age of Pleistocene tephra, derived from eruption of Toba, in central India. Journal of Earth System Science 120:503–530

Williams MAJ, Royce K (1982) Quaternary geology of the middle Son valley, north central India: implications for prehistoric archaeology. Palaegeography Palaeoclimatology Palaeoecology 38: 139–162

Williams MAJ, Ambrose SH, Van der Kaars S, Ruehlemann C, Chattopadhyaya U, Pal J, Chauhan PR (2009) Environmental impact of the 73 ka Toba super-eruption in South Asia. Palaegeography Palaeoclimatology Palaeoecology 284:295–314

Chapter 12
Active Tectonics–Climate Interaction in the Structural Depo-Centres of Central Indian Tectonic Zone (CITZ)

Snigdha Ghatak and Mriganka Ghatak

Abstract The Late Cenozoic period in the Central Indian Tectonic Zone (CITZ) marked by several episodes of crustal adjustments resulted in the development of various landforms, drainage anomalies and pedosedimentary successions. The fault bound tectonic basins viz. Tapi, Purna and Narmada flanking the Satpura ranges hold tell-tale evidence of varying climatic and tectonic regimes. This chapter summarizes data from various literature, the Active Fault Mapping programme in CITZ and the fluvial archives to unravel the role of the tectonics-climate feedback system in shaping landscape architecture in and around the structurally controlled Quaternary River basins of Purna, Tapi and Narmada. The cumulative interactions between the deformational events and ongoing surface processes are best witnessed by the marker horizons, anomalous river courses, basin geometry, hill slopes and elevation characteristics and the way these have influenced its depositional environment. The study suggests that repeated reactivation of older crustal-scale faults under different stress regimes have been a guiding factor in landform evolution in CITZ. While tectonics holds the primary key to the overall landscape development and sedimentation, peninsular climatic or microclimatic variations have equally been significant as palaeo-climatic proxies mimic all India monsoon scenarios.

Keywords Cenozoic period · Sedimentary basins · Soil chronostratigraphy · Drainage anomalies · Climate amelioration

Introduction

The tectonic activity in the Cenozoic period of Earth's history has contributed to the modification of the geography of Asia through the collision between the Indian and Eurasian plates resulting in the formation of the highest (5 km and above msl) mountain chain and most extensive plateau representing the present-day greatest

S. Ghatak (✉) · M. Ghatak
Geological Survey of India, Kolkata, West Bengal, India
e-mail: snigdha.ghatak.gsi@gov.in

N. Khare (ed.), *Science, Policies and Conflicts of Climate Change*, Springer Climate,
https://doi.org/10.1007/978-3-031-16254-1_12

topographic relief – the Himalaya. Scientific investigations suggest that the uplift of the Himalayas and Tibetan Plateau (HTP) caused the intensification of the Asian monsoon through several physical effects (Molnar et al. 1993). For example, physical and chemical weathering induced by this seasonal variation has caused enhanced drawdown of atmospheric CO_2 and resulted in global cooling since ca. 45 Ma in different spells (Raymo 1991). The variations in the intensity of the Indian monsoon (Overpeck et al. 1996) played a very significant role in the sedimentary as well as surface processes in the Indian Subcontinent. These were aided by tectonics driven changes in the source and the sink regions. The evolutionary history of the Asian monsoon has been intensively studied both on land and in marginal seas but attempts to integrate the terrestrial and marine data are still insufficient.

Effects of monsoon proxies preserved in oceanic phytoplanktons, zooplanktons and continental proxies that include tree rings, soils, ice and lake deposits have also been well studied (Kroon et al. 1991; Patnaik et al. 2012). Several site-specific and basin-specific studies were undertaken in the Ganga plain (Srivastava et al. 2003a, b, c, Gibling et al. 2005, Sharma et al. 2004, Singh 2005, Sinha et al. 2007, Williams et al. 2006), sedimentary basins of western Peninsular India (Jain and Tandon 2003; Juyal et al. 2004; Juyal et al. 2006; Chamyal et al. 2003; Raj 2007; Bhandari et al. 2005, south Indian River basins (Thomas et al. 2007; Nagalakshmi and Achyuthan 2004; Resmi and Achyuthan 2021) with an emphasis on the determination of the chronology of various geomorphic surfaces, climate, sedimentation and soil chronostratigraphy, etc.

The Cenozoic Central Indian tectonic basins flanking the ENE-WSW trending Satpuras are the repository of varied palaeoclimatic proxies viz. palaeosols or pedisedimentary sequences and calcretes. Very few workers have addressed the issue of tectonics-climate linkage in these basins. This gains importance as the Late Cenozoic period in the Central Indian Tectonic Zone (CITZ) was marked by several episodes of crustal adjustments. This is reflected in the development of various landforms, drainage anomalies, pedosedimentary successions correlatable to the regional or local tectonics and sedimentation pattern in the fault bound tectonic basins viz. Tapi, Purna and Narmada on either side of the Satpura ranges. However, the basic lithostratigraphic and morphostratigraphic formations of Central Indian River basins and their correlation and chronology (Table 12.1) have been discussed in a few earlier works (Tiwari 1999). Further, a large amount of data on soil climate studies is available for parts of Central and Peninsular India, including the Purna watershed (Pal et al. 1989, 2001). In these regions, climate amelioration is aided by tectonic stability leading to stabilization of sediments, soil profile development and activation of fluvial systems (Singhvi and Kar 2004). Conversely more arid episodes can result in desiccation, decreased vegetation cover and mobilization of sediments (Dhir et al. 2010). Depending on the slope and other dependent factors, flash flood events or tectonic activity can trigger coarser sediment flux into the basins causing termination in pedogenesis and incomplete soil profile development or preservation.

This chapter presents a synoptic view from available literature and data generated in the course of the Active Fault (fault which has been active in Quaternary times,

Table 12.1 Generalized stratigraphy

Narmada valley (250–300 m)	PM	Purna valley (425 m)	PM	Tapti valley (400 m)	PM	Upper Wardha (25–30 m)	Age
Ramnagar Formation		Purna Formation/ Dark grey silt formation		Dark grey silt formation/ Mehun Formations		Mengawadi Formation/ Wardha formation	Holocene
Bauras Formation							Holocene
Hirdepur Formation		Kural Formation/ Light grey silt formation		Light grey silt formation/ Tahakali Formation		Bharaswada / Siri Formation	Lower Holocene to Upper Pleistocene
Baneta Formation (Tephra: Transported and in situ)	BN	Kodori Formation/ Brown Silt Formation (Tephra: in situ)	BN	Brown silt formation/ Kurkhali Formation		Belora/ Jamalpur Formation	Upper Pleistocene
						Simbhora Formation	
Surajkund Formation	BN						Middle Pleistocene
Dhansi Formation	BN-MR	Vaghoi Formation/ Red silt Formation	BN-MR			Red silt formation	Mid. Pleistocene to Lower Pliestocene
Pilikarar Formation		Ferruginised gravels and sands					Lower Pleistocene
Based on the Quaternary successions							
Deccan Traps					Basalt flows with associated inter-trappean		Upper Cretaceous - Eocene
Lameta Formation					Coarse sandstones, impure carbonates		Cretaceous
Upper Gondwana Group					White clay, coarse-grained sandstone		Triassic- Lower Cretaceous
Granite mylonite					Pink foliated granite mylonite		Mesoproterozoic

After Tiwari (1999); *PM* Palaeomagnetic date, *BN* Bruhn's Normal, *MR* Matuyama Reversal.

Michael 2000; Roy and Devarajan 2003) Mapping Programme (Devrajan et al. 2004; Ghatak et al. 2005, 2008; Ghatak and Ghatak 2010) in CITZ and the study of fluvial archives to unravel the role of the tectonics-climate feedback system in shaping landscape architecture in and around the structurally controlled Quaternary River basins of Purna, Tapi and Narmada. The Quaternary deposits in the shallow river basins of Godavari, Wardha, Waiganga, Hasdo, Mahanadi, Son and Chambal, etc. (other than the upper Wardha valley) have been, however, referred to as exclusively climatogenic (Tiwari 1999) and are kept beyond the purview of the discussion.

Regional Geological Framework

Central Indian Tectonic Zone (CITZ) is a mobile belt and a group of crustal-scale lineaments separating the Bundelkhand craton to its north and the Baster craton in the south (Roy and Prasad 2003; Golani et al. 2001). CITZ is represented by a collage of different lithotectonic terranes ranging from Archaean to Recent (Roy and Devarajan 2001). The Satpura ranges situated in the intraplate ENE-WSW trending (CITZ) (Acharyya and Roy 2000) are bound in the north and south by three major seismogenic fault zones, namely Son Narmada South Fault (SNSF) in the north and Gavilgarh Fault Zone (GFZ) and Tapi Fault Zone (TFZ) in the south, respectively (Fig. 12.1). The present response of CITZ to accumulation of stress and attendant seismicity is governed by the structures generated due to the early seismic history of rocks within it (Acharyya and Roy 2000). This area has witnessed several earthquake events in the past and many of these faults on the map scale show unambiguous evidence of fault movement within the Quaternary period (Narula et al. 2000, IMD data). Much of these fault movements have resulted in the upliftment of land mass. However, the intraplate seismicity, fault geometry and kinematics in various active fault zones and their relation to sedimentation have been studied by few and there are significant gaps in the existing information. The primary reasons for this can be attributed to scanty historical and instrumental data on regional seismicity and lack of detailed geological studies and availability of marker horizons. However, the major earthquakes in Peninsular Central India, viz., 1967 Koyna earthquake, 1993 Latur earthquake and the 1997 Jabalpur earthquake have brought into focus the importance of seismotectonic appraisal of the major faults of the region. The area is characterized by low magnitude events with a high return period which is characteristic of the intraplate region. The generalised Quaternary stratigraphy in and around the Tapi, Purna and Narmada basins and basement to Quaternary sediments is at Table 12.1.

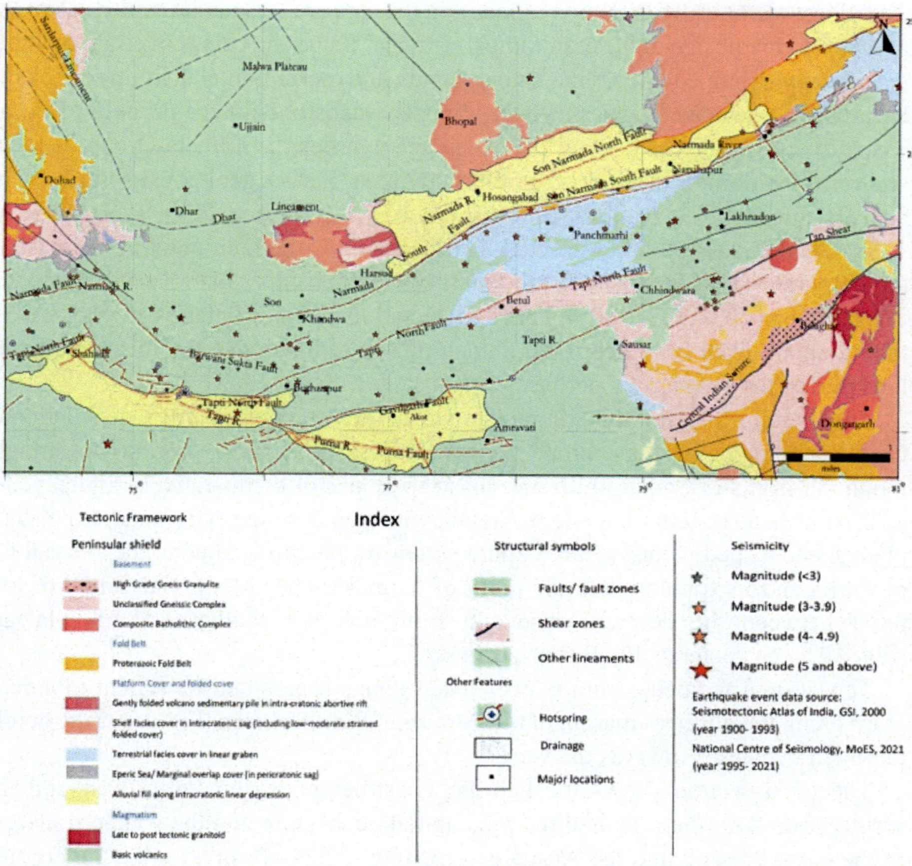

Fig. 12.1 Regional map of the area showing major faults of CITZ, earthquake locations and structurally controlled Quaternary basins. (Modified after Narula et al. 2000, Earthquake data source: NCS, IMD and GSI, up to November 2021)

Gavilgarh Fault Zone Vis-à-Vis Sedimentation in the Purna Basin

Geology-Geomorphology of Purna Basin

The crescent-shaped, semi-arid, Purna drainage basin, covering an area of 18,000 sq. km (approx.), is a fault-controlled depocentre hosting 425 m of supposed Quaternary sediments (Ravi Shanker 1987). Although Purna is part of the Tapi River watershed, the distinctive geological milieu of Purna qualifies for separate discussion here.

Located to the north of Purna valley and as a western extension of the Tan shear, the Gavilgarh Fault Zone along the south Satpura foothills runs for ~230 km exposing an array of rocks ranging in age from Mesoproterozoic to Quaternary.

The oldest tectonically exhumed lithounit in the area are the granite mylonites of the Mesoproteriozoic age in the easternmost part near Salbardi. This is overlain by other tectonically exhumed units, viz., Upper Gondwana rocks, Lameta Group of rocks of Permian to Cretaceous age and inter-trappean sandstone-limestone units. Deccan Traps of the Upper Cretaceous to Eocene age overlies the Lameta Group of rocks and form the dominant lithology in the study area. Earlier geological studies in the area (Blanford 1869; Pradhan 1980; Rajurkar 1982; Saxena 1988; Ray and Rana 1989; Chandradas 1999; Tiwari et al. 2001) described the fault marking the base of the Satpuras variously as Gavilgarh Fault, Ellichpur Fault, Salbardi Fault (Rajurkar 1982; Saxena 1988; GSI 2000). Purna River and its major tributaries emerging from the mountain front have deposited a thick pile of sediments in the form of alluvial fans during the Quaternary period.

The landforms in this intraplate area are predominantly denudational. Geomorphologically, three broad zones: (1) faulting induced sinuous to straight mountain fronts of Satpura hills and its steep slopes of comparatively higher relief at the north, (2) zone of moderate relief with gentle slope, i.e., southern Ajanta plateau and its pediments, south Satpura pediment, piedmont zone of the alluvial fan of Purna and its tributaries in the north of the basin and (3) alluvial valley of low relief between the northern mountain front and the southern Ajanta plateau, (Fig. 12.2) characterize the Purna basin.

Terrain morphometry from a level sliced image brings out the salient mountain scarp along the northern margin of the Purna and Wardha basin (Fig. 12.2) the height of which decreases towards the west.

The cross-section shows the half-graben structure of the Purna basin on the surface with a northern faulted margin, central depocentre and the southern margin of the basin grading into the Ajanta plateau (Fig. 12.2b–d). In a terrain undergoing tectonic uplift, the first geomorphic agents to be affected are the streams as their courses are governed by the natural gradient of the area. Purna River originates from Sawalmedha at the northeastern corner of the basin, flows in an axial course south of the GFZ and finally joins Tapi River in the northwestern corner of the basin (Fig. 12.2a). Higher-order transverse antecedent drainages join the Purna River emanating from the northern mountain front. Many of the major rivers of the area show structurally controlled abrupt changes in their courses, while lower-order streams at the pediment zone are very closely and regularly spaced as the characteristics of the streams are emerging from a tectonically active flank (Talling et al. 1997; Burbank and Anderson 2001; Ghatak et al. 2005). These features add to the fact that the axial Purna River flows along a course distal to the mountain range, which points towards an overall slow tectonic activity in the area (Burbank and Anderson 2001). Further, aspect data displayed as a grey-scale image (Fig. 12.2e) is independent of illumination parameters and helps in picking up the coalesced palaeo-fan surfaces of Purna and other transverse tributaries lying south of GFZ.

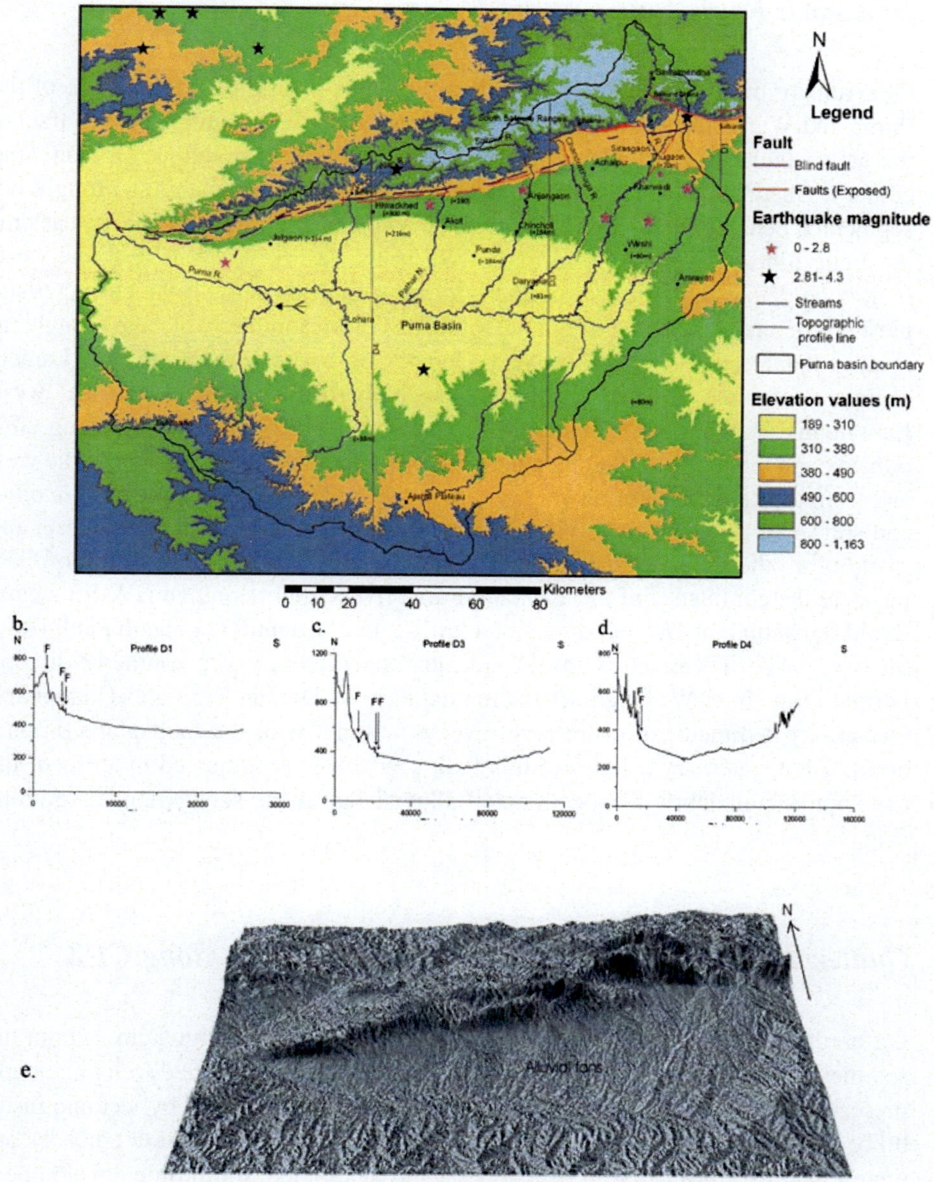

Fig. 12.2 (**a**) SRTM derived level slice map of Purna basin and adjacent areas showing general geomorphology, micro-earthquake epicentres, fault location etc. (**b–d**) Digital cross-section across Purna basin. 'F' denotes the location of faults in the profiles. For elevation classes and locations of profiles refer to (**a**). (**e**) 3D perspective view of the terrain with aspect data draped on the DTM

Gavilgarh Fault Zone

GFZ runs primarily through Deccan Traps and cuts across the water divide of the Purna and Wardha basins (Fig. 12.2a). Salient fault zone signatures help in tracing the active faults in GFZ which otherwise runs through monotonous Deccan Trap country. Ghatak and Ghatak 2010 in the course of active fault mapping brought out segmented behaviours of GFZ along with the gaps, overlaps or offsets. The easternmost curvilinear (swerving from ENE-WSE in the east to WNW-ESE to the west) fault zone defines a vertical to northerly dipping fault plane with the characteristic presence of a tectonic sliver of granite mylonites of Mesoproterozoic age brought up against the Deccan Traps. Cofolded Upper Gondwana sandstones and Lameta sediments against Deccan Traps in the Salbardi-Belkundhana in the west, Behramghat-Belkher in the centre are indicative of reverse reactivation along GFZ (Ghatak and Ghatak 2005) along a northerly dipping fault plane. Further to the west and from the central sector onwards (west of Chandrbhaga River), the fault is blind and shows surface signatures viz. tilting of strata (Fig. 12.3), hogbacks, cuesta and rotation of otherwise vertical columns within basalts. The westernmost tip of GFZ lies near the confluence of the Tapi and Purna rivers after taking an NW-SE swing. Rapid thickening of Quaternary about 4 km. south of Foothill Gavilgarh Fault (Ravi Shanker 1987, Tiwari et al. 2001) indicates the presence of a southerly dipping normal fault (in E-W direction) at the boundary of Deccan Trap and Quaternary. Quaternary sediments also preserve reverse faulting, with the fault plane dipping north. The Quaternary activity of this fault segment is documented in terms of the development of debris flow dominated alluvial fan at the south Satpura foothills (Fig. 12.2e).

Quaternary Sedimentation and Climatic Imprints Along GFZ

Different climatic and tectonic episodes have led to different signatures within the sedimentary record of the Purna basin. The process of soil profile development and preservation of pedosedimentary sequences had been interrupted by tectonic instability followed by an influx of coarser sediments in the basin. These episodes are represented by termination in pedogenesis and incomplete soil profile development or preservation (Ghatak and Ghatak 2010). The sedimentological characters of the fan near the fault are dominantly of debris flow to sheet flow character, while away from the fault zone stream flow and sheet flood facies continue. The terminal part of the fan is represented by the mixing up of the fan and alluvial plain environment (Fig. 12.4). Evidence of fan reworking from the middle fan part is ample. Apart from debris flow dominated proximal fan deposit, sheet flood and stream flow deposit which is even found in the middle part of the fan (almost 10–12 km from the fault line) requires some amount of transport where it is not only controlled by the hill slope processes. This is because hill slopes reach limiting values after certain

Fig. 12.3 Geological and geomorphological expressions of GFZ. (**a**) Tectonic sliver of Meso Proterozoic granite mylonite with steep northernly dip [location: Salbardi]. (**b**) Brecciation along the fault zone [Location: Bhili] (**c**) Butressed Fold in Gondwana against the fault plane of GFZ [Location: Belkher] (**d**) Northerly tilted bole bed [Location: Jambhal Talav] (**e**) Regional hogbacks along with GFZ (**f**) Evidence of compressional tectonics within Quaternary [Location: Wan River Section]. N: north, S: south

distances. A higher rate of bed load transport indicates a frequent event of storms. Well preserved cross-beds in the stream flow require current action and sheet flood overland flow without channelization is called for. All parts of the fan facies bear calcrete nodules/layers and palaeosol horizons (Ghatak and Ghatak 2010; Srivastava

Fig. 12.4 Representative litho packages in the Purna basin. (**a**) Silty clay at Udkhed. (**b**) Red pedosedimentary, debris flow fan deposit at Wishroli (**c**) Sheet Flood Facies south of Wishroli (**d**) Stream flow dominated fanglomerate facies at Wan River Section (**e**) River bank deposit along the present active channel of Purna showing the ball and pillow structure

et al. 2021) except for the present-day deposit, which is restricted to the present day floodplain. The youngest sediment is mainly dark grey silts and clay. Based on the relation with the source area, this fan sequence can be described as a 'mountain front fan' (Harvey 1990). These mountain front alluvial fans are characteristic of tectonically active mountain chains (Harvey 1988; Guzzetti et al. 1997).

Depth distribution of Quaternaries available from borehole data/point resistivity data (Ravi Shanker 1987) shows a sudden increase in sediment thickness (more than 300 m) west of Akot (Fig. 12.2a). Towards the south of the basin near Lohara sediment thickness near the gradational basin boundary located near Ajanta Plateau becomes 38 m. Stratigrahic drilling in the Quaternaries in the middle part of the Purna basin, undertaken as part of Quaternary studies (Tiwari et al. 2001) has established a Quaternary sediment thickness of more than 250 m with a lower horizon attributed to being of limnic origin. Other workers like Ravi Shanker 1987 have mentioned a total sediment thickness of nearly 400 m in the basin. The generation of this huge sediment thickness has so far been attributed to the regional scale Gavilgarh Fault disposed at the base of the Satpuras. Surprisingly, the sudden increase of Quaternary at Deccan-Quaternary contact west of Chandrabhaga was intriguing where not much sediment yield is expected from a blind reverse reactivation of a fault segment to fill in depositional loci for 425 m. Such multi-parametric considerations of fault geometry, sedimentation pattern and other field evidence revealed the buried Purna north fault with a downthrow to the south defining the Deccan/Quaternary contact.

Considering segmented behaviour of the fault zone, three key sections along the GFZ at Udkhed, Wishroli, Wan near the fan head were studied (Fig. 12.5). Broad lithofacies associations are as follows.

Lithofacies Association at Udkhed

Udkhed lies south of the most active mountain front (Dhamandas-Belkund Dhana) in the Wardha basin and east of the Purna cross fault (Ghatak et al. 2009). A 3 m thick mottled, brownish to orange silty clay, with intense clay venations/fractures filled with white clay has yielded an Optically Stimulated Luminescence date of 212 ± 17 Ka (Ghatak et al. 2009). Field observations suggest that this lies on the block defining the northeast basin boundary cross fault along the course of the Purna River in NE-SW direction. Several micro earthquake activities are also concentrated along this fault determined using the MEQ network along GFZ (Ghatak et al. 2009).

Broad Lithofacies Association at Wishroli

Purna River emerges from the mountain front at Wishroli and initiates fan formation. Wishroli exposes a nearly 18.5 m thick fanglomerate sequence which can be divided into three units based on lithofacies assemblage and the nature of their contact (Fig. 12.5a). The basal unit 1 (Red palaeosol-gravel association) overlies eroded Deccan Traps strath terrace is a multistoried gravel sequence (~6 m) characterized by debris flow. The termination of this unit is occasionally marked by bedded calcrete developed over the upper planar and sharp contact. The middle and upper parts of unit 1 have yielded dates of 129 ± 8 Ka and 117 ± 9 Ka, respectively (Ghatak et al. 2009). Unit 2 is middle gravel-sand association with a sharp planar contact that is ~4 m thick. Unit 3 is the upper pebble-sand unit which is nearly 4 m thick and

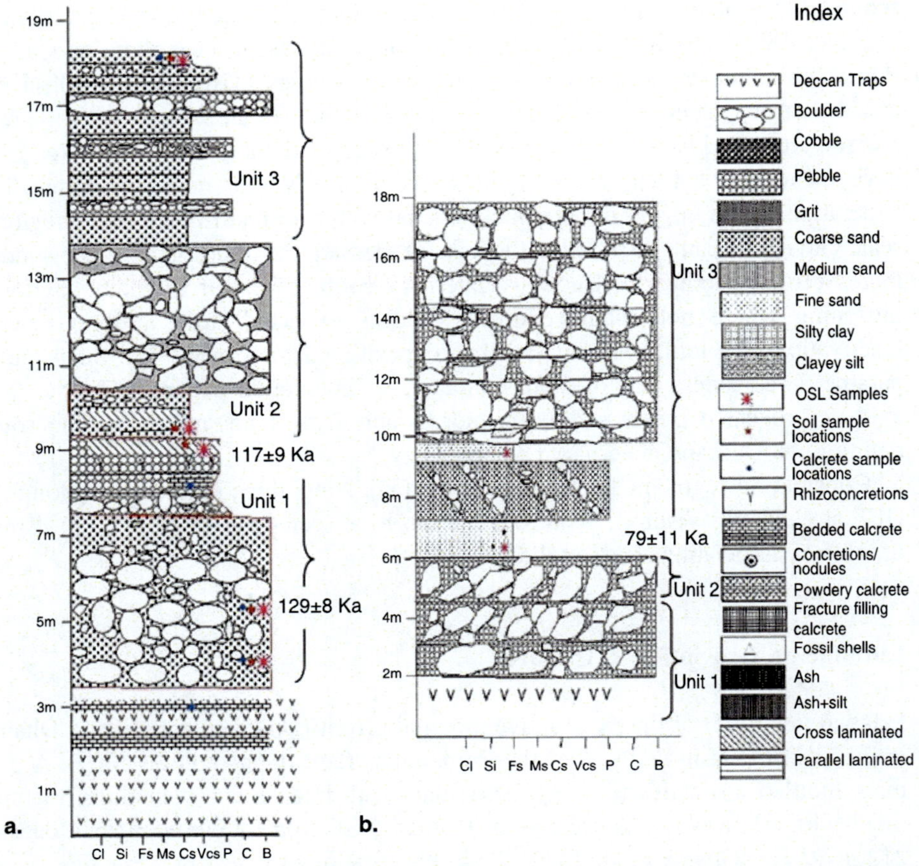

Fig. 12.5 (**a**) The lithology of Purna left bank section at Wishroli. (**b**) The lithology of Wan left bank section near mountain front

comprises alternation of coarse sandy layers and intervening pebbly layers. The contact between units 2 and 3 is very sharp and planar. Unit 3 is also devoid of visible calcrete association. However, as in the next underlying unit, this unit is also indurated due to the cementation of the matrix by carbonate material. Calcretes found in the section are chalky carbonate cemented sand and rhizoliths. Calcrete micromorphology indicates the development of typical alpha fabric and pedogenesis in soil vadose zone with limited plant activity (presence of rhizoliths). The entire section broadly shows an abundance of smectite clay (Ghatak and Ghatak 2010). This is indicative of episodes of high rainfall phases. It is difficult to resolve the formation of large amounts of smectite clay in vertisols with the current semi-arid climates (Pal et al. 1989). Therefore, smectite in the sediments at Wishroli were formed during an earlier humid climate and preserved in the non-leaching environment of arid and semiarid climate.

Lithofacies Association at Wan

Wan is a tributary of Purna, flowing in its upper reaches through a strath terrace carved on Deccan Traps near Wan village. Wan River exposes a 19 m thick fan sequence (Figs. 12.4d and 12.5b) divided into three sub units based on lithofacies assemblage and contact relationship. All the units have an abrupt and planar contact relationship with the overlying or underlying units. Unit 1 represents a gravel-boulder bed that directly overlies the weathered Deccan Traps and represents an unsorted boulder bed with a crude parallel orientation of boulders. The unit thickness is 4 m. Unit 2 is a fine-grained buff coloured palaeosol/sand of 1.2 m thickness overlying the Gravel-boulder bed with sharp contact. It is thoroughly pedogenised for which the sedimentary structures are not well preserved. Unit 3 is a cross to parallel laminated sand, bedded pebble-boulder association. This unit is highly calcretised and calcretes occur mostly along the bed boundaries and across bed boundaries. This unit is overlain by fine-grained parallel laminated sand. This sub unit yielded a date of 79 ± 11 Ka. This in turn is again overlain by a parallel bedded/ imbricated boulder bed. It is friable and white. The initiation of this Wan fan with the absence of the typical red soil or pedosedimentary sequence at its base defines a younger sedimentary package than at Wishroli and sediment influx is triggered by the central fault segment of GFZ which got reactivated later in comparison to the eastern fault segment of Gavilgarh. A shift in climatic condition from unit 1(debris flow) to unit 2 (stream flow) indicates a shift in water availability or predominance of humid conditions in the later phase. The Holocene unit is not well developed near the fan apex. Evidence of fan reworking and proximal fan incision are ample. All of these point towards a destructive phase of landform development.

Tapi Fault Zone Vis-à-Vis Sedimentation in the Tapi Basin

Geology-Geomorphology of Tapi Basin

The E-W stretching Tapi watershed is one of the most spectacular faults bound basins in Central India located at the southern foothills of the Satpura Mountains. Tapi Fault Zone (TFZ) is famous for its geothermal springs. Previous literature regarding TFZ (Ravi Shanker and Dubey 1984; Ravi Shanker 1987; Tiwari 1996; Tiwari 1999; Sinha and Tiwari 1984) had emphasized the neotectonic activity with special reference to geothermal and broad Quaternary morphostratigraphic classification around Unabdeo, Ramtalab, Nazardeo and Sunabdeo hot springs in Jalgaon and Dhule districts, Maharashtra. The ENE-WSW trending Tapi lineament of CITZ cuts across the entire width of the Indian Peninsula affecting the Traps, Gondwanas, Vindhyans and Archaean groups of rocks (Ghatak et al. 2009b). In the area lying between Palasur in the east and Nhavi in the west in the Tapi basin, the oldest exposed lithounit are the Lameta Group of rocks of Upper Cretaceous age (variegated arenaceous facies and impure carbonates). These occur as tectonic slivers in limited stretches along the foothills of the Satpuras and are one of the markers for the Tapi North Fault (TNF; Ghatak et al. 2009). The dominant lithology of the area is the Deccan Traps of Upper Cretaceous to Palaeocene age overlying the Lametas. The Tapi River and its major tributaries emerging from the mountain front have contributed to the deposition of a thick pile of sediments of Quaternary age in the fault bound Tapi basin at the foothills of the Satpura ranges (Fig. 12.6a–f).

Regionally, three geomorphologic units – the middle/lower level plateau ending with a prominent scarp defining TNF, a pediment zone at its base and the vast expanse of the Quaternary alluvial valley filling south of it and Ajanta Plateau in the south characterize the Tapi Basin (Fig. 12.7).

Tapi is one of the major rivers in central India with a length of 724 km covering an area of 65,145 sq km. The river originates in Multai in the eastern Satpura Range of Madhya Pradesh. It then flows westwards running in Madhya Pradesh's Nimar region, Maharashtra's Khandesh and east Vidarbha regions in the northwest corner of the Deccan Plateau and empties into the Gulf of Cambay of the Arabian Sea. The Tapi is predominantly a bed rock channel (Kale 2005). The climate of the basin is humid tropical ranging from sub-humid to semi-arid. Precipitation is more in the hinterland compared to the basinal part.

Geomorphologically, the westerly flowing Tapi River originates in the middle-level plateau, and passes through an intermontane valley near Palasur and west of its confluence and its major tributary Purna flows axial to the basin. The most striking geomorphic feature here is the spectacular fault line scarp marking the southern limit of the Satpura ranges. Asirgarh ranges (~867 m) have the highest elevation in the area, while the lowest point lies on the river course/Tapi plain. The southern part of the area is dissected Ajanta/Buldana plateau, reaching up to an elevation of nearly 600 m. Terrain morphometry using a hill shade map shows the tectonic half-graben structure of the Tapi basin.

Tapi Fault Zone

Tapi basin is traversed by various discontinuous segments of ENE-WSW trending faults and cross faults in the N-S direction (Ghatak et al. 2009b). Tapi North Fault (TNF) is a master normal fault with a downthrow to the south, separating the mountain range from the Tapi basin fill. In exposed sections, the fault plane is mostly sub-vertical to vertical. The regional scale ENE-WSW trending TNF has caused secondary back tilting of a Lameta Group of rocks (Fig. 12.6) and Deccan Traps, sills and dykes. The undisturbed geological formations in the study area are horizontal to very gently dipping (1°–6°) while the secondary dip exhibited by them varies from 19° to 48° towards the north. The accommodation space provided by the southern downthrown block of TNF has acted as a depocentre of a thick pile of Quaternary sediments. The maximum width of this fault zone is ~100 m. Several hot and cold springs, brecciated horizons, presence of a high pressure-temperature zone

Fig. 12.6 Fault zone signatures in the Tapi basin. (**a**) Tapi North fault separating Deccan Trap and Quaternary. (**b**) A sliver of coarse-grained Lametta sandstone against Deccan Traps vertical contact [Location: Bholana]; (**c**) Doleritic dyke intruded into Deccan Traps along the course of the Tapi River following TRF [Location: Tapi sluice] (**d**) Northerly tilted Lameta carbonates [Location: Near Chinchala] (**e**) Gushing spring in TFZ [Location: Bholana] (**f**) Quaternary sediments over degraded fault plane in Lametas [Location: Kala Chabutra]

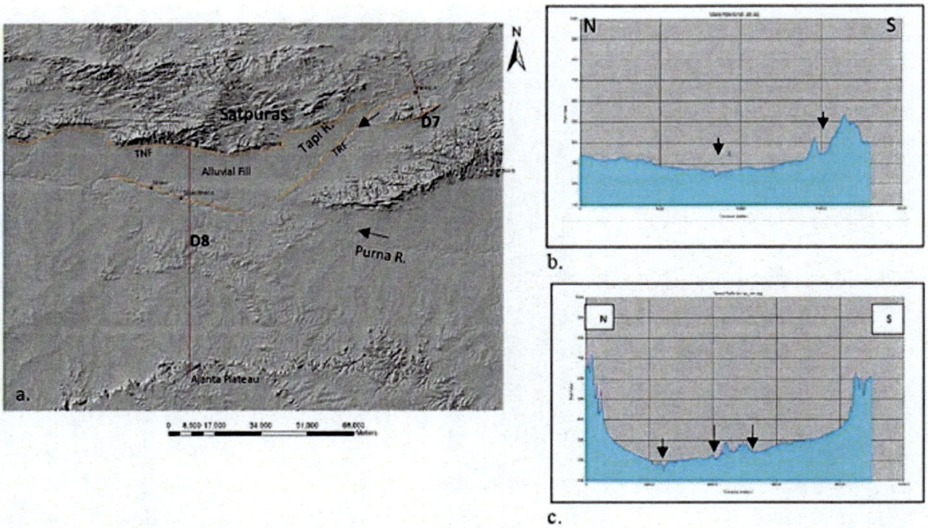

Fig. 12.7 (**a**) Hill shade map of the area showing major fault locations. (**b**) and (**c**) profile along with D7 and D8. Arrows show the position of major faults

(yielding hot water in bore wells with an average temperature of 40 °C) bears indirect evidence of the active nature of TNF. Tapi River Fault (TRF), synthetic to TNF also occurs in discrete segments and mostly follows the trend of the major trunk channel. Many synthetic faults to the major Tapi Fault zone and cross faults in the mountain front and basinal area striking NW-SE are evidenced in the field; they have displaced interflow contacts, red bole horizons and at places, Quaternary sediments (Ghatak et al. 2009).

Uplift along the range bounding TNF has altered the base level of the streams and the effort of maintaining a balance between uplift and erosion has caused intense vertical degradation even in the first order nalas along the mountain front within Quaternary. Numerous first-order streams with very close and regular spacing emerge from the mountain front. This is characteristic of tectonically active mountain fronts (Burbank and Anderson 2001). Owing to various sets of active faults present in the basin, river captures and drainage anomalies are frequent phenomena. Reverse faults with northerly dipping fault planes developed within Quaternary sediments are also present (Fig. 12.8a–f).

Quaternary Sedimentation Vis-S-Vis Climatic Imprints in TFZ

The sediments occur mainly in the form of overbank/channel deposits between Palasur in the east to Nhavi in the west. Fanglomerate facies are restricted to a few of the transverse tributaries of Tapi. Two key sections at Palasur in the eastern middle-level plateau along with TRF and at Takadkheda in the central part of the

Fig. 12.8 Representative lithopackages at Tapi Basin. (**a**) Disposition of Red and buff coloured palaeosol and associated ash on either side of Tapi River, terraces are unpaired [Location: Palasur]. (**b**) Ash layer and branching rhizocretes in Unit 1 [Location: Palasur]. (**c**) Calcified gritty sand on top of buff-coloured palaeosol [Location: west of Palasur] (**d**) Multiple bedded calcrete profile [Location: Takadkheda] (**e**) Deccan Trap riding over Quaternary (reverse faulting), [Location: west of Takarkheda] (**f**) Holocene flood facies rich in plant debris [Location: south of Takarkheda]

basin along TRF has different geomorphic setting and Quaternary horizons of different paleoclimatic influences exposed along the main course of the Tapi River and its major transverse tributaries.

The sediments in the Tapi basin are dominantly fluviatile and characterized by the presence of intercalated palaeosol and calcretes of varying morphology formed during periods of non-deposition or sediment starvation. They include pedogenic, groundwater, mixed groundwater pedogenic, valley calcretes, etc. (Ghatak et al. 2010; Sinha et al. 2006) and are found in states ranging from powdery to nodular to

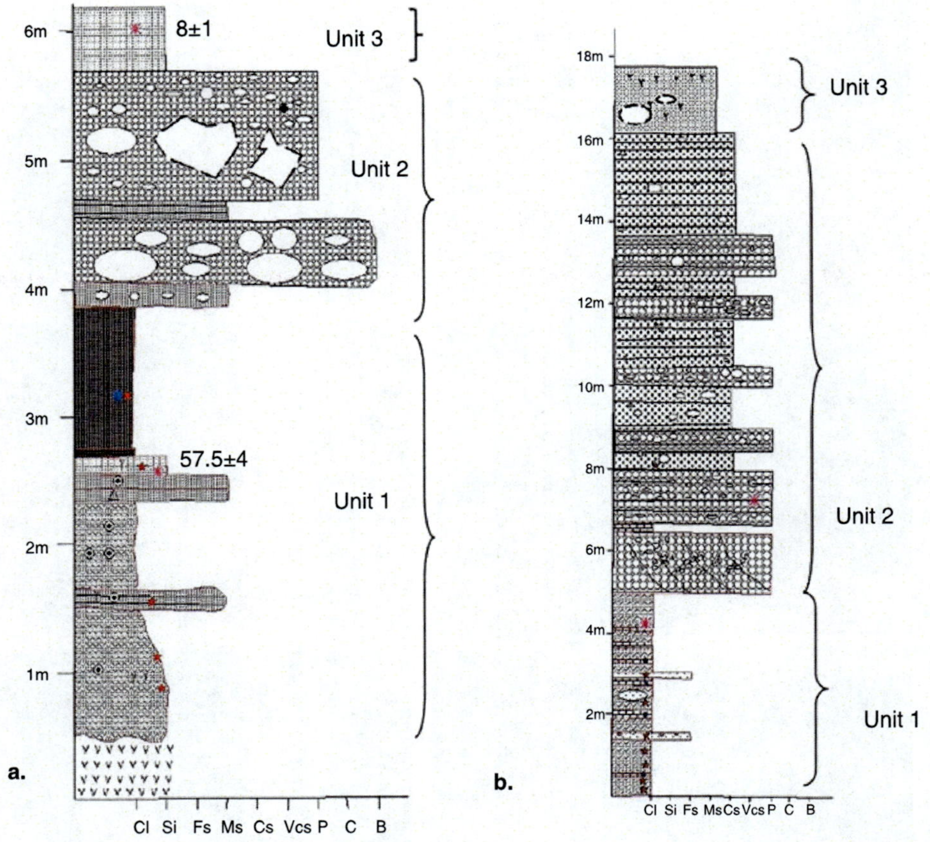

Fig. 12.9 Lithologs from (**a**) Palasur section (**b**) Takadkheda section

highly indurated (Goudie 1983) form. The term calcrete has been applied here to all of these types, and commonly denotes secondary carbonate introduced into the surface and near-surface soil, sediment and rock (in the phreatic and vadose zones) through replacive and displacive processes and/or passive cementation. In pedogenic calcretes, alpha and beta fabrics are considered petrographic (micromorphological) end members, with many calcretes exhibiting mixtures of the two (Wright and Tucker 1991; Bedelean 2004). Palaeoseismites are also found in abundance in the Quaternary sediments that owe their origin to fault-controlled active structural setup in the Tapi basin.

Considering segmented behaviour of the fault zone that provides the varying set-up for sediment deposition, broad litho facies association at two key sections, i.e., at Palasur and Takadkheda are described below (Fig. 12.9).

Lithofacies Association at Palasur

The Tapi River section at Palasur has some of the best-preserved evidence of Quaternary tectonic history and the palaeoclimate. Here, Tapi traverses through an intermontane valley in the lower/middle level plateau of Satpura. TRF displaces the Deccan/Quaternary contact on either bank by an elevation difference of nearly 10 m (Ghatak et al. 2009, Ghatak and Ghatak 2010, Fig. 12.8a) forming an unpaired terrace. The entire section has been divided into three units based on lithofacies assemblages and contact relationships (Fig. 12.9a).

Unit 1 represents the basal gravel-palaeosol (vertisol)- ash bed association and is the oldest depositional unit of the Tapi basin developed over the southern bank strath terrace. The entire unit has attained a very high degree of pedogenesis and is brick red. The primary structure of this part has been lost due to pedogenesis. The entire unit has preserved calcrete nodules, rhizoconcretions and fossil bivalve. This unit also contains an ash layer. This is the first report of ash from this part of the Tapi basin (Ghatak et al. 2009). Its preservation points towards a period of quiescence in this part of the basin (Fig. 12.8b). The sediment below the ash layer has been dated as 57.5± 4 Ka. The upper part of this silt/ash has well preserved B horizon. Calcretes from unit 1 are powdery, nodular (average diameter 5 cm) and rhizolith bearing showing variable sizes and shapes. The calcrete nodules in this horizon coalesce as they may have received enough time for growth. They occur isolated or in clusters (coalesced) indicating stage II maturity of the soil profile (Gile et al. 1965; Gile et al. 1966; Candy et al. 2003; Candy and Black 2009).

Unit 2 is the pebble sand buff palaeosol (vertisol) association which is present on the northern bank of the river on the upthrown unpaired terrace, north of TNF. This unit also bears calcrete with rhizoliths at the base, powdery calcretes throughout and an indurated, carbonate cemented pebbly grit bed capping this unit. The upper part of unit 2 is capped by cross-bedded carbonate cemented gritty sand bed. At the base and top of the buff coloured palaeosol, the paleo-flood deposit has yielded dates of 27 ± 2Ka and 17.6 ± 2 Ka, respectively. Morphological characteristics of the calcretes indicate biogenic pedocalcic pedogenesis (Ghatak et al. 2009).

Unit 3 is dark grey thick pebbly sand with occasional boulders (crudely imbricated) fine to medium-grained sand with well-preserved parallel laminations, thick coarse-grained pebbly sand which contains meter scale co-seismic bank collapse structures, with angular blocks of the ash belonging to unit 1. Unit 3 ends at Palasur with thick dark humus rich silty clay that has well defined A, B and C horizons. Unit 3 dates back to 8 ± 1 Ka.

Ghatak and Ghatak 2010 have reported the predominance of smectite over other mineral phases in the total sediment package along with shrink-swell properties of these soils atypical of vertisols of semiarid regions. Similar to the Purna basin, the predominance of smectite indicates an earlier phase of humid climate. Second, the mineral assemblage of red palaeosol and top of the buff palaeosol indicates flood plain aggradation on which pedogenesis has taken place (smectite to kaolinite alteration or feldspar to kaolinite alteration).

Lithofacies Association at Takarkheda

Before reaching Takarkheda, the Tapi River descends onto the plains. Deccan/Quaternary contact is not observed at exposed sections and the observed thickness at Takarkheda is ~18 m. Based on lithofacies assemblages and contact relationships broadly three units can be separated (Fig. 12.9b).

Unit 1 is basal red palaeosol and multiple bedded calcrete association. Six bedded calcrete units are interspersed within the red palaeosol or silty/clay unit. The physical appearance of the calcrete is brownish grey, and flaky comprising coalesced pisoliths of compact nodules of irregular shape and size. Chalky veins penetrate the nodules and it also contains clasts of red clay pellets derived from host sediments.

The hiatus in the depository cycle has been followed by rejuvenation in the stream power paving way for a major cycle of sedimentation, identified as unit 2 which is pebble sand and calcrete association. This unit begins with the deposition of cross-bedded channel sand over a scoured base of the red pedosedimentary sequence. It is dominantly clast supported indicating winnowing of the sediments. An arid phase represented by a bedded calcrete overlies this channel sand. Repeated cycles of pebble beds and parallel laminated coarse-grained sand have followed this transient dry phase. The uniform parallel laminated sand with pebble beds in the upper part of unit 2 indicates riverbank flooding.

Coalesced rhizoliths and nodules from the bedded calcretes indicate biogenic control on calacrete development at least at the lower part of unit 1. As calcretes develop in easily identifiable stages, time significance has been assigned to them (Machette 1985). Time estimates for the development of mature (stage 4) calcrete profiles range from thousands to one million years (Wright and Tucker 1991), although these estimates are poorly constrained. This is corroborated by the fact that calcretes at Takarkheda are older compared to the Palsur section. OSL estimates from the upper part of this red palaeosol unit have yielded an age of 139 Ka.

Unit 2 is followed upward by a 1.5 m thick unit dominantly comprising fine-grained parallel laminated sand deposited in the inset terraces carved in units 1 and 2. This contains lamina rich in plant debris indicating a phase of high rainfall in the basin during the upper Holocene.

The predominance of smectite in this sequence indicates a high rainfall scenario. Preservation of feldspar indicates flood plain aggradation events (Ghatak and Ghatak 2010). It is because under sub-humid to arid climates weathering of primary minerals contributes very little towards the formation of smectite (Srivastava et al. 2002). It is thus difficult to reconcile the retention of smectite in vertisols of a tropical humid climate and also its formation in sub-humid to arid climatic conditions. These smectites were formed in a humid climatic condition during ~139 Ka, a humid phase that was witnessed by the entire Indian subcontinent and got preserved with minor alteration to Kaolinite in the later semi-arid to sub-humid climate of the Tapi basin (present-day annual average range fall of 800 mm).

Son Narmada South Fault Zone Vis-à-Vis Sedimentation in the Narmada Basin

Geology-Geomorphology of the Narmada Basin

Narmada basin is bound by Son Narmada North Fault (SNNF) in the north and Son Narmada South Fault (SNSF) in the south. Further south of SNSF, the geomorphology of the area is defined by dissected Deccan plateau, structural hills and ridges developed on the Palaeoproterozoic Mahakoshal supracrustals, tors and inselbergs of intrusive granites and residual hills developed over the Gondwana Supergroup capped by Deccan Traps at places (Roy and Devarajan 2001). The oldest litho-succession in this area belongs to the Mahakoshal Group and is largely confined between SNNF and SNSF. There has been the intrusion of granites along the margin of Mahakoshal rocks. Later reactivation of the SNSF zone under brittle to brittle-ductile conditions during Phanerozoic controlled the deposition of Phanerozoic sequences such as Gondwana Supergroup and Lameta Formation, outpouring of Deccan Traps and the deposition of the Quaternary sediments. Gondwana, Lameta and Deccan volcanic cover large parts of the CITZ, especially in the western parts. Jabalpur Formation, an alternating sequence of sandstone, shale and clay, is the dominant Gondwana sequence in this area. Jabalpur Formation is overlain by the Lameta Formation, a sequence of impure carbonates and sandstone, also known as infra-trappean. Deccan Trap flows of Upper Cretaceous to Lower Eocene overlie the Jabalpur Formation and the Lameta Formation and occur extensively in the southern and south-eastern parts of the basin. Quaternary sediments of Upper Pleistocene to Holocene extensively occur between SNNF and SNSF (Devarajan et al. 2004; Acharyya and Roy 2000).

A north-south view across the SNNF-Quaternary valley fill-SNSF shows a major steep scarp in the north which defines the SNNF separating the Vindhyans against the Mahakoshals (Fig. 12.10a, b). The Hiran River, a tributary of Narmada, passes at its base. The quaternary basin in this part is nearly 35 km wide. The southern margin of the basin is marked by SNSF beyond which the dissected Satpura ranges ascend (Fig. 12.10c). Anomaly in river courses as a response to tectonism is a constant feature in both margins of Satpura and is seen in Narmada and its tributary rivers. The streams have experienced a rejuvenation in their energy as evidenced by signatures of large-scale bank collapse structures within Quaternary sediments. The rivers are engaged in the deep incisions of their beds and at places, nearly 45–50 m thick sections are seen (Devarajan et al. 2004).

Active Fault Mapping along with SNSF (Devrajan et al. 2004) has revealed various phases of reactivation along with SNSF during different geological periods. Overall, SNSF exhibits a series of ENE-WSW trending faults from north to south with successive downthrows towards the south (Figs. 12.11a, b). The presence of seismites within different Quaternary horizons in the basin points towards repeated seismic events in the basin.

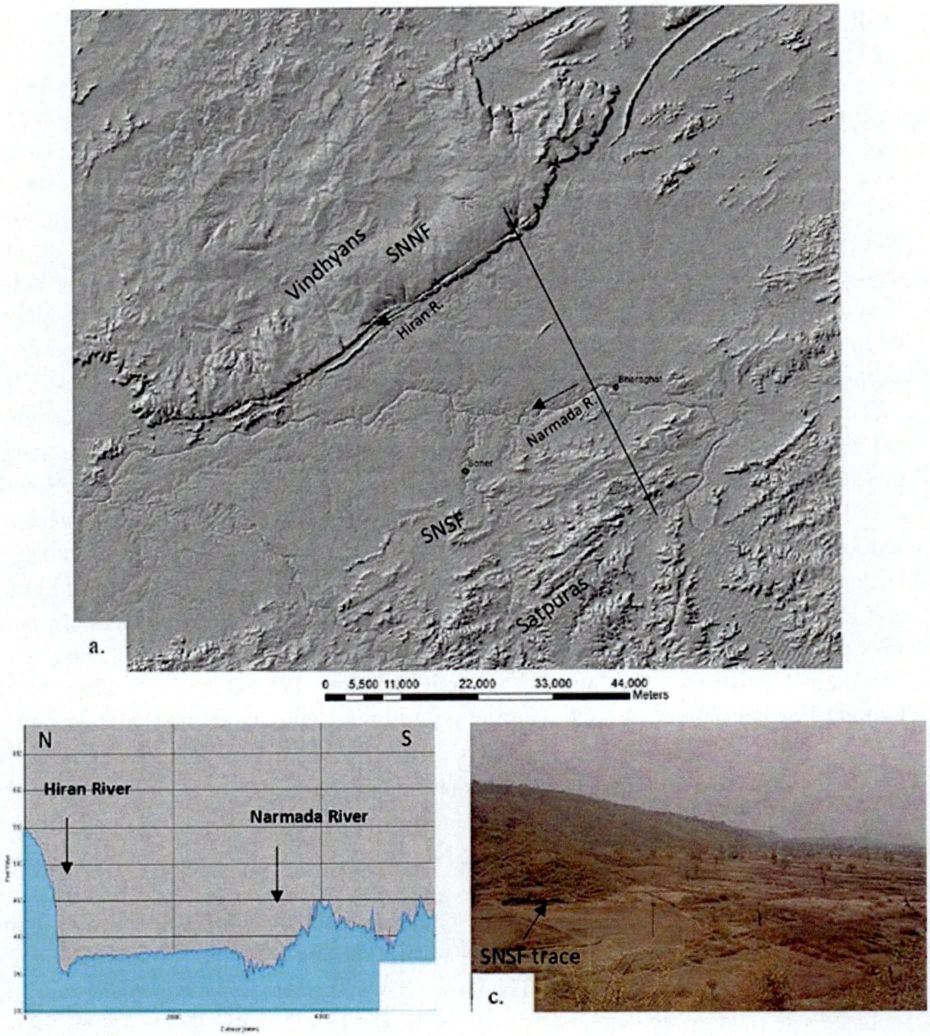

Fig. 12.10 (**a**) SRTM (90 m) derived hill shade map of part of the Narmada basin showing the disposition of major faults. (**b**) Topographic cross-section showing surface expression of a graben. (**c**) Dissected Deccan Traps at the northern margin of Satpura ranges. A viewer looking towards NW. SNSF is located at the foothills, marked with the red line

Quaternary Sedimentation Vis-S-Vis Climatic Imprints from the Narmada Basin

The Quaternary succession in the Narmada basin has been divided into seven formations (Table 12.1). Pilikarar Formation, the oldest in the sequence, is a reddish ferricrete gravel bed and rests over the eroded surface (Tiwari and Bhai 1997) of the basement. The youngest Ramnagar Formation is mostly deposited in inset terraces and has preserved vegetal remains indicative of a climate favourable for plant growth

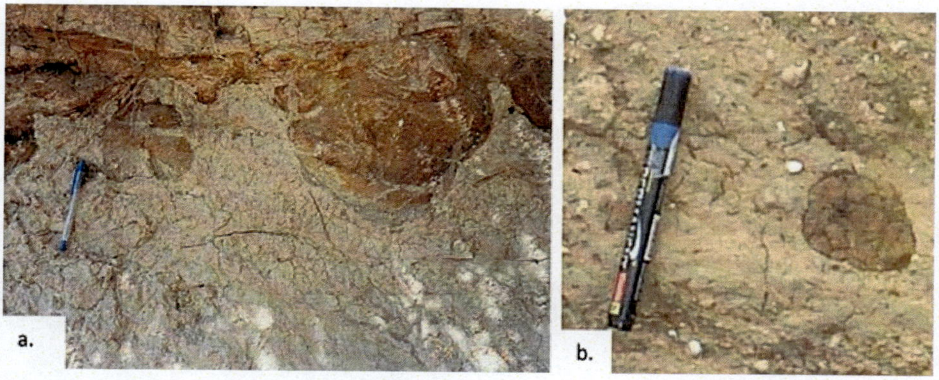

Fig. 12.11 Palaeo-seismites (**a**) Ball and pillow structure at the basal part of Hirdepur Formation. (**b**) Rip-up clasts in the upper part of Baneta Formation

(Devarajan et al. 2004; Ghatak and Ghatak 2010). Paleosols and calcretes are present throughout the sediment archive. One of the litho-sections with maximum nos. of established formation and chronometric data control (C-14 dates, GSI, Kolkata lab) is described below.

Lithofacies Association and Calcrete Development at Soner

Soner Nadi, a tributary of Narmada, flows in a south to north course and meets Narmada in its middle reaches west of Jabalpur. The deposits contain evidence of tectonic activity in terms of preservation of paleoseismic features and Quaternary faults. The total exposed section is nearly 20 m thick and is divided into four units based on lithofacies assemblage and their contacts (Fig. 12.12a).

The basal red palaeosol association is developed on top of the Baneta Formation of the middle to upper Pleistocene age (Tiwari and Bhai 1997). It comprises red silty clay; internal structures are not observed within this paleosol horizon. Fine sediment size is attributed to low energy conditions in the basin which had later experienced stability and high rainfall condition leading to pedogenesis and oxidation of the sediments. The paleosol horizon has bedded calcrete which has been displaced subsequently (Fig. 12.12b).

Unit 2 is 1 m thick and has a pebble-boulder association, separated by an erosional unconformity. It heralds the onset of Hirdepur Formation sedimentation in the basin. The unit contains rip-up clasts of the red palaeosol belonging to unit 1 (Fig. 12.12b). These clasts are rounded and indicate a fair degree of transport. The upper conglomerate bed of this unit is highly calcified. Unit 3 is the sand-silty clay association. Thinner, discontinuous sand layers are present within the dominant clay. This unit has well-preserved seismites formed due to a major seismic episode during that period. The seismites are in the form of disrupted bedding, convolute laminations, folds and cycloid-column association (Fig. 12.12a). The sand unit has yielded

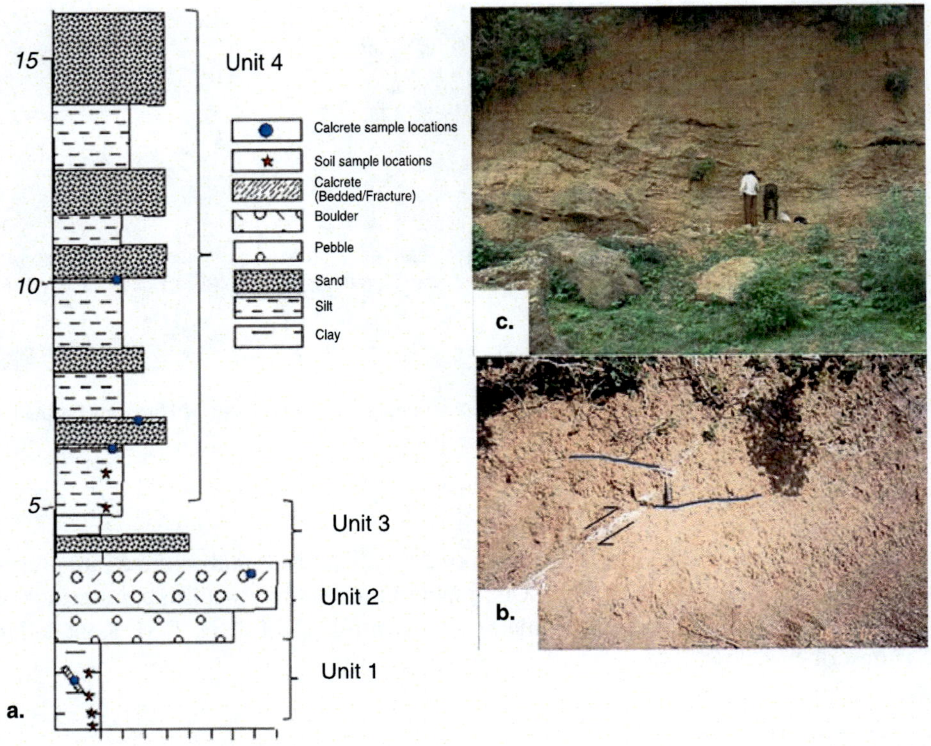

Fig. 12.12 (**a**) Litho-succession at Soner River, (**b**) Bedded calcrete from unit 4 (**c**) displaced bedded calcrete in paleosol horizon (unit 1). There has been subsequent development of calcrete along the fault plane

a radiocarbon age of 28,891 ± 294 Ka (Devarajan et al. 2004). Unit 4 is a sand-silt association, a thick sequence of alternate medium to fine-grained sand and silt. Few bedded calcrete horizons are present in this unit. Unit 1 at the Soner Nadi section comprises fracture filling calcrete and disseminated rhizocretes. Units 2 and 3 do not contain noticeable calcrete development. However, the upper part of unit 4 contains equally spaced bedded calcrete horizons. Macroscopic observation in general points towards groundwater origin.

Discussion and Conclusions

Looking through the lens of intermediate time scale tectonic deformation and (Burbank and Anderson 2001) climatic interaction, landscape responses during Quaternary times for Purna, Tapi and Narmada basins bear certain basic similarities. Repeated reactivation of older crustal-scale faults under different stress regimes has been a guiding factor in landform evolution in CITZ. The cumulative interactions between the deformational events and ongoing surface processes are best witnessed by the marker horizons, anomalous river courses, basin geometry, hill slopes and elevation characteristics (Chen et al. 2003) and the way these have influenced its depositional environment (Ghatak and Ghatak 2005).

Segmented behaviour of GFZ with salient transfer zones (especially near Belkund Dhana and Chandrabhaga) have been the probable sites of stress release. Other cross faults viz. Purna River fault have also contributed to the sediment flux to the basin. The Foothill Gavilgarh Fault (Fig. 12.2a) has behaved as a basin margin fault during the deposition of the Gondwana and Lameta sediments to the north of the fault under an extensional regime. Post Deccan tectonic inversion or during an episode of shortening, this northerly dipping fault had provided the necessary buttressing for the folding of cover sediments viz. Gondwana (Fig. 12.3c) and Lameta. The fault plane is well exposed in the field and testimony towards a sub-vertical to northerly dipping fault plane (Fig. 12.3b) as against the earlier understanding of this part of the fault zone (Ravi Shanker 1987). Post Deccan and possibly in the pre-Quaternary/ early Quaternary period (prior to deposition of the Purna alluvium, as the oldest dated sediment is 212 ± 17 Ka) an intermittent phase of extensional stress regime had set in/ or extension took place locally under overall compressional regime resulting in normal faulting marked by Deccan-Quaternary boundary (Purna North Fault) with the down throw to the south forming the depositional loci for Purna alluvium. However, the compressional stress regime continued into Quaternary. Reverse/thrust faults within Quaternary sediments with northerly dipping fault plane (Fig. 12.3f) bear testimony to this. Sediments dated on the strath terraces of Purna (129 ± 8 Ka) and Wan River (79 ± 11 Ka) refer to different fault segment reactivation. Various cross faults also affect the sediment distribution pattern in the basin. The alluvial fans and other geomorphic landforms are presently in the destructive phase, engaged in the deep incision of paleofan sequences. The thick gravelly facies of sub angular to sub rounded pebbles and boulders occurring as cohesive debris-flow deposits in proximal parts of the fan represent the syntectonic growth of the gravelly facies in response to the reactivation and upheaval along the GFZ. Thick pedosedimetary sequences ranging from lower Pliestocene to Holocene, traced from the fan head to the distal part of the Purna fan characterize the dominant formative processes over the fans. While debris flow to sheet flood deposit characterizes the proximal part of the fan, stream flow and reworked fan sediments derived through fan head trenching and distal aggradation dominate in the terminal part. Three different exposed palaeosol/pedosedimetary units (red, buff and grey colour, respectively) with varying degrees of pedocalcic pedogenesis (nodular and rhizogenic calcretes) from the fan head to the distal part suggest a period of tectonic stability and soil formation interspersed with high energy sediment flux due to tectonic events and/or climatic control. The basal red pedosedimentary sequence indicates a high rainfall phase with tectonics providing the orographic control for precipitation during the lower Pleistocene. A suspected semi-arid phase during the middle Pleistocene witnessed aridity induced flash flood events resulting in fan head trenching and distal aggradation. A comparatively stable landscape was prevalent during the upper Quaternary with discrete periods of soil formation and pedogenic calcrete development (Goudie, 1996). The palaeosol facies belonging to Holocene is devoid of many calcretes. The present study reveals, that rapid aggradation of the gravelly facies and less frequent overbank deposition marked by tectonic activity or flash flood events possibly allowed only weak to moderate pedogenesis of the overbank and fanglomerate facies. Eventually, distal fan aggradation and continuous calcrete development throughout the entire fan surface, led to the ultimate fan surface induration, controlling subsequent landscape evolution. The same situation holds for the Wan River fan except for the fact that the Wan fan is younger probably of middle to upper Pleistocene compared to the Purna fan with preservation of one palaeosol unit. The rest of the pedosedimentary fan sequence is composed of debris flow to stream flow character. Another important aspect of the Purna basin lies in the fact that the drainage divide between Purna and Wadha near the mountain front preserves the oldest datable sediments at Udkhed. This zone is seismically active and records low magnitude earthquakes. Purna River Cross fault with upthrown eastern block possibly aided the formation of the drainage divide of Purna and Wardha in this sector. The stratigraphic drilling at Akot also indicates the presence of clay beyond 60 m from the surface. The sediment flux, slope condition and surface processes also differ spatially and temporally. From east to west, the scarp height of the Gavilgarh fault decreases and this has caused micro climatic variation along with the fault trend. The present observations suggest that the faulting induced south Satpura mountain front creates first-order control on sedimentation, microclimatic variation and orographic precipitation in the Purna valley during Quaternary.

The Tapi basin also demonstrates similar reactivation history under different stress regimes. The regional scale normal fault at the foothills of Satpura opened up the basin during or prior to the Quaternary period. Sediment influx was from fault segments separated spatially and temporally. Sedimentation in the Tapi basin had started subsequent to the Purna basin (Ghatak and Ghatak 2010). The Quaternary sediments in the Tapi basin also demonstrate features that are akin to the compressional stress regime continuing during Quaternary times (Deccan overriding Quaternary near-fault zone, close-spaced first-order stream on the pediment slope). Initiation of sedimentation was during different periods; sedimentation along western segments of TFZ had started before in the eastern part as evident from the age of Quaternary sediments in the Tapi basin. The level of maturity of calcretes supports this observation. Tectonic uplift and influx of sediments in the basin were followed by stability in the basin during which paleosols with multiple calcrete horizons with different levels of maturity had developed. Sediments of different age on either bank at Palasur establish Tapi occupying a structural course. The humid and arid conditions interspersed between as observed in the paleosol and calcrete horizons match fairly well with the south-Asian monsoon trends (Ghatak and Ghatak 2010). Subsequent to deposition, the river is engaged in the deep incision of earlier sediments, intense scouring of strath terraces, and base cutting leading to the collapse of river banks (*op.cit.*). Kilometre scale formation of carbonate-cemented sand over buff coloured palaeosol (upper horizon dates back to ~17 Ka) indicates a period of aridity which is coeval with the Last Glacial Maxima (LGM). Holocene deposits are limited in occurrence in the Tapi basin, at least in the upper to the middle part of the basin. In both Purna and Tapi basins, the youngest Holocene deposits show abundant vegetal remains to indicate a high rainfall scenario supporting plant growth.

The alluvial succession of the Narmada valley had been well-constrained morpho/litho stratigraphically by earlier workers (Tiwari 1999). A sediment thickness of 250–300 m has been postulated for the central part of the Narmada basin. However, chornostatigraphic control in sedimentation is limited. The sediment package is dominated by fines and occasionally contains pebble bearing horizon. The overall energy condition is controlled by slope parameters, nature of faulting and the wide watershed area of the basin. The pulsating reactivation of the faults along the basin margin has resulted in high energy sediment flux in the basin. The predominance of fines in the studied sections along with the presence of palaeosols and calcretes indicate wide distributed periods of climatic amelioration and pseudogenization. The underlying red palaeosol from Baneta formation indicates a period of high rainfall, i.e. humid climate, tectonic stability and proliferation of biota. Calcretes from this unit indicate pedogenic origin. The climatic shift from humid to sub-humid is witnessed in the next younger Hirdepur Formation. The basal part of this unit has witnessed tectonic instability with the preservation of palaeoseismites. Both of these phenomena aided shallowing of the local groundwater table, evaporation and development of groundwater channel calcrete, bedded calcrete with evidence of weak pedogenesis. Although SNSF has been the most active fault system in CITZ, a minor microclimatic variation can be perceived due to the fault movements or orographic variation. The key factor for sedimentation in this basin is technogenic while following all Indian monsoon scenarios.

While the fault bound basins in the south of Satpura have preserved fans, there is no formation of fans in SNSF bound Narmada basin. The reasons behind this may be attributed to the lesser number of transverse drainages across the regional fault and lesser height of mountain scarp. Field observation and calcrete micromorphology from part of the Narmada basin adjacent to SNSF suggest that alluvial succession in Bheraghat and Soner sections is restricted to the middle to upper Pleistocene. Figure 12.13 displays the intensity of the Indian monsoon plotted against the last 150 Ka period and how various pedo-sedimentary sequences from the studied Quaternary basins mimic the monsoonal highs.

Fault segment controlled sedimentation varies spatially and temporally aided by climatic or microclimatic variation and hence the overlapping period or the hiatus in sedimentation are to be considered while carrying out morpho-stratigraphic classification. This aided with a precise chronostratigraphic approach based on high-resolution dating has helped to identify the tectonics-climate interaction in these basins. Since the Cenozoic tectonic basins from Central India preserve excellent evidence of palaeoclimatic episodes of microclimatic and regional scale climatic fluctuation, further research in this line will open up the avenues to correlate the tectonic events and palaeoclimatic interaction at the subcontinental scale drawing inferences from the Himalayas and Central India.

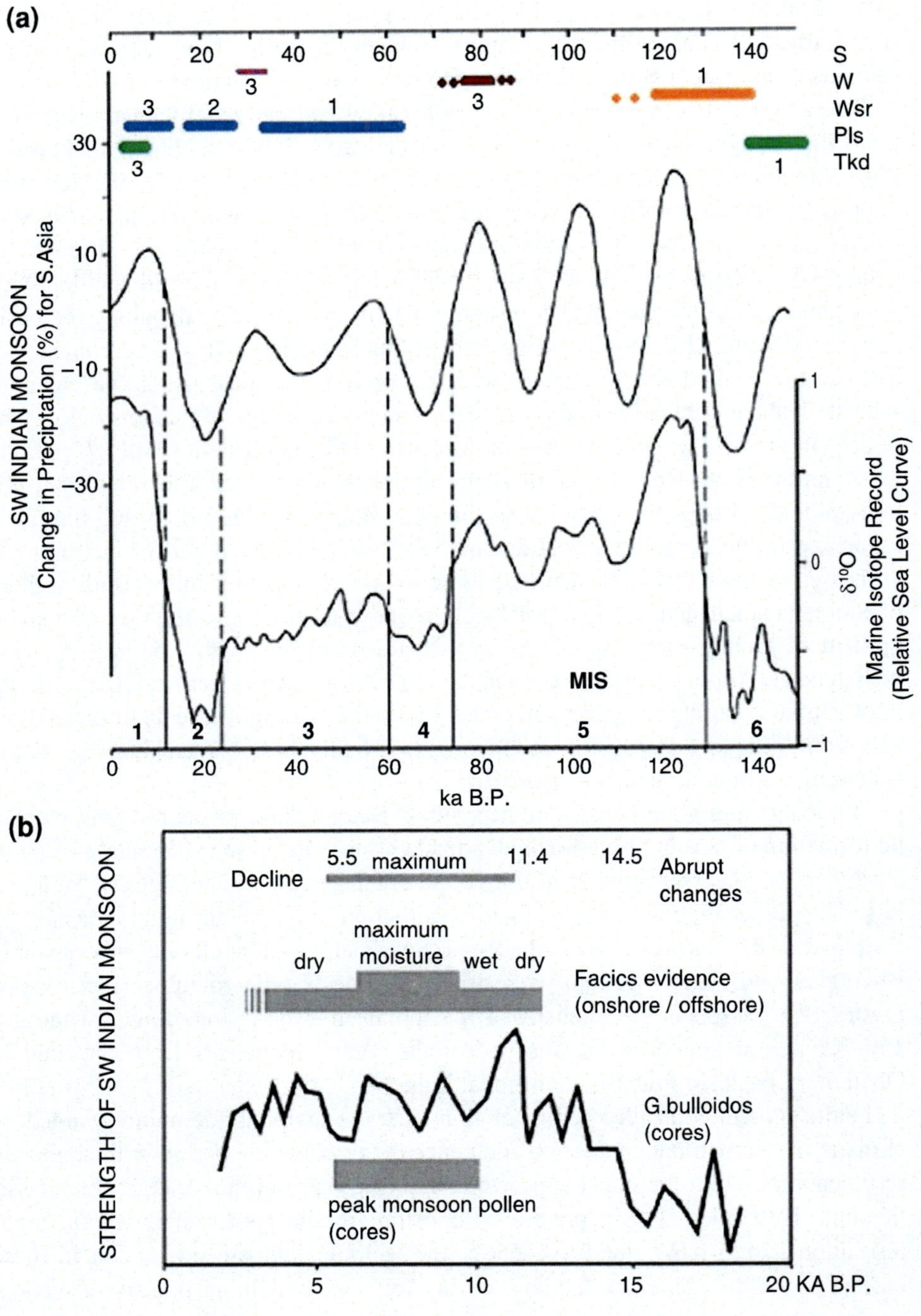

Fig. 12.13 Variation in precipitation associated with Southwest Indian Monsoon. (**a**) Modelling results for the past 150 ka, from Prell and Kutzbach (1987), shown in relation to proxy eustatic curve (Waelbroeck et al. 2002). MIS, Marine Isotope Stage. (**b**) Major trends in precipitation for the past 20 ka, from Overpeck et al. (1996). Geological information includes facies evidence from dated sections in the Arabian Sea and onshore in Asia, and the abundance of Globigerina bulloides and African pollen (linked to wind strength and upwelling) from Arabian Sea cores. The chronology of dated sediments from Purna (W: Wan, Wsr: Wishroli), Tapi (Pls: Palasur, Tkd: Takadkheda), Narmada (S: Soner river section) are superposed on the curve with its corresponding unit numbers

Acknowledgements The paper is the result of a decade of work by the authors on the characterization of the active faults and paleoclimate of inland basins flanking the Satpura ranges as a part of the annual programme of the Geological Survey of India. Dr. Abhinaba Roy, Sr. DDG (Retired), GSI has always been an overbearing influence as a mentor, and his contributions can never be repaid. Dr. Anjan Kumar Chatterjee, ADG (Retired), GSI had stood stoically with the authors for undertaking this project and is remembered with gratitude. The authors thankfully acknowledge Dr. M. K. Devarajan for his major contribution to mapping SNSF and GFZ. Dr. D.K. Pal, Head (Retired), Soil Sciences Division, NBSS & LUP is thanked for the stimulating discussions the authors had with him on the various aspects of soil micromorphological studies. The authors are also indebted to Dr. K.S. Misra, Sr. DDG (Retired), GSI for his encouragement. Dr. O.P. Mishra, Scientist 'G' and Director, NCS, Ministry of Earth Sciences has been a continued inspiration over the years; he has shared the earthquake record for this paper. Lastly, the authors place on record their gratitude for Dr. Neloy Khare, Scientist 'G', Ministry of Earth Sciences for his invitation to contribute to this volume and for his continuous encouragement. The Director-General, GSI is thanked for his kind encouragement to publish the chapter.

References

Acharyya SK, Roy A (2000) Tectonothermal history of the central Indian tectonics zone and reactivation of major faults/shear zone. J Geol Soc India 55:239–256

Bedelean H (2004) Study on the diagenetic calcareous accumulations in a soil profile from Floreşti (Cluj county, Romania). Studia universitatis babeş-bolyai, geologia, xlix 1:75–85

Bhandari S, Maurya DM, Chamyal LS (2005) Late Pleistocene alluvial plain sedimentation in lower Narmada valley, western India: Palaeoenvironmental implications. Jour Asian Earth Sci 24: 433–444

Blanford WT (1869) Geology of the Tapti and lower Narbada valley and some adjoining districts. Mem Geol Surv India VI:182–189

Burbank D, Anderson RS (2001) Tectonic geomorphology. Blackwell Science:1–274

Candy I, Black S (2009) The timing of quaternary calcrete development in semi-arid Southeast Spain: investigating the role of climate on calcrete genesis. Sedi Geol 218:6–15

Candy I, Black S, Sellwood BW, Rowan JS (2003) Calcrete profile development in quaternary alluvial sequences, Southeast Spain: implications for using calcretes as a basis for landform chronologies. Earth Surf Process Landf 28(2):169–185

Chamyal LS, Maurya DM, Raj R (2003) Fluvial systems of the drylands of western India: a synthesis of late quaternary environmental and tectonic changes. Quat Int 104:69–86

Chandradas M (1999) Final report on stratigraphic drilling to establish Gondwana stratigraphy below the Deccan trap to the north of Purna. GSI unpublished report

Chen YC, Seng Q, Cheng KY (2003) Along with strike variations of morphotectonic features in the western foothills of Taiwan: tectonic implications based on stream- gradient and hypsometric analysis. Geomorphology 56:109–137

Devarajan MK, Ghatak M, Chakrabarti U, Siddaiah S, Punekar DV, Shrivastava RK, Chaurasia AH, Naik KK, Guha D (2004) Report on the mapping of the son-Narmada south fault zone between Jabalpur and Chargawan, Jabalpur District. Madhya Pradesh. Unpublished GSI report

Dhir RP, Singhvi AK, Andrews JE, Kar A, Sareen BK, Tandon SK, Kailath A, Thomas JV (2010) Multiple episodes of aggradation and calcrete formation in late quaternary Aeolian sands, central Thar desert, Rajasthan, India. J Asian Earth Sci 37:10–16

Ghatak S, Ghatak M (2005) Morphotectonic evaluation of active tectonic domain from Central India: a case study from Gavilgarh Fault Zone, vol 85. Contributions to Kangra Earthquake Centenary Seminar 2005, Geological Survey of India Special Publication, pp 237–248

Ghatak S, Ghatak M (2008) Active faulting and it's bearing on the geology, drainage and landscape in Tapi Fault Zone. Gond Geol Magz 23(1):31–37
Ghatak S, Ghatak M (2010) Active tectonics-climate feedback in Purna, Tapi and Narmada basins of Central India. Unpublished GSI report
Ghatak S, Ghatak M, Devarajan MK, Naik KK, Livingstone D, Siddaiah S, Ramamurthy V, Gaonkar SG (2009) Report on active fault studies along the Gavilgarh Fault Zone in Purna valley Quaternary basin. Unpublished GSI report
Ghatak S, Ghatak M, Chandradas M (2009b) Report on active fault studies along Tapi North Fault. Unpublished GSI report
Ghatak S, Ghatak M, Chatterjee AK, Mujtaba SA, Bhattacharyya S (2010) Palaeoclimatic and sedimentation history inferred from multiple bedded Calcrete profile of Takarkheda section. Gondwana Geological Magazine, Special 12(2010):329–332
Gibling MR, Tandon SK, Sinha R, Jain M (2005) Discontinuity-bounded alluvial sequences of the southern Gangetic plains, India: aggradation and degradation in response to monsoonal strength. J Sedi Res 75:369–385
Gile LH, Petersen FF, Grossman RB (1965) The K horizon: a master soil horizon of carbonate accumulation. Soil Sci 99:74–82
Gile LH, Petersen FF, Grossman RB (1966) Morphological and genetic sequences of carbonate accumulation in desert soils. Soil Sci 101:347–360
Golani PR, Bandopadhyay BK, Gupta A (2001) Gavilgarh – Tan shear: a prominent ductile shear zone in Central India with multiple reactivation history. Geo Surv Ind Spl Pub 64:265–272
Goudie AS (1983) Calcrete. In: Goudie AS, Pye K (eds) Chemical sediments and geomorphology. London academic Press 439p, pp 93–131
Goudie AS (1996) Organic agency in calcrete development. J Arid Environ 32:103–110
GSI (2000) Seismotectonic atlas of India.
Guzzetti F, Marchetti M, Reichenbach P (1997) Large alluvial fans in the north-central Po plain. Geomorphology 18:119–136
Harvey AM (1988) Controls of alluvial fan development: the alluvial fans of the Sierra de Carrascoy, Murcia, Spain. Catena 13:123–137
Harvey AM (1990) Factors affecting alluvial fan development in Southeast Spain. In: Rachocki AH, Church M (eds) Alluvial fans: a field approach. Wiley, pp 247–269
Jain M, Tandon SK (2003) Quaternary alluvial stratigraphy and palaeoclimatic reconstruction at the Thar margin. Curr Sci 84:1048–1055
Juyal N, Chamyal LS, Bhandari S, Maurya DM, Singhvi AK (2004) Environmental changes during late Pleistocene in the Orsang River basin, western India. Jour Geol Soc Ind 64:471–479
Juyal N, Chamyal LS, Bhandari S, Bhusan S, Singhvi AK (2006) Continental record of the southwest monsoon during the last 130 ka: evidence from the southern margin of the Thar desert. India Quaternary Sci Rev 25:2632–2650
Kale VS (2005) The sinuous bedrock channel of Tapi River, Central India, its form and processes. Geomorphology 70:296–310
Kroon D, Steens TNF, Troelstra SR (1991) Onset of monsoonal related upwelling in the western Arabian Sea as revealed by planktonic foraminifers. Proc ODP Sci Results 117:257–263
Machette MN (1985) Calcic soils of the southwestern United States. In: Weide DL (ed) Soils and Quaternary geology of the southwestern United States, vol 203. Geological Society of America Special Paper, pp 1–21
Michael NM (2000) Active, capable, and potentially active faults — a paleoseismic perspective. J Geodyn 29(3):387–392. https://doi.org/10.1016/S0264-3707(99)00060-5
Molnar P, England P, Martiod J (1993) Mantle dynamics, uplift of the Tibetan plateau and the Indian monsoon development. Rev Geophys 34:357–396
Nagalakshmi T, Achyuthan H (2004) Radiocarbon dating and Holocene episodes of alluvial sedimentation in the Koratallaiyar-Cooum River basin, Chennai, South India. Jour Geol Soc India 64:461–469

Narula PL, Acharyya SK, Banerjee J (eds) (2000) Seismotectonic Atlas of India and its environs. GSI publication

Overpeck JT, Anderson DM, Trumbore S, Prell WL (1996) The southwest monsoon over the last 18,000 years. Clim Dyn 12:213–225

Pal DK, Deshpande SB, Venugopal KR, Kalbande AR (1989) Formation of di and trioctahedral smectite as evidence for paleoclimatic changes in southern and central Peninsular India. Geoderma 45:175–184

Pal DK, Balpande SS, Srivastava P (2001) Polygenetic Vertisols of the Purna valley of central India. Catena 43:231–249

Patnaik R, Gupta AK, Naidu PD, Yadav R, Bhattacharyya A, Kumar M (2012) Indian monsoon variability at different time scales: marine and, terrestrial proxy records. Proc Indian Natn Sci Acad 78(3):535–547

Pradhan SM (1980) GSI News September 1980. II(9):4

Prell WL, Kutzbach JE (1987) Monsoon variability over the past 150,000 years. J Geophysical Lett Atmos 92(D7). https://doi.org/10.1029/JD092iD07p08411

Raj R (2007) Late Pleistocene fluvial sedimentary facies, the Dhandar River basin, Western India. Quat Int 159:93–101

Rajurkar ST (1982) Photogeological interpretation of Upper Wardha project and surrounding areas, Wardha. Amravati and Betul districts. GSI unpublished report

Ray A, Rana NS (1989) Geology of Buldana and Amravati districts. Maharashtra. Unpublished GSI report

Raymo ME (1991) Geochemical evidence supporting T. C Chamberlin's theory of glaciation. Geology 19(4):344–347. https://doi.org/10.1130/0091-7613(1991)019<0344:GESTCC>2.3.CO;2

Resmi MR, Achyuthan H (2021) The interplay of mid to late Holocene tectonism and northeast monsoon on the landscape evolution. In: Quaternary climate change over the Indian sub-continent. Edited by Neloy Khare, CRC Press, pp 109–139

Roy A, Devarajan MK (2001) Occurrence of seismites from the quaternary deposits of Narmada Valley. Geological Survey of India News 18:15–16

Roy A, Devarajan MK (2003) Tectonics and seismicity of Central India: a review. In: Roy A, Mohabey DM (eds) Seismicity of Central India, vol 5. Gondwana Geological Magazine Special Publications, pp 23–44

Roy A, Prasad MH (2003) Tectonothermal events in Central Indian Tectonic Zone (CITZ) and its implications in Rodinian crustal assembly. J Asian Earth Sci 22:115129

Saxena RK (1988) Geothermal investigation around Salbardi hot springs area, district Betul (M.P.) and Amravati (Maharashtra). GSI unpublished report

Shanker R (1987) Neotectonic activity along the Tapi-Satpura lineament in Central India. Indian Minerals 41(1):19–30

Shanker R, Dubey R (1984) Landform, neotectonism and geothermal activity in the part of Tapi basin, district Jalgaon, Maharashtra. Rec Geol Surv India 123(6):70–85

Sharma S, Joachimiski M, Tobschall HJ, Singh IB, Sharma C, Chauhan MS, Morgenroth G (2004) Late-glacial and Holocene environmental changes in Ganga plain, northern India. Quaternary Sci Rev 23:145–159

Singh IB (2005) Quaternary palaeoenvironments of the Ganga plain and anthropogenic activity. Man and Environ 30:1–35

Singhvi AK, Kar A (2004) The aeolian sedimentation record of the Thar desert. Proc Indian Acad Sci (Earth Planet Sci) 113(3):371–401

Sinha M, Tiwari MP (1984) Geomorphology and geology of the Quaternary deposits in part of Tapti basin, Jalgaon district, Maharashtra. Unpublished GSI report

Sinha R, Tandon SK, Sanyal P, Gibling MR, Stuben D, Berner Z, Ghazanfari P (2006) Calcretes from a late quaternary interfluve in the Ganga Plains, India: carbonate types and isotopic systems in a monsoonal setting. Palaeogeogr Palaeoclimatol Palaeoecol 242:214–239

Sinha R, Bhattacharjee P, Sangode SJ, Gibling MR, Tandon SK, Jain M, Godfry D (2007) Valley and interfluve sediments in the southern Ganga plains, India: exploring facies and magnetic signatures. Sed Geol 201:386–411

Srivastava P, Bhattacharyya T, Pal DK (2002) Significance of the formation of calcium carbonate minerals in the pedogenesis and management of cracking clay soils (vertisols) of India. Clays and Clay Minerals 50(1):111–126. 1 February 2002

Srivastava P, Singh IB, Sharma M, Singhvi AK (2003a) Luminescence chronometry and late quaternary geomorphic history of the ganga plain, India. Palaeogeogr Palaeoclimatol Palaeoecol 197:15–41

Srivastava P, Sharma M, Singhvi AK (2003b) Luminescence chronology of incision and channel pattern changes in the River Ganga, India. Geomorphology 51:259–268

Srivastava P, Sharma S, Shukla UK, Singh IB, Singhvi AK (2003c) Late Pleistocene-Holocene hydrological changes in the interfluve areas of central Ganga plains, India. Geomorphology 54: 279–292

Srivastava AK, Bansoda, Mangla N, Khare N (2021) Calcretes from the quaternary alluvial deposit of Purna basin, Central India: lithological and climatic controls. Rhizosphere 18:100343

Talling PJ, Stewart MD, Stark CP, Gupta S, Vincent SJ (1997) The regular spacing of drainage outlets from linear fault blocks. Basin Res 9:275–302

Thomas PJ, Juyal N, Kale VS, Singhvi AK (2007) Luminescence chronology of late Holocene extreme hydrological events in the upper Penner River basin, South India. J Quaternary Sci 22: 747–753. ISSN 0267–8179

Tiwari MP (1996) Neotectonism in Tapti-Purna valleys and its probable correlation with geothermal activity. In: Pitale UL, Padhi RN (eds) Geothermal energy in India, Geol. Surv. India Special Publ, vol 45, pp 325–332

Tiwari MP (1999) Quaternary Lithostratigraphic formations of Central Indian River Basins; their correlation and chronology. Gondwaana, National Symposium of Quaternary of India, Geol Magz Spl 4:17–31

Tiwari MP, Bhai HY (1997) Quaternary stratigraphy of the Narmada Valley. Geol Surv India Spec Publ 46:33–63

Tiwari MP, Bhai HY, Bandyopadhyay KP (2001) Quaternary geology of the Purna valley, Amravati, Akola and Buldhana districts, Maharashtra. Unpublished GSI Report

Waelbroeck C, Labeyrie L, Michel E, Duplessy JC, McManus JF, Lambeck K, Balbon E, Labracherie M (2002) Sea-level and deep water temperature changes derived from benthic foraminifera isotopic records. Quat Sci Rev 21(1–3):295–305. https://doi.org/10.1016/S0277-3791(01)00101-9

Williams MAJ, Pal JN, Jaiswal M, Singhvi AK (2006) River response to quaternary climatic fluctuations: evidence from the Son and Belan valleys, north-central India. Quaternary Sci Rev 25:2619–2631

Wright VP, Tucker ME (1991) Calcretes: an introduction. In: Wright VP, Tucker ME (eds) Calcretes. Blackwell Scientific. 352, Oxford, pp 1–22

Chapter 13
Application of Cosmogenic Radionuclides in the Quaternary Sciences Using Accelerator Mass Spectrometry

Atul Kumar Singh and Pankaj Kumar

Abstract The Quaternary Period begins at 2.58 Ma and was defined by the Global boundary Stratotype Section and Point (GSSP) in 2009. The Quaternary comprises only 0.4% of Earth's history but is crucial for humans because of glacial and interglacial events. The oxygen isotope studies in the microfossils of the deep marine sediments have revealed the glacial and interglacial cycles at 100 ka, 43 ka, 24 ka and 19 ka for the past 700 ka due to Milankovitch cycles. Therefore, it is important to reconstruct the past climate and environmental changes and model them to predict such events which may happen in future. Accelerator Mass Spectrometry (AMS) is an ultrasensitive technique to handle a very small sample size and provides isotopic ratios for many elements which help understand the evolution of Quaternary landforms and climate. In this chapter, we have discussed the working principle of AMS and different cosmogenic radionuclides (CRN) which help reconstruct past climate and tectonic events. Further, different geomorphological features and landforms such as loess, lakes, river terraces, ice cores, etc., are also discussed, where cosmogenic radionuclide proxy can be applied to understand the past changes in the climate on a global or regional scales. Denudation rates are an important proxy to understand the palaeo-humidity and temperature conditions, and the role of CRNs to estimate accurate palaeo-denudation rates has also been discussed in the chapter.

Keywords Holocene epoch · Atmospheric circulations · Microfossils · Accelerator mass spectrometer

Introduction: The Quaternary Period

The Quaternary is the most recent geological record of the Earth's history. Along with the Neogene and Palaeogene Periods, it forms the Cenozoic Era. The boundary of the Quaternary System/Period was defined by the Global boundary Stratotype

---

A. K. Singh · P. Kumar (✉)
Inter-University Accelerator Centre, New Delhi, India
e-mail: pkb@iuac.res.in

N. Khare (ed.), *Science, Policies and Conflicts of Climate Change*, Springer Climate,
https://doi.org/10.1007/978-3-031-16254-1_13

Section and Point (GSSP) in 2009 at 2.58 Ma which matches with the Marine Isotope Age (MIS) 103 (Gibbard and Head 2010; Head 2019). The Quaternary System/Period is divided into the Pleistocene and Holocene Series/Epochs. The Pleistocene Epoch begins with the Quaternary at 2.58 Ma and ends at 0.0117 Ma or 11.7 ka. This marks the beginning of the Holocene Epoch which continues till date. The term Quaternary was introduced in 1759 by Giovanni Arduino to define a fourth-order (*quarto ordine)* or stage of alluvial sediments of the River Po from the Venetian and Tuscan regions of Italy (Arduino 1759; Vaccari 2006; Lowe and Walker 2014; Head 2019). However, technically he never used the term Quaternary (Gibbard 2019); the term for the first time was used by the French geologist Jules Desnoyers in 1829.

Quaternary has been the most active period in terms of environmental changes in the last 60 Ma. $\Delta^{18}O$ is a proxy used for reconstructing deep-water temperature and global ice cover. The ice cover started developing around 34 Ma ago during the Cenozoic, which was most extensive in the Pleistocene (Zachos et al. 2001; Bradley 2015). The Quaternary comprises only 0.4% of Earth's history and is known for its glacial cycles. This change can be caused by a combination of factors, which are changes in long-term positions of landmass, mountain-building processes which control the atmospheric circulations and atmospheric composition through erosion and weathering (Zachos et al. 2001; An Yin 2006) and cyclic variations in Earth's orbital position (Hays et al. 1976). However, the mechanism through which these ice sheets advance and recede is not very well understood (Muller and MacDonald 1997). The beginning of the Quaternary was marked by the development of substantial ice sheets. Milankovitch (1941) proposed three astronomical processes which controlled the insolation and thus the climate of the Earth: (a) eccentricity of the orbit, the shape of the Earth's orbit around the sun changes from circular to elliptical due to planetary gravitational influences between 95 and 136 ka, this is referred to as 100 ka eccentricity cycle, (b) obliquity of the ecliptic, the tilt of the Earth's axis varies from 21°39′ to 24°36′ over a time of ~41 ka and (c) precession of the equinoxes, change in the direction of the rotation of the Earth axis to a fixed star (wobbling on its axis like a top) is caused by the gravitational pull of the Sun and the Moon at an interval of ~23 and 19 ka. It is widely accepted that these astronomical changes in the eccentricity, obliquity and precession of the Earth around the Sun are the main driving forces of climate shifts through time (e.g. Berger and Loutre 1991; Clement et al. 2001; Leuschner and Sirocko 2003). The oxygen isotope studies in the microfossils of the deep marine sediments have revealed the cyclicity of glacial/ interglacial cycles at 100 ka, 43 ka, 24 ka and 19 ka for the past 700 ka (Hays et al. 1976; Lowe and Walker 2014). However, the earth has witnessed changes in the climate over a shorter time duration during the Holocene such as the Medieval Warming Period (MWP), Little Ice Age (LIA), etc., with a smaller period of a few hundred years. The Milankovitch theory does not explain changes at these smaller time intervals and a large dataset is required to understand these processes. The phenomenon such as mountain building processes, which control the atmospheric circulations pattern affects land-ocean thermodynamic relations and affect the climate at a regional scale (An et al. 2001; Wünnemann et al. 2007). For example, in the

Indian subcontinent, the climate is strongly influenced by the formation of the Himalayas due to the convergence of the Indian and the Eurasian plates. The uplift of the landmass results in exposure of the fresh rock to the exogenic forces which lead to the weathering and erosion of these rocks. The chemical weathering of silica-rich rocks consumes the atmospheric CO_2 (Ebelmen 1845; Urey 1952; Lupker et al. 2016), thus decreasing the greenhouse gas from the atmosphere and bringing down the global temperatures.

Accelerator Mass Spectrometry

Accelerator Mass Spectrometry (AMS) is a mass spectrometry of high energy with an accelerator. Since accelerators provide energy to the ions, the AMS can be termed as high-energy mass spectrometry as well. By making the accelerator an integral part of mass spectrometry, the limitations of conventional mass spectrometers can be considerably reduced and the sensitivity of AMS measurements can be improved to 10^{-14}–10^{-16}. Therefore, AMS is the best suitable technique for the measurement of cosmogenic produced long-lived radionuclides present at a very low level in nature. In AMS, ions are produced and accelerated to high energy (in the range of tens of MeV) by using a tandem accelerator. At this energy, several important techniques (such as nuclear detectors, electric and magnetic filters, etc.) become more effective and help in the suppression of unwanted interferences to a very low level.

A typical tandem accelerator-based AMS system is composed of an ion source, an injector system with dipole magnets, a tandem accelerator, a post acceleration analysis system and a detector. The ion source produces pre-accelerated negative ions, which undergo a selection procedure by the injector system through ME/q^2 filtering. These selected negative ions are accelerated and converted into positive ions with the help of an accelerator. Tandem accelerators have a stripper system in the middle of their tank, which converts negative ions to positive and also helps in reducing the molecular background by breaking the molecule. Due to the high energy achieved by the accelerator (order of MeV) atom counting is possible using particle detectors. High energy also helps in removing isobaric and isotopic interferences. Furthermore, various other filters like velocity filter (E/M selector), electrostatic analyzer (E/q selector), gas-filled magnet, etc., are used in AMS to reduce background. The use of a high-energy accelerator in combination with various filters increases the sensitivity of AMS to 10^{-14}–10^{-16} and makes it a versatile and powerful method to accurately measure ultra-trace elements. Figure 13.1 shows the schematic diagram of the AMS facility at the Inter-University Accelerator Centre (IUAC), New Delhi for the measurement of ^{14}C, ^{10}Be, and ^{26}Al radionuclides.

Usually, AMS determines the ratio of radioisotope to a stable isotope, as absolute counting of a particular nuclide is difficult. However, two isotopes may not be ionized or transmitted through the system in an exactly similar manner and, therefore, the measured ratio of radioisotope to stable isotope may not reflect the true ratio in the unknown sample. To solve this problem, a standard sample having a known

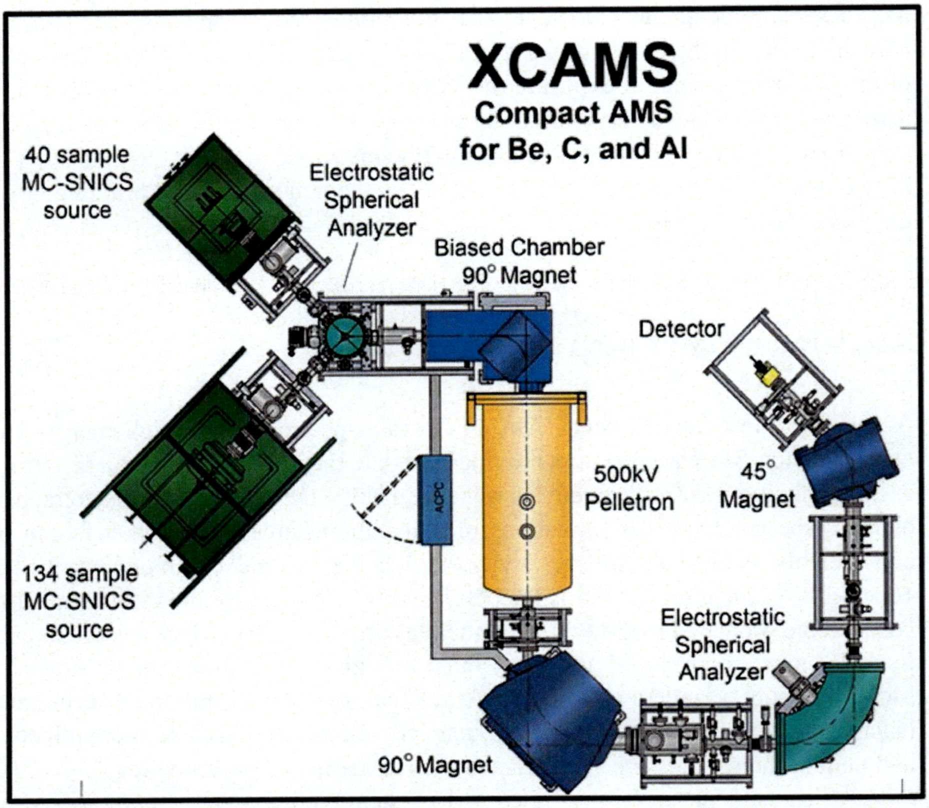

Fig. 13.1 Schematic diagram of accelerator mass spectrometry facility at IUAC, New Delhi (after Sharma et al. 2019)

ratio of radioisotope to stable isotope is measured along with the unknown sample to normalize the results.

The sample material chosen for the AMS studies may contain several other unwanted elements along with the element of interest. Sometimes an element of interest stays inside a mineral, e.g., ^{10}Be and ^{26}Al formed by the in situ method stay in the quartz mineral. Physical and chemical processing of samples needs to be done before the AMS measurement to remove these unwanted elements to the maximum possible level and to extract the element of interest from the mineral (Singh et al. 2022). During the chemical processing, the sample material is converted into a chemical form suitable to produce a high yield of negative ions from the ion source. In addition to this, chemical separation reduces the concentration of isobar by several orders of magnitude. All these physical and chemical processes are performed in a controlled environment to reduce the background level. Any AMS facility is usually accompanied by suitable physical and chemical processing laboratories.

Another method of detecting cosmogenic radionuclides is decay counting utilizing their decay property. Estimating a radioisotope is possible by simply counting the radiation (α, β, or γ) emitted during the decay of the radioisotope. The radiations emitted during decay are proportional to the number of radioisotope atoms present at that time and the half-life of the radioisotope. The natural abundance of long-lived radioisotopes is usually in the range of 10^{-10} to 10^{-16}. Therefore, decay counting takes a very long counting time and needs a large amount of sample material for counting long-lived radioisotopes to achieve considerable statistics. Decay counting is best suited for short-lived radioisotopes. AMS directly counts the long-lived radionuclides without waiting for their feeble radioactive decay, and therefore AMS is preferred over decay counting in the case of the cosmogenic radionuclides having a long half-life targeting geological research.

Cosmogenic Radionuclides

Earth's atmosphere is constantly bombarded by high-energy particles called cosmic rays that originate from the Sun, from different solar systems within our galaxy, and also from other galaxies. Cosmic rays can be classified as primary and secondary cosmic rays. Primary cosmic rays are the ones that arrive from the Sun and outside of the solar system. On entering the Earth's atmosphere, primary cosmic rays produce a cascade of lower energy secondary particles called secondary cosmic rays. Primary cosmic rays are composed mainly of protons and alpha particles with few electrons, positrons, gamma rays, and antiprotons. Based on origin, primary cosmic rays can be classified as galactic cosmic rays (GCR), solar cosmic rays (SCR), and anomalous cosmic rays (ACR). GCR is generated outside of the solar system with an average energy of 2 GeV in a range of 0.1 to 10 GeV (Tuniz et al. 1998). SCR originate from the Sun with an energy range of 1 to 100 MeV and their intensity increases manifold during solar flares. ACR is thought to originate from the ionization of neutral interstellar atoms either by solar radiation or solar winds (Jokipii and Giacalone 1998; Tuniz et al. 1998).

Cosmogenic radionuclides are produced by the interaction of high-energy cosmic rays with the stable nuclei of atoms present in the atmosphere or on the Earth's surface radioisotopes are produced. For example, when neutrons present in the cosmic rays spectrum interact with nitrogen present in the atmosphere ^{14}C ($T_{1/2} = 5730$ years) is produced following the nuclear spallation reaction $^{14}N(n,p)\,^{14}C$. Thus formed ^{14}C is quickly oxidized into $^{14}CO_2$ and equilibrates with volatile carbon species. The $^{14}C/C$ ratio in the atmosphere is homogeneous with a natural value of about 1.2×10^{-12}. The distribution of atmospheric ^{14}C in the hydrosphere and biosphere is done by gaseous exchange and by photosynthesis, respectively, which is a part of the carbon cycle. Once ^{14}C is mixed, it presents everywhere like in plants, animals, the hydrosphere, and the biosphere. As long as plants or animals are alive, they stay in equilibrium with nature and the natural ^{14}C value remains constant. However, when they are dead, the natural value starts decreasing and the older the sample, the fewer is the number of ^{14}C atoms in the sample. Therefore, ^{14}C can act as a natural clock present in the natural archive.

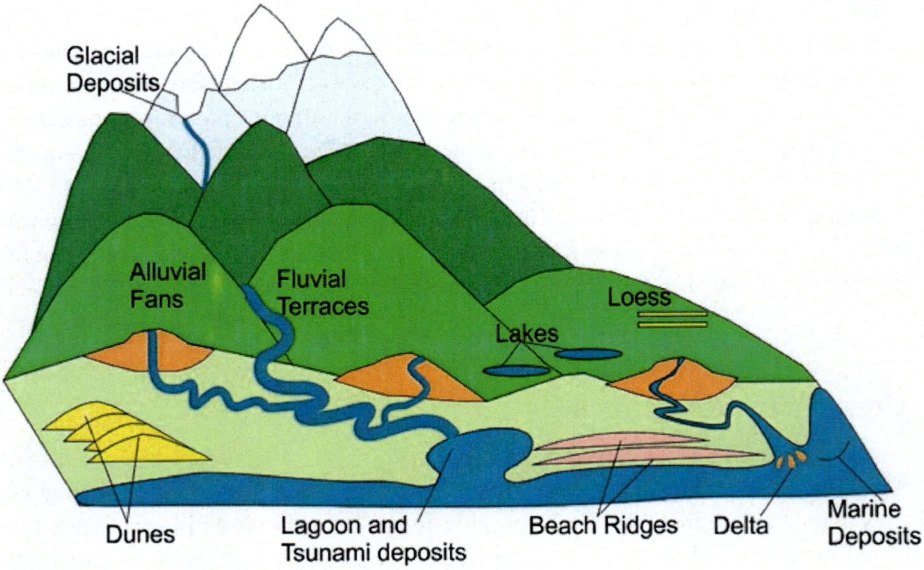

Fig. 13.2 The figure shows different geological landforms where cosmogenic radionuclides can be applied. (Modified after Rhodes 2011)

The cosmogenic radionuclides have wide application in Earth and atmospheric sciences. These could be used for dating, palaeo-climate reconstruction, denudation rates, solar activity, geomagnetic variations, etc. The cosmogenic radionuclides help understand the geological evolution of different landforms (Fig. 13.2), and these are briefly discussed here.

Loess

Loess is aeolian derived silt that covers about 10% of the Earth's land surface (Busacca and Sweeney 2005; Bradley 2015). Extensive loess covers are present in China, Central Asia, Russia, Europe, Argentina, New Zealand, and North America (Muhs 2014). In India, a thick loess cover is found in the Karewa basin of Kashmir. Loess can be of glacial and non-glacial origin, a classical approach is that the silt-sized particles are produced by glacial grinding of rocks, deposited as till, reworked by the fluvial processes, and finally picked up, transported, and deposited by the wind (Roberts et al. 2007; Muhs 2014). This leads to the view that the loess deposits are markers of glacial periods. However, some authors have suggested that the loess may be of desert origin and the topic is under debate (Smalley 1995; Wright 2001; Muhs and Bettis 2003). The classical approach is well proven in regions close to glaciers, whether they are continental or mountainous glaciers, e.g., in North America (Ruhe 1969), Europe (Frechen 2003), and Asia (Dodonov 1991). The loess deposits are usually unaltered sediments with intercalated paleosol layers. The paleosol layers form when the loess activity is minimal and the layers become stabilized under relatively moist conditions. Therefore, the interplay between glacial and interglacial periods leads to formations of loess and paleosols, respectively. Thus, these can be used as an indicator of climate change.

In a recent study, Meenakshi et al. (2018) worked on the loess-paleosol deposits of the Dilpur Formation, Karewa basin, Kashmir using the AMS facility at IUAC and showed that the Last Glacial Maxima (LGM) was between 18.6 ka to 22.3 ka. They further extrapolated the results to show two more glacial periods between 67–71 ka and 124–127 ka.

Lakes

Lakes accumulate terrestrial sediments from the surroundings and thus preserve records of any environmental change (Cohen 2003; Jeziorski et al. 2008; Bradley 2015; Shah et al. 2020). Lake sediments are composed of two types of materials, one which generates within the lake basin (autochthonous) while the other is transported to the lake by rivers, streams, overland flow, etc. (allochthonous material). A wide variety of proxies can be used to reconstruct the past climate from the lake sediments. These include sedimentology, inorganic geochemistry, varves, pollen, macrofossils, phytoliths, ostracods, diatoms, stable isotopes, and organic biomarkers. It is worth mentioning that lake sediments provide a plethora of information about past climate using above-mentioned proxies but this information will be limited in absence of proper chronology. This is where the role of AMS becomes important because the lake sediments are rich in organic carbon and can be dated using ^{14}C.

The sedimentary facies are the indicator of depositional condition (Miall 1985; Das et al. 2017) and can be used to get basic information about palaeo-climates. Similarly, magnetic susceptibility data can be used to interpret the amount of clastic in the lake sediments (Polissar et al. 2006). Geochemical data can be useful in interpreting the palaeo-climate, e.g., the Ti concentration in the lake sediment tells about the dominant erosional agent in the region (Bakke et al. 2009; Yancheva et al. 2007). Varves are an important proxy for palaeo-climate reconstruction; these are annually deposited laminated sediments in lakes (Anderson and Dean 1988). Varves are difficult to preserve and are usually found in glacial and paraglacial environments. In some cases, varves can be used as a scale to calibrate the radiocarbon timescale (Staff et al. 2013). Pollen, macrofossils, and phytoliths are other important proxies that provide a plethora of information on the palaeo-environment and palaeo-climate. Pollen has been widely used to verify the general circulation model simulations (Bennett and Willis 2001). Plant macrofossils such as leaves, fruits, seeds, etc., are helpful in vegetation reconstruction (Jackson et al. 1997).

River Terraces

River terraces are the remnants of the flood plains that run parallel to the present-day river (Leopold et al. 1964; Schumm 1977; Ritter et al. 2002; Pazzaglia 2013). The terraces based on their relative elevation on both the banks of the river may be classified as paired and unpaired terraces. The paired terraces occur on opposite sides of the valley at similar elevations (when not eroded) and represent the former longitudinal profile of the stream when connected (Bull 2007 p. 43). The unpaired terraces are more common and are at different elevations on both sides of the valley. These terraces form when a vertically incising river laterally migrates in its flood-plain. ^{14}C and Optically Stimulated Luminescence (OSL) dating are well-established dating methods for detrital organic materials and sediments (quartz and feldspar), respectively, that are commonly found in terrace deposits (Wegmann and Pazzaglia 2002; Duller 2004; Pazzaglia 2013; Singh et al. 2016, 2017). Crustal deformation and rock uplift associated with the tectonic forcing are one of the primary reasons for the formation of the fluvial and alluvial terraces. Climate change has played an important role in the formation of the fluvial, alluvial, and marine terraces. Thus, terraces are important geographic markers that can be studied to interpret the past climate and tectonics.

Glacial and Periglacial Landforms

The cryosphere consists of continental and mountainous glaciers, seasonal snow cover on land, and sea ice. The fluctuations in the glaciers result from changes in net accumulation and net ablation. An increase in net accumulation will result in positive mass balance and will lead to the advance of the glacier snout and similarly, an increase in the net ablation will result in the retreat of the snout (Bradley 2015). The variation in glacier snout is a complex process combining short- and long-term climatic fluctuations. The smaller glaciers have a shorter response time, whereas the large glaciers would have a longer response time to climate change. It is interesting to note that despite rising global temperatures some glaciers in the Himalayas have managed to advance; this shows the influence of micro-climate zones and topography on glacial dynamics (Scherler et al. 2011).

The fluctuations in the glaciers can be recorded using the glacial moraines which are left behind when the glaciers retreat (Nesje et al. 2000). The recent advance in glaciers may overwrite the signatures of previous advances and recession, thus complicating the reconstructions. However, a proper stratigraphic study may reveal buried soils or some erosional contacts which were buried by recent glacial moraine (Röthlisberger 1976; Bradley 2015). The biggest hurdle in using glaciers as a palaeo-climate proxy is dating them. The glaciers usually lack organic material, and even if present will provide minimum ages of the glacial advance because there is a time lag in the retreat of glaciers and the development of soil (Bradley 2015). The recent developments in surface exposure dating using cosmogenic radionuclides have opened a new era in the reconstruction of glacial studies.

Ice Cores

Ice cores have immerged as a very important proxy in palaeo-climate reconstruction because the natural oscillations in the greenhouse gases are preserved in the ice caps. Ice cores trap these fossil gases such as CO_2, CH_4, and N_2O; ice cores also provide a plethora of information such as stable isotope composition of ice matrix and trapped molecular oxygen, cosmogenic isotopes, dust, and electrical conductivity (Bales and Wolff 1995; Cronin 1999). The measurement of cosmogenic nuclides such as ^{14}C, ^{10}Be, and ^{36}Cl provides not only dating information but also changes in the production rate, geomagnetic field, and solar activity (Lal and Peters 1967; Beer et al. 1988; Stuiver and Braziunas 1989; Raisbeck et al. 1992; Cronin 1999; Masarik and Beer 1999; Horiuchi et al. 2008). The variations in the past solar activity, especially for the Holocene epoch, are very well recorded by the ^{14}C tree-ring record and Be ice core record (Edouard et al. 1997; Solanki et al. 2004). The changes in the solar irradiance are directly related to the changes in the solar magnetic activity on an 11-year timescale (Willson and Hudson 1988; Radick et al. 1990; Field et al. 2006). The increased solar magnetic activity (which regulates GCR) results in greater deflection and thus Earth's exposure to the GCR is reduced resulting in reduced production of the cosmogenic radionuclide. Thus, an inverse relationship exists between solar activity and cosmogenic radionuclide production. There are different uncertainties associated with these topics and have been addressed by Lean et al. (2002); McCracken (2004) and McCracken et al. 2004; Horiuchi et al. (2008), who collected ice core from the Dome Fuji (Antarctica) and have shown that the concentration and the flux of the ^{10}Be increased at known solar activity minimums at 1010–1050 year CE, 1280–1340 year CE, 1420–1540 year CE, 1654–1715 year CE and 1795–1820 year CE.

The Earth's magnetic field is very dynamic and varies on a time scale of less than a second to millions of years; these could be palaeosecular variations to geomagnetic excursions and reversals (Simon et al. 2020). The geomagnetic excursion is defined as a deviation of more than 40–45° in the virtual geomagnetic pole (VGP) from the geographic pole (Merrill and McFadden 1994) and may occur on timescales of a few thousand years (Roberts 2008). The excursions can also be understood as failed reversals of Earth's magnetic field (Simon et al. 2020) and are associated with weak magnetic palaeo-intensities in marine and lacustrine sediments (e.g. Thouveny et al. 1993; Nilsson et al. 2011; Channell et al. 2017; Lund et al. 2005, 2017). The Laschamp excursion (41 ± 1 ka) (Laj and Channell 2015) is one such extensively studied excursion. The palaeomagnetic data of directions and palaeo-intensities has been extensively used to study these excursions but these datasets suffer from some inherent problems such as low dipole field intensities during different time intervals and discontinuous terrestrial sedimentary records (Roberts 2008). There is an inverse relationship between Geomagnetic dipole moment (GDM) and cosmogenic nuclide production rates (Elsasser et al. 1956; Lingenfelter 1963; Lal and Peters 1967; O'Brien 1979; Simon et al. 2020).

Denudation Rates

The rate of erosion or denudation has been a central theme of research for geomorphologists and geologists. The conventional method is to measure the average mass of material leaving a drainage basin (Milliman and Meade 1983), but the measurements have high uncertainty (Bierman and Steig 1996). Basin-scale denudation rates are important to explain the geomorphic evolution of mountain belts and drainage systems and also to establish the relationship between erosion, climate and tectonics (Molnar and England 1990; Raymo et al. 1988; Dash et al. 2020).

In recent times, the ratio of cosmogenic ^{10}Be to the stable isotope ^{9}Be has emerged as a very successful proxy for basin-scale denudation rate calculations (von Blanckenburg et al. 2012; Wittmann et al. 2015; Rahaman et al. 2017). This proxy is not only useful in determining the present-day denudation rates but also palaeo-denudation rates. The meteoric variety of ^{10}Be is produced in the atmosphere and gets attached to dust and aerosols and is removed from the atmosphere due to precipitation (wet or dry). This meteoric ^{10}Be gets adsorbed onto the soil and sediments and has several advantages over the in situ produced ^{10}Be such as higher concentration, smaller sample size and applications to quartz free lithologies (von Blanckenburg et al. 2012). However, the meteoric or garden variety of ^{10}Be suffers from some inherent problems such as its adsorption dependence on the pH of the solution and the grain size of the sediments. To circumvent this problem, the concentration of ^{10}Be should be normalized with the second element having identical chemical properties. ^{9}Be is the stable isotope of Be and will show similar partitioning behaviour as ^{10}Be. $^{10}Be/^{9}Be$ has emerged as a robust proxy for denudation rates using bulk soil and river sediments which have ^{10}Be adsorbed onto rock surfaces and precipitated during secondary weathering (von Blanckenburg et al. 2012). Denudation rates can be determined using

$$\left(\frac{^{10}\mathrm{Be}}{^{9}\mathrm{Be}}\right)_{\mathrm{reac}} = \left(\frac{^{10}\mathrm{Be}}{^{9}\mathrm{Be}}\right)_{\mathrm{diss}} = \frac{\left(\frac{F_{\mathrm{met}}^{^{10}\mathrm{Be}}}{D*[^{9}\mathrm{Be}]_{\mathrm{parent}}}\left(\frac{[^{9}\mathrm{Be}]_{\mathrm{min}}}{[^{9}\mathrm{Be}]_{\mathrm{reac}}}+1\right)\right)}{\left(1+\frac{Q}{K_d}*\frac{[^{9}\mathrm{Be}]_{\mathrm{min}}}{D*[^{9}\mathrm{Be}]_{\mathrm{parent}}}\right)}$$

where $(^{10}Be/^{9}Be)_{reac}$ = ratio of ^{10}Be and ^{9}Be in reactive phases; $(^{10}Be/^{9}Be)_{diss}$ = ratio of ^{10}Be and ^{9}Be in river water, $^{9}Be_{parent}$ = concentration of ^{9}Be in bedrock; $^{9}Be_{reac}$ = concentration of ^{9}Be in the reactive phase; $^{9}Be_{min}$ = concentration of ^{9}Be after removal of dissolved and reactive phases; K_d = Partition coefficient of *Be* between dissolved and reactive phases.

F_{met}^{10Be} = Flux of meteoric ^{10}Be to the Earth's surface (in atoms/m^2/year); D = Denudation rate (in T/km^2/year); Q = Water flux out of system (in m/year).

If this condition is followed:

$$\frac{Q}{D} \gg K_d * \frac{[^9\text{Be}]_{\text{parent}}}{[^9\text{Be}]_{\text{min}}}$$

Then, the above equation can be modified to:

$$\left(\frac{^{10}\text{Be}}{^9\text{Be}}\right)_{\text{reac}} = \left(\frac{^{10}\text{Be}}{^9\text{Be}}\right)_{\text{diss}} = \left(\frac{F_{\text{met}}^{10\text{Be}}}{D * [^9\text{Be}]_{\text{parent}}}\left(\frac{[^9\text{Be}]_{\text{min}}}{[^9\text{Be}]_{\text{reac}}} + 1\right)\right)$$

Dhal et al. (2018) estimated present and past denudation rates for the Cauvery river catchment based on a ^{10}Be (meteoric) study of sediment samples from two cores drilled at Uttrangudi and Valangaiman sites in the Cauvery delta, which vary from 18.7 ± 1.6 to 48.1 ± 8.3 T/km^2 /yr during the Holocene.

Dash et al. (2020) estimated the denudation rate using the ^{10}Be$_{(\text{meteoric})}$/^{9}Be ratio in monsoon dominated regions and found that a high/low denudation rates were observed during the strong/weak phases of ISM, respectively, and increase in catchment erosion is associated with intensified ISM.

Reservoir Effect

The atmosphere, biosphere and oceans work as active reservoirs of radiocarbon. A biogeochemical cycle in which carbon is exchanged between different reservoirs is known as a carbon cycle (Craig 1957). Carbon dating has an inbuilt assumption that the radiogenic carbon (^{14}C) in the atmosphere is in a state of equilibrium with all the living organisms. However, there are certain localities in which this condition is not fulfilled, e.g. the plants which grow in close vicinity of volcanoes will have a lot of older carbon emancipated from the volcanoes. Similarly, the organisms living close to the industrial hubs will also inhale air that is enriched in older carbon being generated by burning of the fossil fuels.

In addition to this, there are reservoir effects (marine and freshwater) that result in older radiocarbon ages. The cosmogenic radiocarbon present in the atmosphere is dissolved in oceans in the form of CO_2 and at the same moment is consumed by the plants through photosynthesis and enters food chains. The content of ^{14}C is not the same at the surface of the ocean and in the deep ocean because the exchange between the ocean surface and atmosphere is comparatively faster than between the surface and the deep ocean. Therefore, the marine organisms of the same age will have different radiocarbon content at the same time, which will result in anomalous radiocarbon ages. In general, it has been observed that the marine radiocarbon ages are 400 years older than the terrestrial radiocarbon ages (Fairbanks 1989;

Timothy Jull et al. 2013). However, these corrections are site-dependent and depend on coastline shape, ocean topography, ocean currents, climate and season, and wind directions because these factors affect the upwelling which controls the mixing of ocean water.

Similarly, freshwater reservoir effects in carbon dating have also been seen. As discussed earlier the living organism is in equilibrium with the atmospheric ^{14}C, but when an organism obtains its carbon when other reservoirs have a lesser content of ^{14}C, it will result in an overestimation of radiocarbon ages. The most common cause of this effect is the presence of dissolved carbonates that contain old carbon, also known as the hard water effect. The hard water effect may result in errors of up to 6000 years (Clark and Fritz 1997; Bente 2013). Deep freshwater lakes may also have delayed mixing of surface water and deep lake waters resulting in older ages similar to marine reservoir effects. In addition, the dating of riverine sediments is also tricky because the sediments are derived from different sources and may contain organic carbon of different ages. The fluvial sediments thus may contain a lot of older carbon resulting in overestimation of the ages.

References

Anderson RY, Dean WE (1988) Lacustrine varve formation through time. Palaeogeogr Palaeoclimatol Palaeoecol 62(1–4):215–235. 0031018288900557. https://doi.org/10.1016/0031-0182(88)90055-7

Arduino G (1759) Sopra varie sue Osservazionifatte in diverse parti del Territorio di Vicenza, ed altrove, appartenentialla Teoria Terrestre, ed allaMineralogia. Letter to Prof. Antonio Vallisnieri, dated 30th March

Bakke J, Lie O, Heegaard E, Dokken T, Birks HH, Haug GH, Dulski P (2009) Rapid oceanic and atmospheric changes during the Younger Dryas cold period. Nat Geosci 2(3):202–205. https://doi.org/10.1038/ngeo439

Bales RC, Wolff EW (1995) Interpreting natural climate signals in ice cores. Eos Trans Am Geophys Union 76(47):477–477. https://doi.org/10.1029/95EO00293

Beer J, Siegenthaler U, Bonani G, Finkel RC, Oeschger H, Suter M, Wölfli W (1988) Information on past solar activity and geomagnetism from 10Be in the Camp Century ice core. Nature 331(6158):675–679. https://doi.org/10.1038/331675a0

Bennett KD, Willis KJ (2001) Pollen. In: Smol JP, Birks HJB, Last WM (eds) Tracking environmental change using lake sediments, vol 3. Kluwer Academic, Dordrecht

Bente P (2013) The freshwater reservoir effect in radiocarbon dating. Herit Sci 1(1):24. https://doi.org/10.1186/2050-7445-1-24

Berger A, Loutre MF (1991) Insolation values for the climate of the last 10 million years. Quat Sci Rev 10(4):297–317. 027737919190033Q. https://doi.org/10.1016/0277-3791(91)90033-Q

Bierman P, Steig EJ (1996) Estimating rates of denudation using cosmogenic isotope abundances in sediment. Earth Surf Process Landf 21(2):125–139

Bradley RS (2015) Chapter 7-Loess. Palaeoclimatology, 3rd edn. Academic Press, San Diego, pp 279–290

Bull WB (2007) Tectonic geomorphology of mountains. Blackwell Publishing, Malden, MA

Busacca AJ, Sweeney MR (2005) LOESS. In: Hillel D, Hatfield JL (eds) Encyclopedia of soils in the environment, vol 3. Elsevier, Amsterdam

Channell JET, Riveiros NV, Gottschalk J, Waelbroeck C, Skinner LC (2017) Age and duration of Laschamp and Iceland Basin geomagnetic excursions in the South Atlantic Ocean. Quat Sci Rev 167:1–13 S0277379117302202. https://doi.org/10.1016/j.quascirev.2017.04.020

Clark I, Fritz P (1997) Environmental isotopes in hydrogeology, Stein J, Starkweather A (eds). Lewis, Boca Raton, New York

Clement AC, Cane MA, Seager R (2001) An orbitally driven tropical source for abrupt climate change. J Climate 14(11):2369–2375

Cohen AS (2003) Paleolimnology: the history and evolution of lake systems. Oxford University Press

Craig H (1957) The natural distribution of radiocarbon and the exchange time of carbon dioxide between atmosphere and sea. Tellus 9(1):1–17

Cronin TM (1999) Principles of paleoclimatology. Columbia University Press

Das SS, Singh A, Singh N (2017) Provenance of bar-top sediments at Tons river near Allahabad. J Geol Soc India 89(5):503–510. https://doi.org/10.1007/s12594-017-0638-3

Dash C, Dhal SP, Kumar P, Pati P, Chopra S (2020) Climate-induced denudational changes during the Little Ice Age inferred from ^{10}Be (meteoric)/^{9}Be ratio: a case study from the core monsoon zone of India. Quat Int

Dhal SP, Balakrishnan S, Kumar P, Singh P, Sharan A, Chopra S (2018) 10 Be/^{9}Be ratios of Cauvery river delta sediments, southern India. Curr Sci 115(9):1770–1781

Dodonov AE (1991) Loess of Central Asia. GeoJ 24(2). https://doi.org/10.1007/BF00186015

Duller GAT (2004) Luminescence dating of quaternary sediments: recent advances. J Quat Sci 19(2):183–192. https://doi.org/10.1002/jqs.809

Ebelmen JJ (1845) Sur les produits de la décomposition des espèces minérales de la famille des silicates. Annales des Mines 7(3):66

Edouard B, Raisbeck GM, Yiou F, Jouzel J (1997) Solar modulation of cosmogenic nuclide production over the last millennium: comparison between 14C and 10Be records. Earth Planet Sci Lett 150(3–4):453–462. S0012821X97000824. https://doi.org/10.1016/S0012-821X(97)00082-4

Elsasser W, Ney EP, Winckler JR (1956) Cosmic-ray intensity and geomagnetism. Nature 178(4544):1226–1227

Fairbanks RG (1989) A 17,000-year glacio-eustatic sea level record: influence of glacial melting rates on the Younger Dryas event and deep-ocean circulation. Nature 342(6250):637–642

Field CV, Schmidt GA, Koch D, Salyk C (2006) J Geophys Res 111(D15):D15107. https://doi.org/10.1029/2005JD006410

Frechen M (2003) Loess in Europe—mass accumulation rates during the Last Glacial Period. Quat Sci Rev 22(18–19):1835–1857. https://doi.org/10.1016/S0277-3791(03)00183-5

Gibbard PL (2019) Giovanni Arduino-the man who invented the Quaternary. Quat Int 500:11–19

Gibbard PL, Head MJ (2010) The newly-ratified definition of the quaternary system/period and redefinition of the Pleistocene series/epoch and comparison of proposals advanced prior to formal ratification. Episodes 33(3):152–158

Hays JD, Imbrie J, Shackleton NJ (1976) Variations in the Earth's orbit: pacemaker of the ice ages. Science 194(4270):1121–1132

Head MJ (2019) Formal subdivision of the quaternary system/period: present status and future directions. Quat Int 500:32–51

Horiuchi K, Uchida T, Sakamoto Y, Ohta A, Matsuzaki H, Shibata Y, Motoyama H (2008) Ice core record of 10Be over the past millennium from Dome Fuji, Antarctica: a new proxy record of past solar activity and a powerful tool for stratigraphic dating. Quat Geochronol 3(3):253–261

Jackson ST, Overpeck JT, Webb T, Keattch SE, Anderson KH (1997) Mapped plantmacrofossil and pollen records of late quaternary vegetation change in Eastern North America. Quat Sci Rev 16(1):1–70. S0277379196000479. https://doi.org/10.1016/S0277-3791(96)00047-9

Jeziorski A, Yan ND, Paterson AM, DeSellas AM, Turner MA, Jeffries DS, Keller B, Weeber RC, McNicol DK, Palmer ME, McIver K (2008) The widespread threat of calcium decline in fresh waters. Science 322(5906):1374–1377

Jokipii JR, Giacalone J (1998) Space Sci Rev 83(1/2):123–136. https://doi.org/10.1023/A:1005077629875
Laj C, Channell JET (2015) Geomagnetic excursions. In: Treatise on geophysics, Geomagnetism, vol 5, 2nd edn. Elsevier, Amsterdam, pp 343–383
Lal D, Peters B (1967) Cosmic ray produced radioactivity on the Earth. In: Kosmische Strahlung II/Cosmic Rays II. Springer, Berlin/Heidelberg, pp 551–612
Lean JL, Wang Y-M, Sheeley NR (2002) The effect of increasing solar activity on the Sun's total and open magnetic flux during multiple cycles: Implications for solar forcing of climate. Geophys Res Lett 29(24):77-1–77-2. https://doi.org/10.1029/2002GL015880
Leopold LB, Wolman MG, Miller JP (1964) Fluvial processes in geomorphology. W. H. Freeman & Sons, San Francisco
Leuschner DC, Sirocko F (2003) Orbital insolation forcing of the Indian monsoon – a motor for global climate changes? Palaeogeogr Palaeoclimatol Palaeoecol 197(1–2):83–95
Lingenfelter RE (1963) Production of carbon 14 by cosmic-ray neutrons. Rev Geophys 1(1):35. https://doi.org/10.1029/RG001i001p00035
Lowe JJ, Walker MJ (2014) Reconstructing quaternary environments. Routledge
Lund SP, Schwartz M, Keigwin L, Johnson T (2005) Deep-sea sediment records of the Laschamp geomagnetic field excursion (~41000 calendar years before present). J Geophys Res Solid Earth 110(B4). https://doi.org/10.1029/2003JB002943
Lund S, Schwartz M, Stott L (2017) Long-term palaeomagnetic secular variation and excursions from the western Equatorial Pacific Ocean (MIS2-4). Geophys J Int 209(2):587–596
Lupker M, France-Lanord C, Lartiges B (2016) Impact of sediment–seawater cation exchange on Himalayan chemical weathering fluxes. Earth Surf Dyn 4(3):675–684. https://doi.org/10.5194/esurf-4-675-2016
Masarik J, Beer J (1999) Simulation of particle fluxes and cosmogenic nuclide production in the Earth's atmosphere. J Geophys Res Atmos 104(D10):12099–12111. https://doi.org/10.1029/1998JD200091
McCracken KG (2004) Geomagnetic and atmospheric effects upon the cosmogenic 10Be observed in polar ice. J Geophys Res Space Phys 109(A4):A04101
McCracken KG, McDonald FB, Beer J, Raisbeck G, Yiou F (2004) A phenomenological study of the long-term cosmic ray modulation, 850–1958 AD. J Geophys Res Space Phys 109(A12): A12103
Meenakshi PK, Shrivastava JP, Chandra R, Sundeep Chopra GS, Roonwal RS (2018) High resolution 14C AMS ages (~50 ka) of organic matter associated with the loess-palaeosol Holocene-Late Pleistocene (8–130 ka) sediments of Dilpur Formation Karewa Group Kashmir India. Quat Geochronol 47:170–179 S1871101417301474. https://doi.org/10.1016/j.quageo.2018.06.004
Merrill RT, McFadden PL (1994) Geomagnetic field stability: reversal events and excursions. Earth Planet Sci Lett 121(1–2):57–69
Miall AD (1985) Architectural-element analysis: a new method of facies analysis applied to fluvial deposits. Earth-Sci Rev 22(4):261–308. 0012825285900017. https://doi.org/10.1016/0012-8252(85)90001-7
Milankovitch MM (1941) Canon of insolation and the iceage problem. Koniglich Serbische Akademice Beograd Special Publication, 132
Milliman JD, Meade RH (1983) World-wide delivery of river sediment to the oceans. J Geol 91(1): 1–21
Molnar P, England P (1990) Late Cenozoic uplift of mountain ranges and global climate change: chicken or egg? Nature 346(6279):29–34. https://doi.org/10.1038/346029a0
Muhs DR (2014) Overview of Paleosols and wind blown sediments. Elsevier, Amsterdam
Muhs DR, Bettis EA (2003) Quaternary loess-paleosol sequences as examples of climate-driven sedimentary extremes. Geological Society of America special paper 370, Boulder, CO, pp 53–74
Muller RA, MacDonald GJ (1997) Glacial cycles and astronomical forcing. Science 277(5323): 215–218

Nesje A, Lie Ø, Dahl SO (2000) Is the North Atlantic Oscillation reflected in Scandinavian glacier mass balance records? J Quat Sci 15(6):587–601. Published for the Quaternary Research Association

Nilsson A, Muscheler R, Snowball I, Aldahan A, Possnert G, Augustinus P, Atkin D, Stephens T (2011) Multi-proxy identification of the Laschamp geomagnetic field excursion in Lake Pupuke, New Zealand. Earth Planet Sci Lett 311(1–2):155–164

O'Brien K (1979) Secular variations in the production of cosmogenic isotopes in the Earth's atmosphere. J Geophys Res 84(A2):423. https://doi.org/10.1029/JA084iA02p00423

Pazzaglia FJ (2013) Fluvial terraces. In: Treatise on geomorphology, vol 9, pp 379–412

Polissar PJ, Abbott MB, Wolfe AP, Bezada M, Rull V, Bradley RS (2006) Solar modulation of Little Ice Age climate in the tropical Andes. Proc Natl Acad Sci 103(24):8937–8942

Radick RR, Lockwood GW, Baliunas SL (1990) Stellar activity and brightness variations: a glimpse at the Sun's history. Science 247(4938):39–44. https://doi.org/10.1126/science.247.4938.39

Rahaman W, Wittmann H, von Blanckenburg F (2017) Denudation rates and the degree of chemical weathering in the Ganga River basin from ratios of meteoric cosmogenic 10Be to stable 9Be. Earth Planet Sci Lett 469:156–169. S0012821X17301802. https://doi.org/10.1016/j.epsl.2017.04.001

Raisbeck GM, Yiou F, Jouzel J, Petit JR, Barkov NI, Bard E (1992) 10Be deposition at Vostok, Antarctica during the last 50,000 years and its relationship to possible cosmogenic production variations during the period. In: NATO advanced research workshop on the last deglaciation: Absolute and radiocarbon chronologies, pp 127–139

Raymo ME, Ruddiman WF, Froelich PN (1988) Influence of late Cenozoic mountain building on ocean geochemical cycles. Geology 16(7):649–653

Rhodes EJ (2011) Optically stimulated luminescence dating of sediments over the past 200,000 years. Annu Rev Earth Planet Sci 39(1):461–488. https://doi.org/10.1146/annurev-earth-040610-133425

Ritter DF, Kochel RC, Miller JR (2002) Process geomorphology. Waveland Press, Long Grove

Roberts AP (2008) Geomagnetic excursions: knowns and unknowns. Geophys Res Lett 35(17): L17307

Roberts HM, Muhs DR, Bettis Iii EA (2007) Loess records| North America. In: Encyclopedia of quaternary science, vol 2, pp 1456–1466

Röthlisberger H (1976) Thermal consequences of the pressure fluctuations in intra-and subglacial water drainage channels. J Glaciol 16(74):309–310

Ruhe RV (1969) Quaternary landscapes in Iowa. Iowa State University Press, Ames. 255 pp

Scherler D, Bookhagen B, Strecker MR (2011) Spatially variable response of Himalayan glaciers to climate change affected by debris cover. Nat Geosci 4(3):156–159. https://doi.org/10.1038/ngeo1068

Schumm SA (1977) The fluvial system. Wiley, New York, NY

Shah RA, Achyuthan H, Lone AM, Kumar S, Kumar P, Sharma R, Amir M, Singh AK, Dash C (2020) Holocene palaeoenvironmental records from the high-altitude Wular Lake Western Himalayas. Holocene 30(5):733–743. https://doi.org/10.1177/0959683619895592

Sharma R, Umapathy G, Kumar P, Ojha S, Gargari S, Joshi R, Chopra S, Kanjilal D (2019) AMS and upcoming geochronology facility at Inter University Accelerator Centre (IUAC) New Delhi India. Nucl Instrum Methods Phys Res B 438:124–130. S0168583X18304154. https://doi.org/10.1016/j.nimb.2018.07.002

Simon Q, Thouveny N, Bourlès DL, Valet JP, Bassinot F (2020) Cosmogenic 10Be production records reveal dynamics of geomagnetic dipole moment (GDM) over the Laschamp excursion (20–60 ka). Earth Planet Sci Lett 550:116547

Singh AK, Jaiswal MK, Pattanaik JK, Dev M (2016) Luminescence chronology of alluvial fan in North Bengal India: Implications to tectonics and climate. Geochronometria 43(1):102–112. https://doi.org/10.1515/geochr-2015-0037

Singh AK, Pattanaik JK, Gagan, Jaiswal MK (2017) Late Quaternary evolution of Tista River terraces in Darjeeling-Sikkim-Tibet wedge: Implications to climate and tectonics. Quat Int 443: 132–142. S1040618216301070. https://doi.org/10.1016/j.quaint.2016.10.004

Singh AK, Manna I, Kumar P, Dawar A, Kumar P, Murari MK (2022) A new and effective method for quartz-feldspar separation for OSL and CRN dating. Quat Geochronol 72:101315. https://doi.org/10.1016/j.quageo.2022.101315

Smalley IJ (1995) Making the material: the formation of silt-sized primary mineral particles for loess deposits. Quat Sci Rev 14:645–651

Solanki SK, Usoskin IG, Kromer B, Schüssler M, Beer J (2004) Unusual activity of the Sun during recent decades compared to the previous 11000 years. Nature 431(7012):1084–1087. https://doi.org/10.1038/nature02995

Staff RA, Schlolaut G, Ramsey CB, Brock F, Bryant CL, Kitagawa H, van der Plicht J, Marshall MH, Brauer A, Lamb HF (2013) Integration of the old and new Lake Suigetsu (Japan) terrestrial radiocarbon calibration data sets. Radiocarbon 55(4):2049–2058. https://doi.org/10.2458/azu_js_rc.v55i2.16339

Stuiver M, Braziunas TF (1989) Atmospheric 14C and century-scale solar oscillations. Nature 338(6214):405–408. https://doi.org/10.1038/338405a0

Thouveny N, Creer KM, Williamson D (1993) Geomagnetic moment variations in the last 70,000 years, impact on production of cosmogenic isotopes. Global Planet Change 7(1–3):157–172

Timothy Jull AJ, Burr GS, Hodgins GWL (2013) Radiocarbon dating reservoir effects and calibration. Quat Int 299:64–71 S1040618212032399. https://doi.org/10.1016/j.quaint.2012.10.028

Tuniz C, Kutschera W, Fink D, Herzog GF, Bird JR (1998) Accelerator mass spectrometry: ultrasensitive analysis for global science. CRC Press, Boca Raton

Urey HC (1952) The planets: their origin and development. Mrs Hepsa Ely Silliman Memorial Lectures

Vaccari E (2006) The "classification" of mountains in eighteenth-century Italy and the lithostratigraphic theory of Giovanni Arduino (1714–1795), vol 411. The Origins of Geology in Italy, p 157

von Blanckenburg F, Bouchez J, Wittmann H (2012) Earth surface erosion and weathering from the ^{10}Be (meteoric)/^{9}Be ratio. Earth Planet Sci Lett 351:295–305

Wegmann KW, Pazzaglia FJ (2002) Holocene strath terraces, climate change, and active tectonics: The Clearwater River basin, Olympic Peninsula, Washington State. Geol Soc Am Bull 114(6): 731–744

Willson RC, Hudson HS (1988) Solar luminosity variations in solar cycle 21. Nature 332(6167): 810–812. https://doi.org/10.1038/332810a0

Wittmann H, von Blanckenburg F, Dannhaus N, Bouchez J, Gaillardet J, Guyot JL, Maurice L, Roig H, Filizola N, Christl M (2015) J Geophys Res Earth Surf 120(12):2498–2528. https://doi.org/10.1002/2015JF003581

Wright JS (2001) 'Desert' loess versus 'glacial' loess: quartz silt formation, source areas and sediment pathways in the formation of loess deposits. Geomorphology 36:231–256

Wünnemann B, Hartmann K, Altmann N, Hambach U, Pachur HJ, Zhang H (2007) Interglacial and glacial fingerprints from lake deposits in the Gobi Desert, NW China. In: Developments in quaternary sciences, vol 7. Elsevier, pp 323–347

Yancheva G, Nowaczyk NR, Mingram J, Dulski P, Schettler G, Negendank JF, Liu J, Sigman DM, Peterson LC, Haug GH (2007) Influence of the intertropical convergence zone on the East Asian monsoon. Nature 445(7123):74–77

Yin A (2006) Cenozoic tectonic evolution of the Himalayan orogen is constrained by along-strike variation of structural geometry, exhumation history, and foreland sedimentation. Earth Sci Rev 76(1–2):1–131

Zachos JC, Pagani M, Sloan L et al (2001) Trends, rhythms and aberrations in global climate 65 Ma to present. Science 292:686–693

Glossary

Alluvial sediments clay or silt or gravel carried by rushing streams and deposited where the stream slows down. synonyms: alluvial deposit, alluvion, alluvium. types: delta. a low triangular area of alluvial deposits where a river divides before entering a larger body of water.

Anomalous Cosmic Rays (ACR) Anomalous cosmic rays are a heliospheric phenomenon in which interstellar neutral atoms stream into the heliosphere, are ionized by either solar radiation or the solar wind, and are subsequently accelerated to very high energies, greater than 1 GeV.

Anthropogenic Intervention Recent studies suggest that anthropogenic modification of land hydrology (e.g. through groundwater mining, dam building, irrigation, deforestation, wetlands drainage, and urbanization) could significantly impact sea level rise, although the magnitude and sign of this effect have been widely debated.

Artificial Neural Network (ANN) Artificial neural networks (ANNs), usually simply called neural networks (NNs), are computing systems inspired by the biological neural networks that constitute animal brains.

Atmospheric Circulations Atmospheric circulation is any atmospheric flow used to refer to the general circulation of the Earth and regional movements of air around areas of high and low pressure. On average, this circulation corresponds to large-scale wind systems arranged in several east-west belts that encircle the Earth.

BedRock Bedrock is the hard, solid rock beneath surface materials such as soil and gravel. Bedrock can be made of most types of rock, such as granite, limestone, or like this piece of bedrock, sandstone.

Calc-silicate Granulite The calc-silicate granulites are characterized by a peculiar mineralogical association made of orthopyroxene–titanite–ilmenite–clinopyroxene–garnet–plagioclase–amphibole, which is ideal for the reconstitution of the P–T path and the metamorphic history of the Gour Oumelalen area.

N. Khare (ed.), *Science, Policies and Conflicts of Climate Change*, Springer Climate,
https://doi.org/10.1007/978-3-031-16254-1

Cenozoic Period Cenozoic (66 million years ago until today) means 'recent life'. During this era, plants and animals look most like those on Earth today. Periods of the Cenozoic Era are split into even smaller parts known as Epochs.

Climate Migrant These are migrants that are led by push factors caused by climate changes such as intense flood, sea level rise, drought or crop failure or water shortage.

Climate It in a narrow sense is usually defined as the 'average weather,' or more rigorously, as the statistical description in terms of the mean and variability of relevant quantities over a period ranging from months to thousands or millions of years.

COP-2 The second Conference of the Parties (COP.2) to the Stockholm Convention on Persistent Organic Pollutants (POPs) was held from 1–5 May 2006, in Geneva, Switzerland. COP-2 considered several reports on activities within the Convention's mandate and adopted 18 decisions on, inter alia, DDT, exemptions, financial resources and mechanisms, implementation plans, technical assistance, synergies and effectiveness evaluation.

Cosmogenic Radionuclides Cosmogenic radionuclide (CRN) dating is based on the rate of accumulation of cosmic rays that stimulate the production and decay of radionuclides such as 14C, 10Be, 26Al, and 36Cl.

Diatom Fossils Diatom fossils are preserved as their original silica cell wall. Other organisms with mineralized structures include bryozoans, which are preserved as their original calcite skeletons in the fossil record. Radiolarians are another type of organism that possess silica structures within their cells.

Disaster Mitigation These measures are those that eliminate or reduce the impacts and risks of hazards through proactive measures taken before an emergency or disaster occurs.

Diurnal Temperature Range (DTR) In meteorology, diurnal temperature variation is the variation between a high air temperature and a low temperature that occurs during the same day.

Dynamic Ordinary Least Square (DOLS) Dynamic OLS (DOLS) is an alternative (parametric) approach in which lags and leads are introduced to cope with the problem irrespectively of the order of integration and the existence or absence of cointegration.

Early Middle Pleistocene The Early-Middle Pleistocene transition (c. 1.2–0.5 Ma), sometimes known as the 'mid-Pleistocene revolution,' represents a major episode in Earth's history. Low-amplitude 41-ka obliquity-forced climate cycles of the earlier Pleistocene were replaced progressively in the later Pleistocene by high-amplitude 100-ka cycles.

Eco-development Eco-development refers to development at regional and local levels, consistent with the potentials of the area involved, with attention given to the adequate and rational use of natural resources, technological styles and organizational forms that respect the natural ecosystems and local social and cultural patterns. The term is also used to describe an integrated approach to environment and development.

Enhanced Thematic Mapper (ETM) The Enhanced Thematic Mapper Plus (ETM+) instrument is a fixed 'whisk-broom,' eight-band, multispectral scanning radiometer capable of providing high-resolution imaging information of the Earth's surface.

Environmental Protection Environmental protection is the practice of protecting the natural environment by individuals, organizations and governments. Its objectives are to conserve natural resources and the existing natural environment and, where possible, to repair damage and reverse trends.

Fluvial Ecosystem These range from the processes operating in a single bend in a river, to the different channel patterns arising from contrasting conditions of water flow, sediment transport and channel gradient, and ultimately to the morphology of entire drainage basins.

Galactic Cosmic Rays (GCR) Galactic Cosmic Rays (GCR) are the slowly varying, highly energetic background source of energetic particles that constantly bombard Earth. GCR originate outside the solar system and are likely formed by explosive events such as a supernova.

Ganges Water Sharing Treaty Indian Prime Minister Indira Gandhi and Bangladesh's founding leader Sheikh Mujibur Rahman signed the wide-ranging Indo-Bangladeshi Treaty of Friendship, Cooperation and Peace on 19 March 1972; as per the treaty, the two nations established a Joint River Commission to work for the common interests and sharing of water.

Geochemical Elements These include Cu, Pb, Zn, Cd, Mo, Hg, Sb, Sn, Tl, Te, As, as well as noble metals. Atmophile elements are those that readily form volatile compounds at relatively low temperatures (<300 K), many of which are preferentially concentrated in planetary atmospheres.

Global Boundary Stratotype Section and Point (GSSP) A Global Boundary Stratotype Section and Point (GSSP) is an internationally agreed-upon reference point on a stratigraphic section that defines the lower boundary of a stage on the geologic time scale. The effort to define GSSPs is conducted by the International Commission on Stratigraphy, a part of the International Union of Geological Sciences. Most, but not all, GSSPs are based on paleontological changes. Hence GSSPs are usually described in terms of transitions between different faunal stages, though far more faunal stages have been described than GSSPs. The GSSP definition effort commenced in 1977. As of 2021, 76 of the 104 stages that need a GSSP have a ratified GSSP.

Global Climate Models (GCM) It is a complex mathematical representation of the major climate system components (atmosphere, land surface, ocean, and sea ice), and their interactions. Earth's energy balance between the four components is the key to long-term climate prediction.

Global Environment Facility (GEF) The Global Environment Facility (GEF) is a financial mechanism that provides grant funds to developing countries for projects covering four 'focal areas': climate change, biodiversity, international waters and ozone depletion.

Global Mean Sea Level (GMSL) Each point in the global mean sea level (GMSL) time series plots is the area-weighted mean of all of the sea surface height anomalies measured by the altimeter in a single, 10-day satellite track repeat cycle (time for the satellite to begin repeating the same ground track).

Green Climate Fund (GCF) The Green Climate Fund's (GCF) aim is to expand collective human action to respond to climate change. The Fund aims to mobilize funding at scale to invest in low-emission and climate-resilient development on our home planet.

Greenhouse Gases (GHGs) It is a gas that absorbs and emits radiant energy within the thermal infrared range, causing the greenhouse effect. The primary greenhouse gases in Earth's atmosphere are water vapour (H_2O), carbon dioxide (CO_2), methane (CH_4), nitrous oxide (N_2O), and ozone (O_3).

Gross Domestic Product (GDP) GDP measures the monetary value of final goods and services – that is, those that are bought by the final user – produced in a country in a given time (say a quarter or a year). It counts all of the output generated within the borders of a country.

Ground Control Points (GCP) Ground Control Points (GCPs) are defined as points on the surface of the earth of a known location used to geo-reference Landsat Level-1 data. GCPs are updated as needed to continually improve Landsat data. GCPs can be downloaded and used as reference data.

Holocene The Holocene is the name given to the last 11,700 years of the Earth's history – the time since the end of the last major glacial epoch, or 'ice age'. Since then, there have been small-scale climate shifts – notably the 'Little Ice Age' between about 1200 and 1700 A.D.

Hydrodynamic condition Hydrodynamic conditions define (bio)fouling properties. Equipment design is a key factor in hydrodynamics and (bio)fouling development. (Bio)fouling formation occurs with higher impact under low hydrodynamics. High shear stresses improve CIP efficiency and (bio)fouling removal.

Indian Summer Monsoon (ISM) The Indian summer monsoon typically lasts from June–September, with large areas of western and central India receiving more than 90% of their total annual precipitation during the period, and southern and northwestern India receiving 50–75% of their total annual rainfall.

Kyoto Protocol The Kyoto Protocol was adopted on 11 December 1997. Owing to a complex ratification process, it entered into force on 16 February 2005. Currently, there are 192 Parties to the Kyoto Protocol. In short, the Kyoto Protocol operationalizes the United Nations Framework Convention on Climate Change by committing industrialized countries and economies in transition to limit and reduce greenhouse gases (GHG) emissions in accordance with agreed individual targets. The Convention itself only asks those countries to adopt policies and measures on mitigation and to report periodically.

Last Glacial Maxima (LCM) The Last Glacial Maximum (LGM), also referred to as the Late Glacial Maximum, was the most recent time during the Last Glacial Period that ice sheets were at their greatest extent. Ice sheets covered much of North America, Northern Europe, and Asia and profoundly affected Earth's climate by causing drought, desertification, and a large drop in sea levels.

Linear Regression (LR) In statistics, linear regression is a linear approach for modelling the relationship between a scalar response and one or more explanatory variables (also known as dependent and independent variables).

Lower Flank The flank is the area on the sides and back of your abdomen, between your lower ribs and your hips. Pain in this area is called flank pain.

Magnetic Susceptibility Magnetic susceptibility indicates whether a material is attracted into or repelled out of a magnetic field. Paramagnetic materials align with the applied field and are attracted to regions of the greater magnetic field. Diamagnetic materials are anti-aligned and are pushed away, towards regions of lower magnetic fields. On top of the applied field, the magnetization of the material adds its magnetic field, causing the field lines to concentrate in paramagnetism, or be excluded in diamagnetism.

Manganese Ore The most important manganese ores are the oxides pyrolusite, romanechite, manganite, hausmannite and the carbonate ore rhodochrosite. Rhodonite and braunite, both silicate ores, are frequently found with oxides. Only ores containing greater than 35 per cent manganese are considered commercially exploitable.

Medieval Period Middle Ages, the period in European history from the collapse of Roman civilization in the 5th-century CE to the period of the Renaissance (variously interpreted as beginning in the 13th, 14th, or 15th century, depending on the region of Europe and other factors).

Medieval Warming Period (MWP) Medieval Warm Period (MWP), also called the medieval warm epoch or little climatic optimum, brief climatic interval that is hypothesized to have occurred from approximately 900 CE to 1300 (roughly coinciding with the Middle Ages in Europe), in which relatively warm conditions are said to have prevailed in various parts of the world, though predominantly in the Northern Hemisphere from Greenland eastward through Europe and parts of Asia.

Natural Ecosystems A natural ecosystem is a community of living and non-living entities that occurs freely in nature. Every component interacts together as a combined unit through physical, chemical and biological processes. The discriminating factor of natural ecosystems from other ecosystems is that they are completely natural.

Ozone depletion Ozone depletion consists of two related events observed since the late 1970s: a steady lowering of about four per cent in the total amount of ozone in Earth's atmosphere, and a much larger springtime decrease in stratospheric ozone around Earth's polar regions. The latter phenomenon is referred to as the ozone hole.

Piezometric Surface It is the imaginary surface to which groundwater rises under hydrostatic pressure in wells or springs.

Plagioclase Index of Alteration (PIA) The Plagioclase Index of Alteration (PIA) was proposed by Fedo et al. (1995) as an alternative to the CIW. Because plagioclase is abundant in silicate and dissolves quickly, the PIA may be used when plagioclase weathering needs to be monitored.

Pleistocene Epoch The Pleistocene epoch is a geological time that includes the last ice age when glaciers covered huge parts of the globe. Also called the Pleistocene era, or simply the Pleistocene, this epoch began about 2.6 million years ago and ended 11,700 years ago, according to the International Commission on Stratigraphy.

Pleistocene The Pleistocene (/ˈplaɪs. təˌsiːn, -toʊ-/ PLYSE-tə-seen, -toh-, often referred to as the Ice Age) is the geological epoch that lasted from about 2,580,000 to 11,700 years ago, spanning the Earth's most recent period of repeated glaciations.

Pre-historic Hominids The earliest fossils that have been proposed as members of the hominin lineage are Sahelanthropus tchadensis dating from 7 million years ago, Orrorin tugenensis dating from 5.7 million years ago, and Ardipithecus kadabba dating to 5.6 million years ago.

Pyroclastic material Pyroclastic materials are classified according to their size, measured in millimetres: dust (less than 0.6 mm [0.02 inch]), ash (fragments between 0.6 and 2 mm [0.02 to 0.08 inch]), cinders (fragments between 2 and 64 mm).

Quaternary Deposit The Quaternary deposits are represented mainly by glacial and water–glacial formations of different ages.

Quaternary Period The Quaternary Period is typically defined by the cyclic growth and decay of continental ice sheets related to the Milankovitch cycles and the associated climate and environmental changes that they caused.

Regional Climate Models (RCM) It is a numerical climate prediction model forced by specified lateral and ocean conditions from a general circulation model (GCM) or observation-based dataset (reanalysis) that simulates atmospheric and land surface processes while accounting for high-resolution topographical.

Representative Concentration Pathways (RCP) It is a greenhouse gas concentration (not emissions) trajectory adopted by the IPCC. Four pathways were used for climate modelling and research for the IPCC Fifth Assessment Report (AR5) in 2014.

River Basin A river drainage basin is an area drained by a river and all of its tributaries. A river basin is made up of many different watersheds. A watershed is a small version of a river basin. Every stream and tributary has its watershed, which drains to a larger stream or wetland.

Sea Level Rise It is an increase in the level of the world's oceans due to the effects of global warming. Burning fossil fuels is one of the causes of global warming because it releases carbon dioxide and other heat-trapping gasses into the atmosphere.

Sediment Geochemistry For a long time, sedimentary geochemistry has been in use to understand the conditions of deposition, climatic variations, tectonic setting, provenance, reservoir characteristics, etc.

Sedimentary Processes Sedimentary processes, namely weathering, erosion, crystallization, deposition, and lithification, create the sedimentary family of rocks.

Soil Erosion Soil erosion is a gradual process that occurs when the impact of water or wind detaches and removes soil particles, causing the soil to deteriorate. Soil deterioration and low water quality due to erosion and surface runoff have become severe problems worldwide.

Solar Cosmic Ray (SCR) Solar cosmic rays are fluxes of high-energy-charged particles accelerated in the corona and chromosphere which appear during solar flares.

Stagnant Water Stagnant water is standing water that is trapped or sits undisturbed for hours/days. Standing or stagnant water is usually a result of floods and water leaks. Basements, crawl spaces, and bathrooms are common areas where you can find standing water.

Standardized Precipitation Index (SPI) It is a relatively new drought index based only on precipitation. It's an index based on the probability of precipitation for any time scale. Some processes are rapidly affected by atmospheric behaviour, such as dryland agriculture, and the relevant time scale is a month or two.

Stratigraphic Horizons Horizon can be characterized by boundaries of only two types: strictly stratigraphic and all stratigraphic. The stratigraphic interval of a horizon can vary from a single chronozone to a stage. Boundaries of neighbouring horizons at the contact between two stages or systems should coincide with the latter.

Sustainable Development Goals (SDGs) The Sustainable Development Goals or Global Goals are a collection of 17 interlinked global goals designed to be a 'blueprint to achieve a better and more sustainable future for all'. The SDGs were set up in 2015 by the United Nations General Assembly and are intended to be achieved by 2030.

Tectonic Stability Shields are considered tectonically stable areas in comparison to the activity occurring at their margins and the boundaries between plates, but their formation required large amounts of tectonic activity and erosion. Initially, a mountain belt is formed at a convergent plate margin.

Index

A
Adjoining lithology, 111
Alluvial fans, 270, 272, 274, 289
Alluvial uplands, 137
Anthropogenic activities, x, xi, 29, 32, 54, 92, 110, 115, 125, 126, 133, 137, 138, 167, 189, 190, 234
Anthropogenic intervention, 133, 313
Atmospheric heat, 55

B
Biophysical approach, 54

C
Cardiovascular diseases (CVD), 29, 37
Chemical weathering, vi, xi, 176, 181, 183, 185, 187, 189–191, 266, 299
Chemical weathering intensity (CWI), 115, 116, 120, 123, 124, 126, 184
Climate oscillations, 31
Climate package, 13, 22
Climate sensitiveness, 56
Climate-resilient community, x
Climatic ecology, 92
Climatic ecosystem, 92
Climatic parameters, 61–66
Coastal erosion, x, 78, 94
Coastal slope, 99, 103, 104
Colloidal influx, xii, 215
Compositional maturity, xi, 187, 191
Convention on Bio-Diversity (CBD), 6
Cosmic rays, 301, 313–315, 319
Cosmogenic radionuclides (CRN), vi, xiv, 297–308, 314
Crop modeling technique, 54
Cross drainage structures, 132

D
Deltaic plains, 93, 104
Depositional environment, xiv, 204, 225, 232, 289
Detrital aluminosilicates, 120
Diatom processing, 112
Dipole magnets, 299
Disaster, 30, 79, 80, 83–86, 106, 132, 246, 314
Disaster mitigation, x, 314
Diurnal temperature range (DTR), ix, 28, 29, 314
Doha Amendment, 10, 11, 20

E
Earth systems, 24, 197, 224
Ecological degradation, 133
Economic development, 4, 8, 56, 85, 105, 223
Environmental challenges, 84

F
Fault zone, 268–289
Flow regime, 150, 155
Fluvial-marine interferences, 93
Fluvio-lacustrine sediments, 247
Food safety, 55, 56

N. Khare (ed.), *Science, Policies and Conflicts of Climate Change*, Springer Climate,
https://doi.org/10.1007/978-3-031-16254-1

G
Geochemical elements, 112, 113, 115, 120, 121, 123, 126, 315
Geochemical mapping, 175, 176, 181, 191
Global element cycles, 176
Global health crisis, 78
Global ice cover, 298
Gradual depletion, 121
Green Climate Fund (GCF), 12, 13, 19, 20, 22, 316
Greenhouse gas (GHGs), vii, viii, 6–13, 15–20, 30, 32–34, 41, 43, 98, 299, 305, 316, 318
Grove casts, 245

H
Heat intensity, 54
High emissions scenario, x
Hydrological cycle, 33, 35
Hydrological drought, 36
Hypolimnetic oxygen, xi, 123, 126

I
Ice cores, xiv, 233, 239, 305–306
Ice sheets, 92, 298, 316, 318
Igneous rocks, 179
Induced polarization (IP), 202–204
Inter-Governmental Panel on Climate Change (IPCC), vii, viii, x, 5, 6, 8, 13, 15–17, 24, 28, 31, 41, 54, 92, 93, 98, 102, 123, 318
Irregular precipitation, 54, 71
Irrigation system, 56

L
Late Cenozoic period, xiii, 266
Leaky capacitor, 202
Low emissions scenario, x

M
Manganese deposits, 198
Manganese ore, xii, 197–201, 204, 205, 208–210, 215–217, 317
Manganese rocks, 197, 216
Marine sediments, xiv, 298
Mass populations, xiii, 238, 248, 250, 251, 257
Meltwater, 80, 84
Mesolithic tools, 226, 228, 230
Meteorological drought, 36
Microclimatic variations, xiv, 290, 291
Micro-climate zones, 304
Migration, xiii, 29, 40, 41, 78, 79, 85, 138, 246, 250, 251, 258
Milankovitch cycles, xiv, 318
Mineralogical maturity, xii, 187, 188, 191
Morphometric analyses, 231

N
Narmada-Son Fault, 241
Natural ecosystems, 28, 41, 314, 317

O
Oldest Toba Tuff (OTT), xiii, 238, 241, 246, 247, 258
Ozone depletion, 5, 315, 317

P
Palaeoenvironment, xiii, 222, 223, 226, 233, 235
Paleoclimatic reconstruction, xxiii, 222
Peninsular climatic, xiv
Plagioclase Index, 176, 181, 317
Post-magmatic process, 247
Precambrian rocks, 137
Proterozoic periods, 197, 216

Q
Quaternary alluvial deposits, 244
Quaternary period, xiii, xiv, 222, 250, 268, 270, 289, 290, 297–299, 318
Quaternary stratigraphy, 222
Quaternary trench, xiii, 226

R
Rare earth elements (REEs), 181
River basins, x, xiii, 32, 36, 58, 78–81, 84–86, 133, 222, 224, 226, 234, 238–241, 244–249, 251, 252, 254–258, 266, 268, 318
River bed morphology, 133
River morphology, 137, 138
Roof pendants, 198

S
Sand laminae, 245
Sea-level rise (SLR), vi, x, 6, 28, 29, 33, 80, 91–106, 318
Seasonal migration, 80
Sediment geochemistry, x, 110, 183, 318

Sediment maturity, xi, 176, 181–183, 187, 189–191
Sedimentary basins, 266
Sedimentary rocks, 179
Shoreline change rate, 94, 96–98, 102
Solar radiation, 28, 29, 69, 301, 313
Solar winds, 301, 313
Stream water velocity, 138
Suitable tectonic evolution, 196
Synthetic fluorinated gases, 32

T
Tectonic basins, xiii, 266, 291
Tectonic stability, 234, 266, 289, 291, 319

U
United Nations Framework Convention on Climate Change (UNFCCC), viii, ix, 6–9, 11, 12, 14, 16, 18, 19, 21–24, 316

V
Virtual geomagnetic pole (VGP), 305
Volcanic ash, 243, 244, 257

W
Western disturbances, 179, 190
Wilderness degradation, 4
World Climate Research Program (WCRP), 5